Food Processes, Biochemistry and Technology

Food Processes, Biochemistry and Technology

Edited by **Jerrod Wesley**

SYRAWOOD
PUBLISHING HOUSE
New York

Published by Syrawood Publishing House,
750 Third Avenue, 9th Floor,
New York, NY 10017, USA
www.syrawoodpublishinghouse.com

Food Processes, Biochemistry and Technology
Edited by Jerrod Wesley

International Standard Book Number: 978-1-68286-133-2 (Hardback)

Contents

Preface

Food science is a multidisciplinary field of study that incorporates principles and concepts of various disciplines like biochemistry, engineering, etc. This book aims to study and analyze various food and biochemical processes and provide significant information to help develop a good understanding of various topics such as food chemistry and physical properties, characterization and profiling of different food products and components, food rheology, etc. The chapters included herein aim to equip students and experts with the advanced topics and upcoming concepts in this area.

The information contained in this book is the result of intensive hard work done by researchers in this field. All due efforts have been made to make this book serve as a complete guiding source for students and researchers. The topics in this book have been comprehensively explained to help readers understand the growing trends in the field.

I would like to thank the entire group of writers who made sincere efforts in this book and my family who supported me in my efforts of working on this book. I take this opportunity to thank all those who have been a guiding force throughout my life.

Editor

Gluten Detection and Speciation by Liquid Chromatography Mass Spectrometry (LC-MS/MS)

Stephen Lock

AB SCIEX, Pheonix House, Centre Park, Warrington, WA1 1RX, UK;
E-Mail: Stephen.lock@absciex.com

Abstract: Liquid chromatography tandem mass spectrometry (LC-MS/MS) has been used historically in proteomics research for over 20 years. However, until recently LC-MS/MS has only been routinely used in food testing for small molecule contaminant detection, for example pesticide and veterinary residue detection, and not as a replacement of microbiological food testing methods, specifically allergen analysis. Over the last couple of years, articles have started to be published which describe the detection of allergens by LC-MS/MS. In this article we will describe how LC-MS/MS can be applied in the area of gluten detection and how it can be used to specifically differentiate the species of gluten used in food, where specific markers for each variety of gluten can be simultaneously acquired and detected at the same time. The article will discuss the effect of variety on the peptide response observed from different wheat grain varieties and will describe the sample preparation protocol which is essential for generating the peptide markers used for speciation.

Keywords: gluten; speciation; LC-MS/MS

1. Introduction

Gluten is known to produce an allergic response, and intolerance to gluten leads to celiac disease. Levels of intolerance to gluten often vary with gluten variety; this is especially relevant to the use of oats which has a low effect on celiac suffers. Most current ELISA methodology, with the exception of assays based on the R5 antibody [1], detect the presence of barley, rye and wheat but also oats. Assays based on the R5 antibody are not sensitive to oats, but this assay still cannot differentiate barley, wheat and rye. Also, all the ELISA methods that detect gluten are based on one section of the gluten protein (for example, in the case of the R5 assay, the peptide sequence glutamine-glutamine-proline-phenylalanine-proline (QQPFP) peptide epitope) and as such are susceptible to false positives and false negatives. Liquid chromatography tandem mass spectrometry (LC-MS/MS) has the ability to detect species based on multiple markers with multiple points of confirmation which makes it far less susceptible to producing false negatives and positives, and gives far more confirmation in detection. Due to its very specific nature, it is also capable of distinguishing species by using multiple peptide markers as shown previously [2–5]. A suitable LC-MS/MS method that could offer the possibility to differentiate between gluten species with a high degree of specificity would be beneficial to both grain producers and consumers.

In addition, legislation is changing with respect to gluten and, in the UK, one of the first allergen limits came into effect on 1 January 2012 [6,7]. On this date the laws governing food labeling were changed such that three different terms can now be used:

1. Gluten-free—is covered by the law and applies only to food which has 20 parts per million (ppm) or less of gluten.
2. Very low gluten—is covered by the law and is for foods which have between 21 and 100 ppm.
3. No gluten-containing ingredients—this is not covered by the law and is for foods that are made with ingredients that do not contain gluten and where cross contamination controls are in place. These foods will have very low levels of gluten, but have not been tested to the same extent as those labeled gluten-free or very low gluten.

These changes in the food labeling law have recently been followed in the US by the FDA [8]. Some initial studies using LC-MS/MS in gluten profiling were presented in 2012 [9]. In these studies trypsin was used as the enzyme for digestion, even though gluten proteins do not undergo a large level of trypsin digestion (due to a low number of lysine and arginine residues) some unique marker peptides were found. Recently, chymotrypsin has also been used as an alternative to trypsin [10] to generate a large number of peptides for the improved detection of wheat gliadin proteins, but this method used longer digestion times and labelling chemistry to better characterise the gluten proteins.

The purpose of this study was a follow up to the original poster presentation in 2012 [9] to investigate a simpler approach to preparing extracts and use recent advances in LC technology to help reach detection limits below the requirements of the current labelling legislation. One of the main purposes of this work was to develop an approach which could analyse a sample in one day using inexpensive available chemicals. In this study single varieties of grain which have not been first milled were ground into flour using a commercial coffee grinder. These samples, together with commercial samples of self-raising flour, gluten-free flour and some gluten and gluten-free foods, were extracted and then the allergenic proteins were reduced, alkylated and digested using trypsin. In this study the extracts produced were simply diluted into 0.1 % formic acid prior to injection and separation by reverse phase chromatography and LC-MS/MS detection. The LC used was a Eksigent ekspert™ microLC 200 UHPLC system (Eksigent, Redwood City, CA, USA) which had been previously evaluated for the detection of egg and milk allergens in wine [11] and had been shown to offer a 5-fold improvement in sensitivity. The mass spectrometry methods utilised *Scheduled* MRM™ (an algorithm which allows the independent monitoring of MRM transitions with a defined window around the expected retentions time for each MRM transition which is available in the Analyst® software version 1.5 and onwards from

AB SCIEX) for multiple peptides for each gluten species, so that presence of allergen can be unambiguously confirmed.

2. Experimental Section

The method described is based on the classic proteomics sequencing approach which involves first the extraction of the protein from a matrix. Once extracted, the proteins are reduced, alkylated and digested. The extracts were finally diluted and analyzed by LC-MS/MS using an AB SCIEX QTRAP® 4500 LC/MS/MS system (AB SCIEX, Warrington, UK).

2.1. Preparation of Tryptic Digests

2.1.1. Extraction of Proteins

Markers proteins from wheat, oats, barley and rye were extracted by placing powdered sample (0.5 g of flour or cookie which had been ground using a commercial coffee grinder) into a falcon tube (15 mL) with extraction buffer [5 mL of a 50:50 mixture of ethanol containing 2 M urea and 50 mM 2-amino-2-hydroxymethyl-propane-1,3-diol (Tris)]. This mixture was shaken by hand (30 s) and then heated and shaken in an orbital water bath (40 °C, 60 min).

2.1.2. Reduction and Alkylation of Proteins

Once extracted the samples were centrifuged (2500 rpm, 5 min, 20 °C). The supernatant (0.5 mL) was then reduced by the addition of TCEP [tris(2-carboxyethyl)phosphine, 0.2 M, 50 µL, 60 °C, 60 min in a thermal mixer] and cooled to room temperature. MMTS (methyl methanethiosulfonate, 0.2 M, 100 µL) was added and the sample left in the dark (30 min) to alkylate the free cysteine residues.

2.1.3. Tryptic Digestion of Proteins

Once the proteins had been alkylated the sample were diluted with buffer (1.35 mL, 0.1 M ammonium bicarbonate solution) and trypsin (80 µL, 0.5 mg/mL, Sigma Aldrich part number 93614) was added. The proteins were then digested for one hour (Eppendorf thermal mixer model number 21516-170, 40 °C, Eppendorf, Stevenage, UK). The digestion was quenched by taking the digest extract (100 µL) and adding 0.1% formic acid (300 µL). The sample was centrifuged (13,000 rpm, 5 min) and then the supernatant was injected into the LC-MS/MS system.

2.2. LC-MS/MS Analysis of Tryptic Digests

All analyses was done using an Eksigent ekspert™ microLC 200 UHPLC system (Eksigent, Redwood City, CA, USA). The extracts (10 µL injection, full loop fill mode) were separated on a reversed-phase Triart C18 column (100 × 0.5 mm, 2.7 µm, YMC, Dinslaken, Germany) at a temperature of 40 °C using the gradient conditions shown in Table 1 where A was water, B was acetonitrile with both phases containing 0.1% formic acid. Micro LC was used as it had previously been shown to improve responses in peptide analysis using electrospray ionization by over 5 fold [11].

Table 1. Gradient elution used for analysis of extracts.

Step	Time (mins)	Flow rate	% A	% B
1	1	25 µL/min	95	5
2	6	25 µL/min	75	25
3	8	25 µL/min	5	95
4	9	25 µL/min	5	95
5	9.2	25 µL/min	95	5
6	12	25 µL/min	96	5

All analyses were performed on an AB SCIEX QTRAP® 4500 LC/MS/MS system (AB SCIEX, Warrington, UK) using electrospray ionization (ESI). The initial method development was carried out using the MIDAS™ workflow (MRM-initiated detection and sequencing [12]) and for microLC analysis the electrode was changed to a microLC hybrid electrode (25 µm ID) designed for microLC [13]. For MIDAS a set of predicted MRM transitions from the known protein sequence were used as a survey scan to trigger the acquisition of EPI spectra (acquired at a scan speed of 10,000 amu/s with dynamic fill time and rolling collision energy active and Q1 resolution set to low) an example of this is shown in Figure 1.

Figure 1. Example of a MIDAS experiment where the top pane shows a MRM trace for a wheat peptide from a flour extract and the bottom pane shows the triggered enhanced product ion (EPI) spectra for the peptide which contains sequence data confirming its identity.

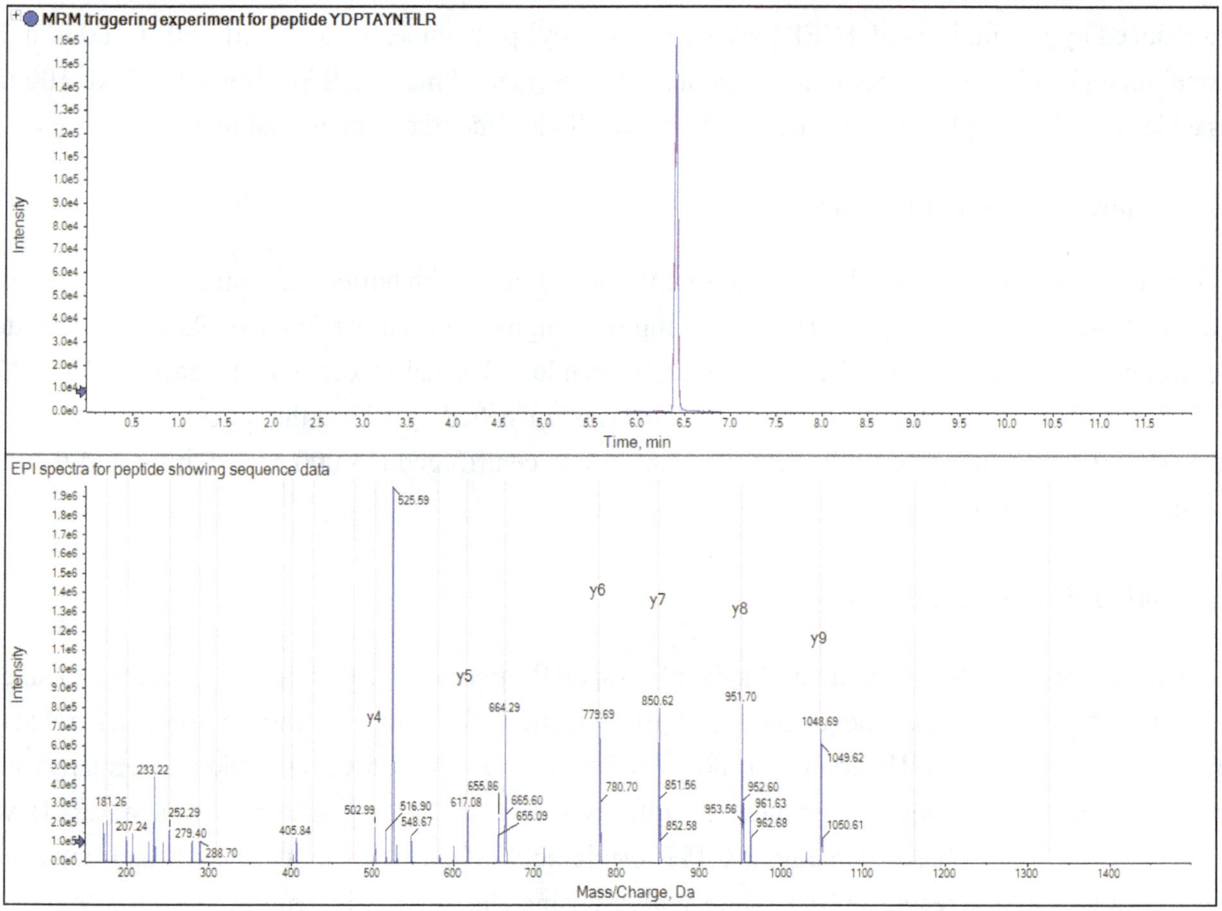

This MIDAS data was submitted to a database search engine for confirmation of peptide identification and to test the feasibility of the MRM transitions for gluten and species identification. With this workflow MRM transitions were designed without the need for synthetic peptides. In the final micro LC method the Turbo V™ source conditions used were gas 1, gas 2 and the Curtain Gas™ interface set to 30 psi, the temperature of the source was set at 350 °C and the IS voltage was 5500 V. The peptides were analyzed using the *Scheduled* MRM™ algorithm with an MRM detection window of 60 s and a target scan time of 0.30 s. Q1 resolution was set to low and Q3 resolution was set to unit. MRM transitions shown in Table 2 were evaluated for rye, oats, wheat and barley, and each MRM transition used the same declustering voltage (80 V) and entrance potential (10 V). These MRM transitions corresponded to the peptides shown in Table 3.

Table 2. MRM transitions used for triggers for generating EPI spectra and peptide detection.

Peptide ID	Q1 mass (amu)	Q3 Mass 1 (amu)	Q3 Mass 2 (amu)	Q3 Mass 3 (amu)	RT (mins)	CE (V)	CXP (V)
Wheat 1	557.3	886.5	548.7	787.4	4.2	32	12
Wheat 2	579.4	897.6	711.5		7.5	33	12
Wheat 3	458.8	730.4	560.3	458.8	3.9	28	12
Wheat 4	594.8	792.4	978.5	538.9	7.1	34	12
Wheat 5	663.8	850.5	779.4	951.7	6.4	37	12
Wheat 6	538.3	547.3	776.4	705.4	4.1	34	12
Barley 1	835.4	947.5	1096.5	1227.6	7.4	47	10
Barley 2	336.2	515.3	554.3	497.3	5.5	13	10
Barley 3	820.4	1096.5	548.3	713.4	7.4	43	10
Barley 4	499.3	785.5	575.3	393.3	6.3	24	10
Barley 5	855.6	980.5	642.4		7.3	38	10
Oats 1	989	997.6	1084.7	1233.7	7.5	54	10
Oats 2	777.4	984.4	1112.5	1225.6	6.1	38	10
Oats 3	627.3	642.4	1012.5		5.8	33	10
Oats 4	365.1	601.2	473.1		3.8	23	10
Rye 1	937	1177.6			7.5	46	10
Rye 1	625	941.6			7.5	24	10
Rye 2	851.7	1199.7	1071.6	1210.2	7.3	43	10
Rye 3	997	1225.6			7.4	46	10
Rye 3	665	1225.6	1128.5		7.4	29	10
Rye 4	988	1197.7	1100.6		7.5	45	10
Rye 4	659	1197.7			7.5	29	10

Table 3. Marker peptides and sequence information used for gluten species markers (the peptide information was taken from searches of the Swiss-Prot database [14]).

Species	Peptide	Protein	Entry number	Peptide sequence
Hordeum vulgare (barley)	1	B1-hordein	P06470	TLPMMCSVNVPLYR
	2	B1-hordein	P06470	GVGPSVGV
	3	B3-hordein	P06471	TLPTMCSVNVPLYR
	4	B3-hordein	P06471	IVPLAIDTR
	5	B3-hordein	P06471	SQMLQQSSCHVLQQ QCCQQLPQIPEQLR
Avena sativa (oats)	1	Avenin-3	P80356	QFLVQQCSPVAVVPFLR
	2	Avenin-3	P80356	SQILQQSSCQVMR
	3	Avenin-3	P80356	QLEQIPEQLR
	4	Avenin-3	P80356	QQCCR
Secale cereale (rye)	1	75k gamma secalin	E5KZQ3	NVLLQQCSPVALVSSLR
	2	75k gamma secalin	E5KZQ4	EGVQILLPQSHQQHVGQGAL AQVQGIIQPQQLSQLEVVR
	3	75k gamma secalin	E5KZQ5	SLVLQNLPTMCNVYVPR
	4	75k gamma secalin	E5KZQ5	QCSTIQAPFASIVTGIVGH
Triticum aestivum (wheat)	1	Glutenin, subunit DY10	P10387	QVVDQQLAGR
	2	Glutenin, subunit PW212	P08489	IFWGIPALLK
	3	Glutenin, subunit DY10	P10387	SVAVSQVAR
	4	Glutenin, subunit DY10	P10387	LPWSTGLQMR
	5	Beta-amylase	P93594	YDPTAYNTILR
	6	Alpha-amylase inhibitor 0.19	P01085	EHGAQEGQAGTGAFPR

3. Results

To test this approach, several samples of grain from single varieties of wheat, rye, barley and oats, together with commercial samples of gluten-free flour, oats cookies, gluten free cookies, wheat cookies and a self-raising flour (from a local supermarket in the UK) were collected. Each sample of grain was milled in a commercially available coffee bean grinder to make single variety flour, and all these samples were extracted and analyzed using the described method. Figure 2 shows the comparisons of the four different grain flours using a beta amylase marker peptide.

Figure 2. The comparison of separate extracts from barley, wheat, rye, oats and gluten free flour. Here the chromatograms for three MRM transitions for a specific marker peptide from beta amylase (wheat 5 in Table 3) have been shown for each species.

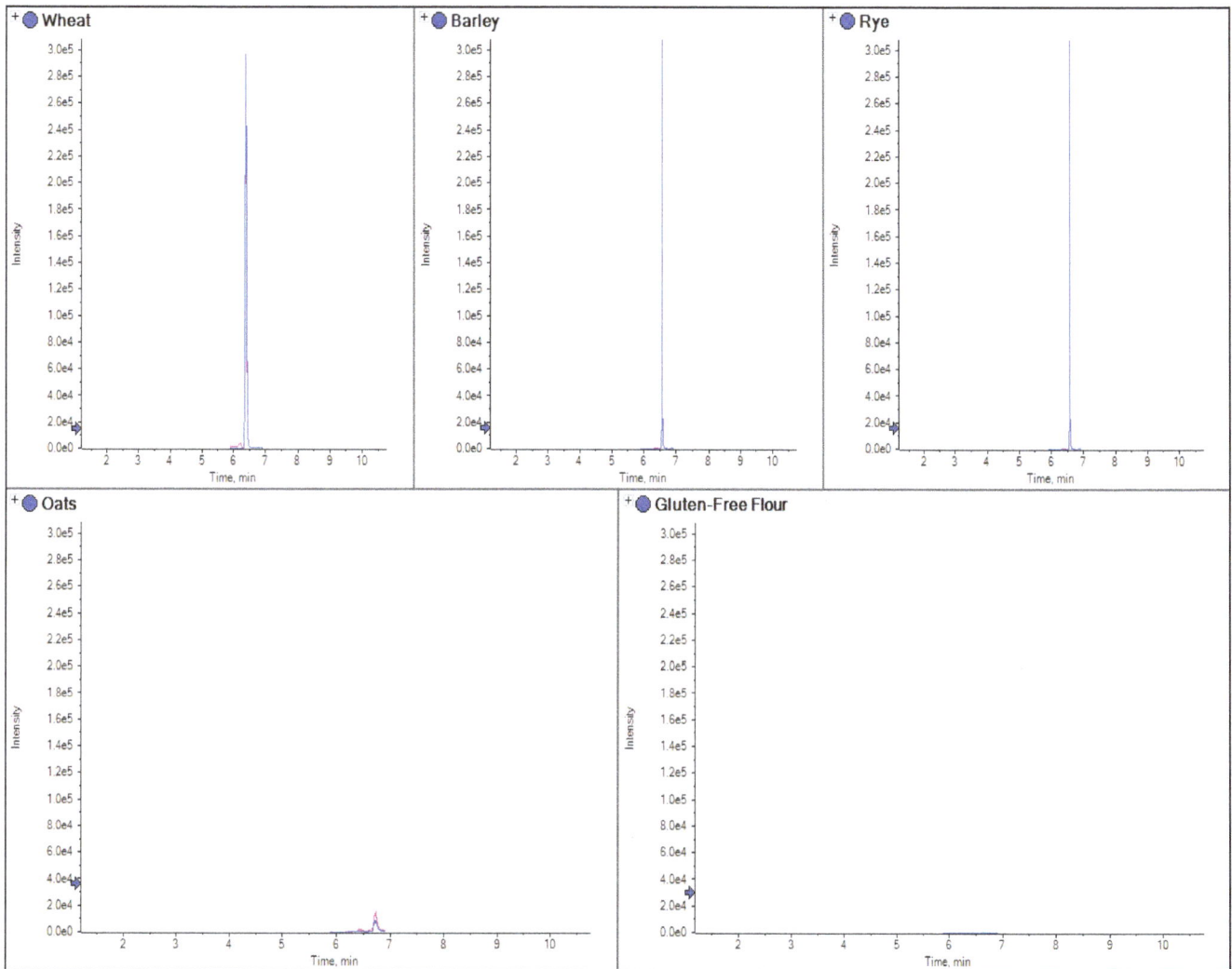

Separate marker peptides were also tested for oats, barley, wheat and rye, which were specific for each species; these are shown in Figures 3–6.

Figure 3. The comparison of separate extracts from barley, wheat, rye, oats and gluten free flour. Here the overlaid chromatograms for oats marker peptides (obtained from the theoretical digestion of avenin) have been shown for flour extract for each sample.

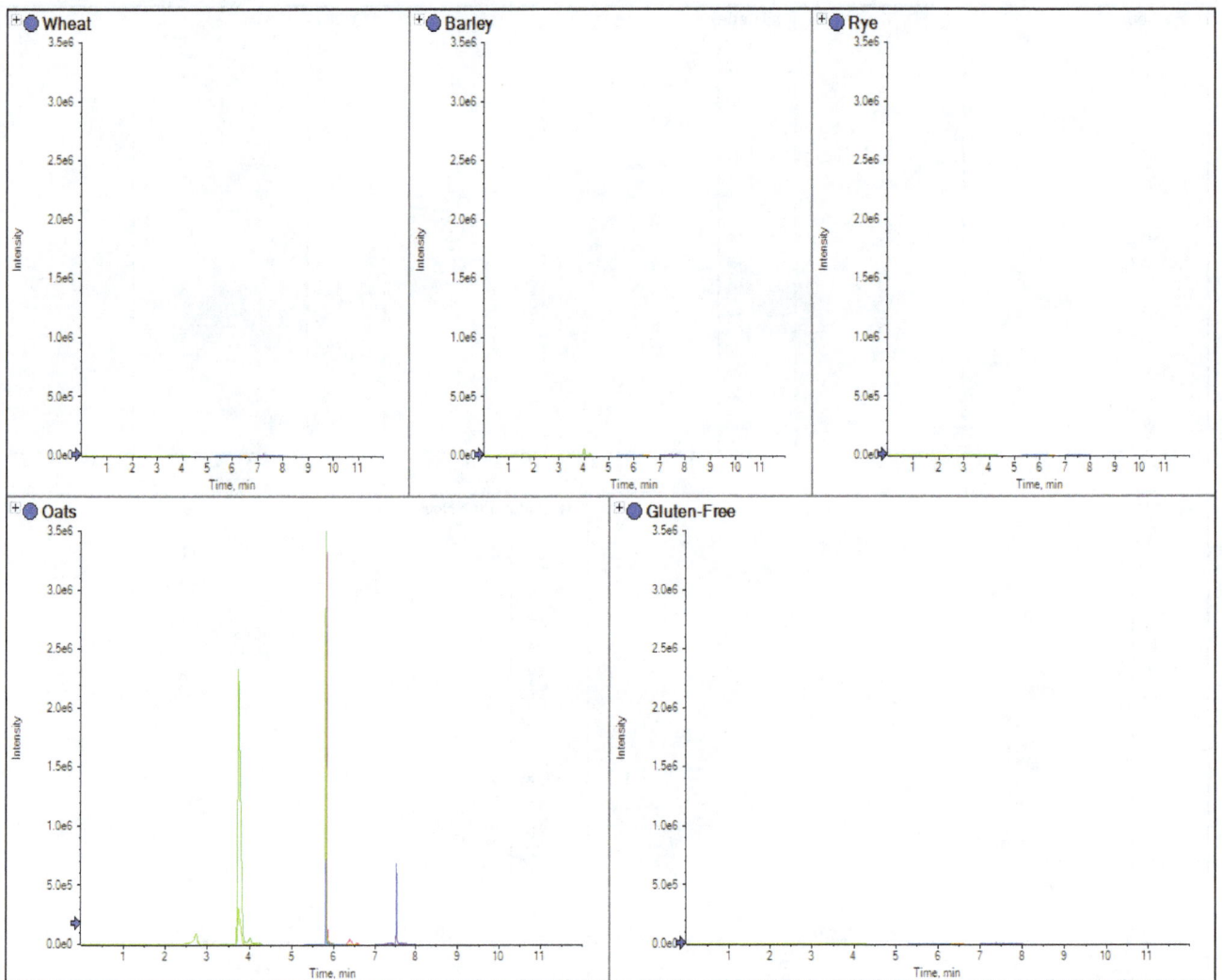

Figure 4. The comparison of separate extracts from barley, wheat, rye and oats flour. Here the overlaid chromatograms for barley marker peptides (obtained from the theoretical digestion of hordein) have been shown for flour extracts for each sample.

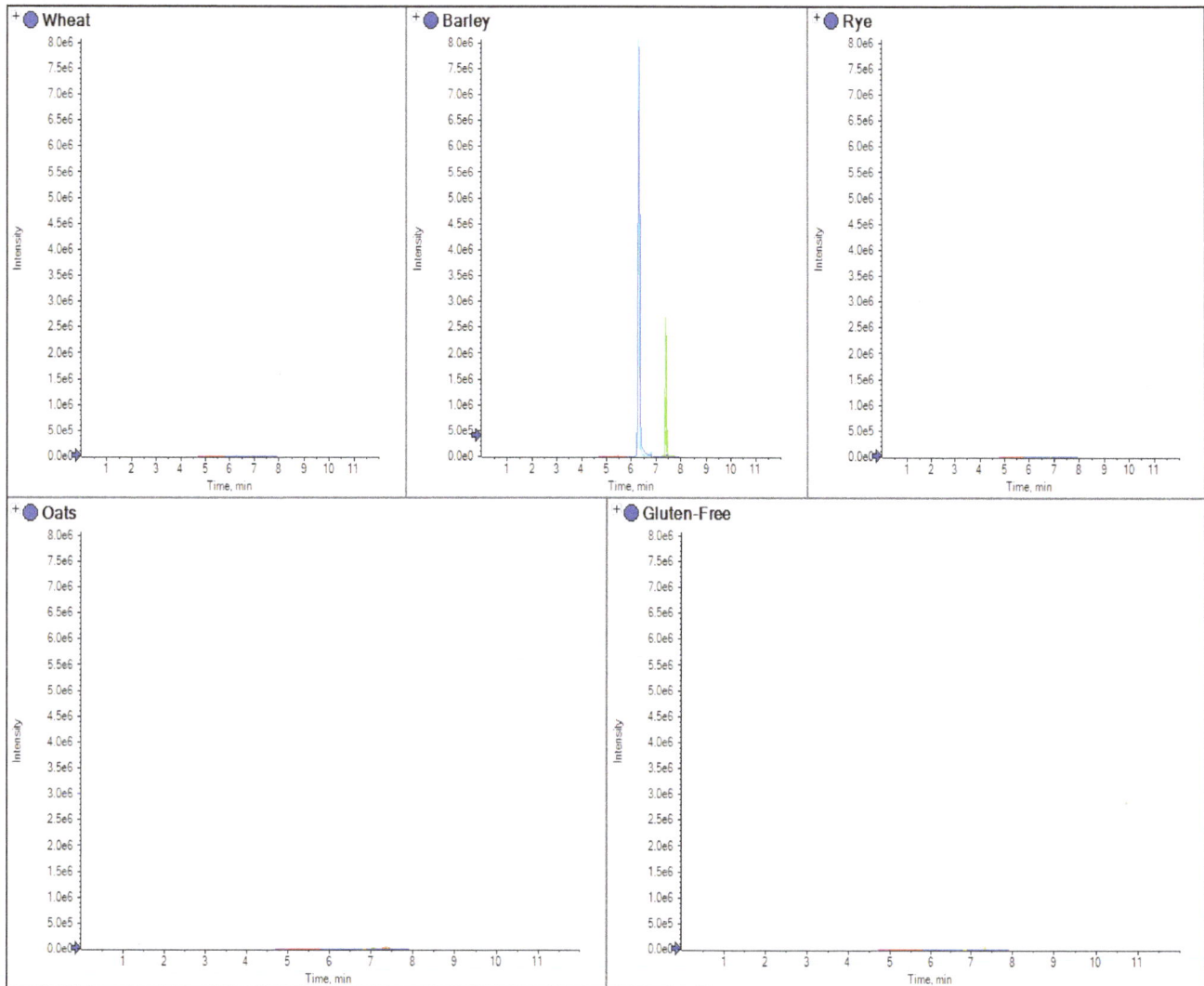

Figure 5. The comparison of separate extracts from barley, wheat, rye and oats flour. Here the overlaid chromatograms for rye marker peptides (obtained from the theoretical digestion of secalin) have been shown for flour extracts for each sample.

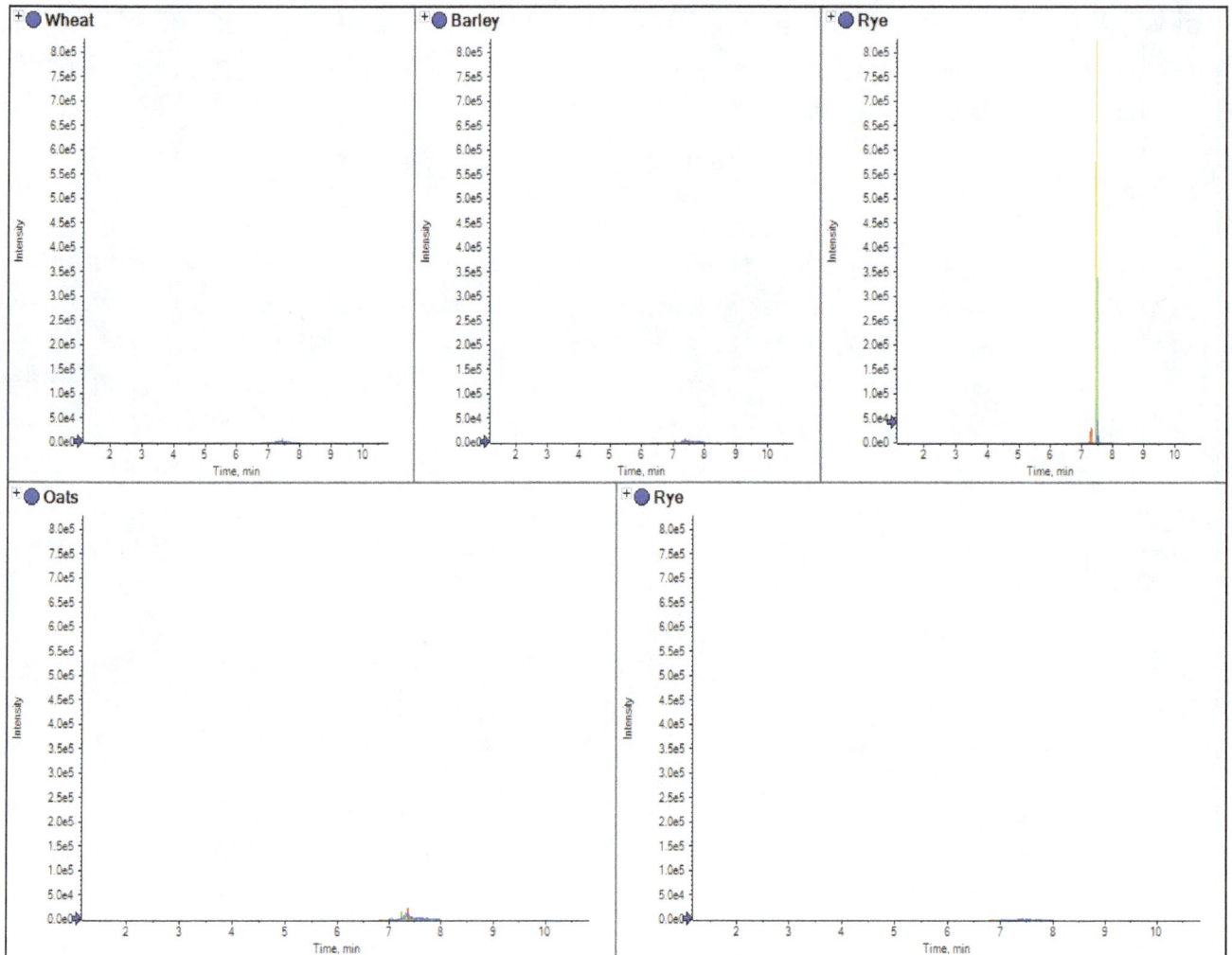

Figure 6. The comparison of separate extracts from barley, wheat, rye, oats and gluten-free flour. Here four separate peptide chromatograms for wheat marker peptides (obtained from the theoretical digestion of glutenin) have been shown for the extracts of each sample.

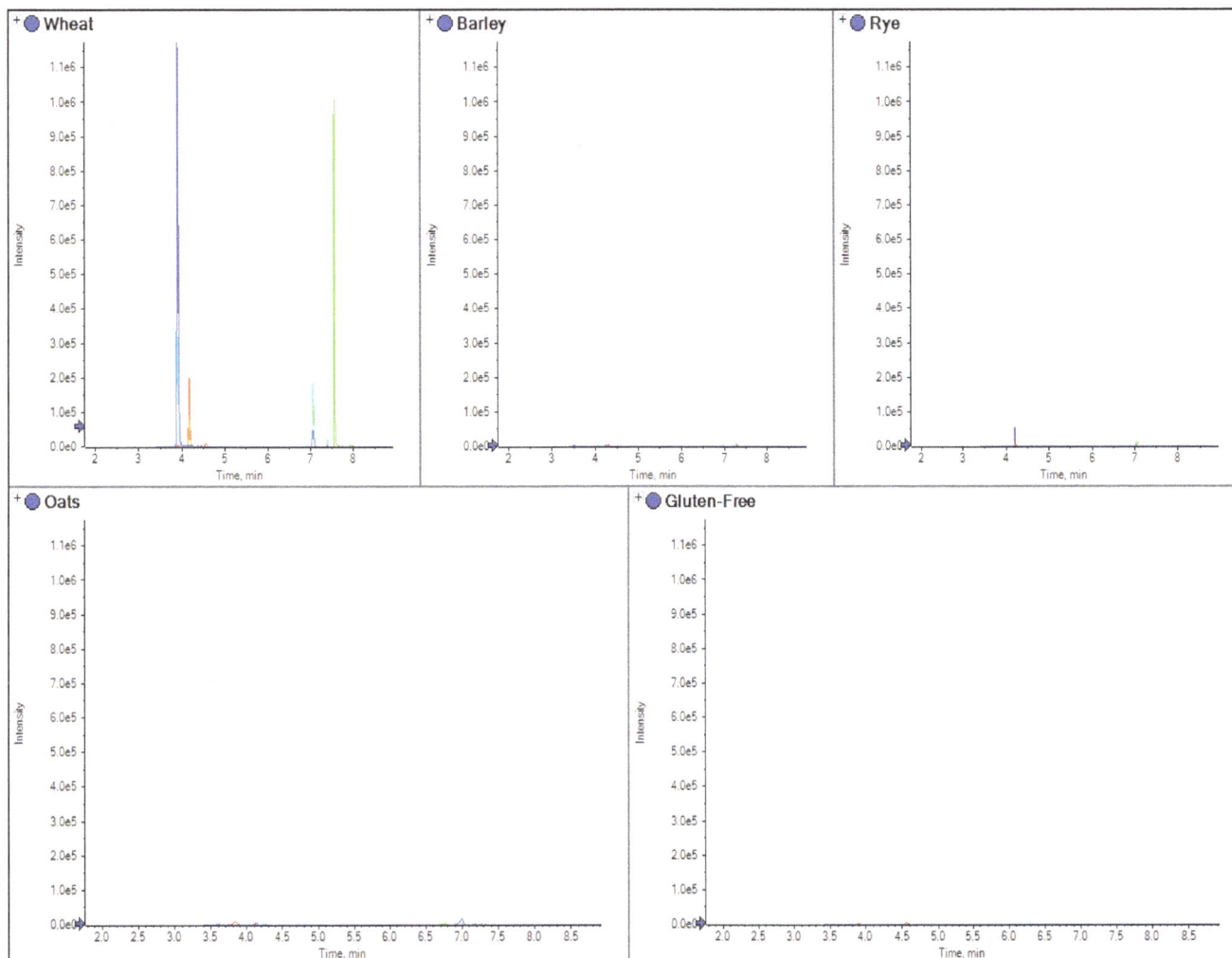

To further evaluate this approach, three samples of single varieties of wheat grain were obtained and extracted and compared with an extraction of gluten-free flour (a mix of tapioca, buckwheat, rice, maize and potato flour), as well as a sample of self-raising flour obtained from a local supermarket (Figure 7). The method was further evaluated by applying the same extraction and analysis method to a sample of gluten-free cake mix and cookies, as well as samples of oats and wheat cookies, to see if it could be applied to processed food (Figure 8). To assess linearity and sensitivity, samples of gluten-free flour were spiked at different levels with gliadin protein from wheat, which had been purchased from Sigma Aldrich (part number G3375, Figure 9).

Figure 7. The comparison of separate extracts of several samples of wheat obtained from single variety grain samples, as well as a sample of gluten-free flour and self-raising flour obtained from a local supermarket using the wheat peptides in Table 3.

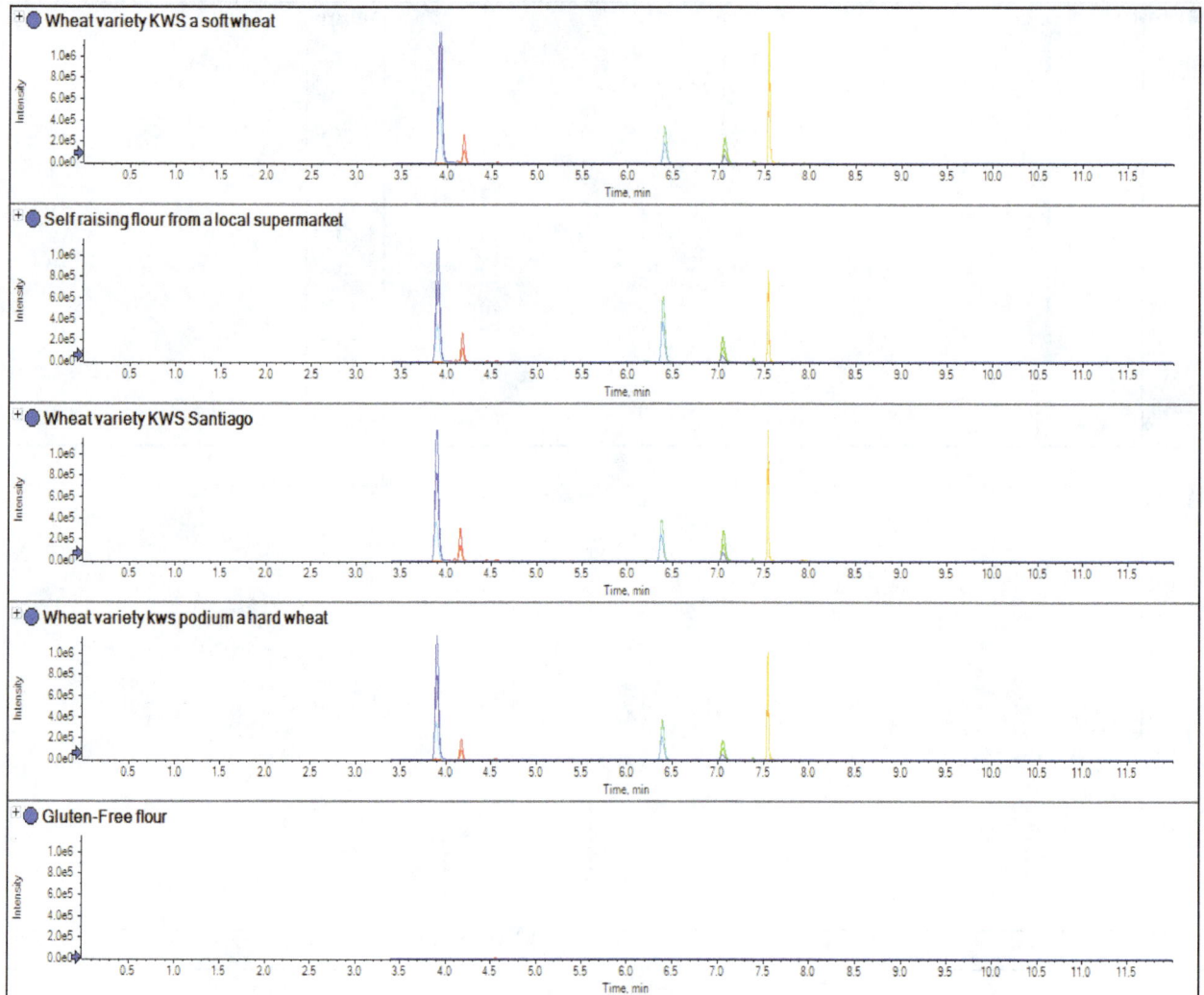

Figure 8. The comparison of extracts from several samples of food collected from a local supermarket and analyzed for gluten markers for oats and wheat.

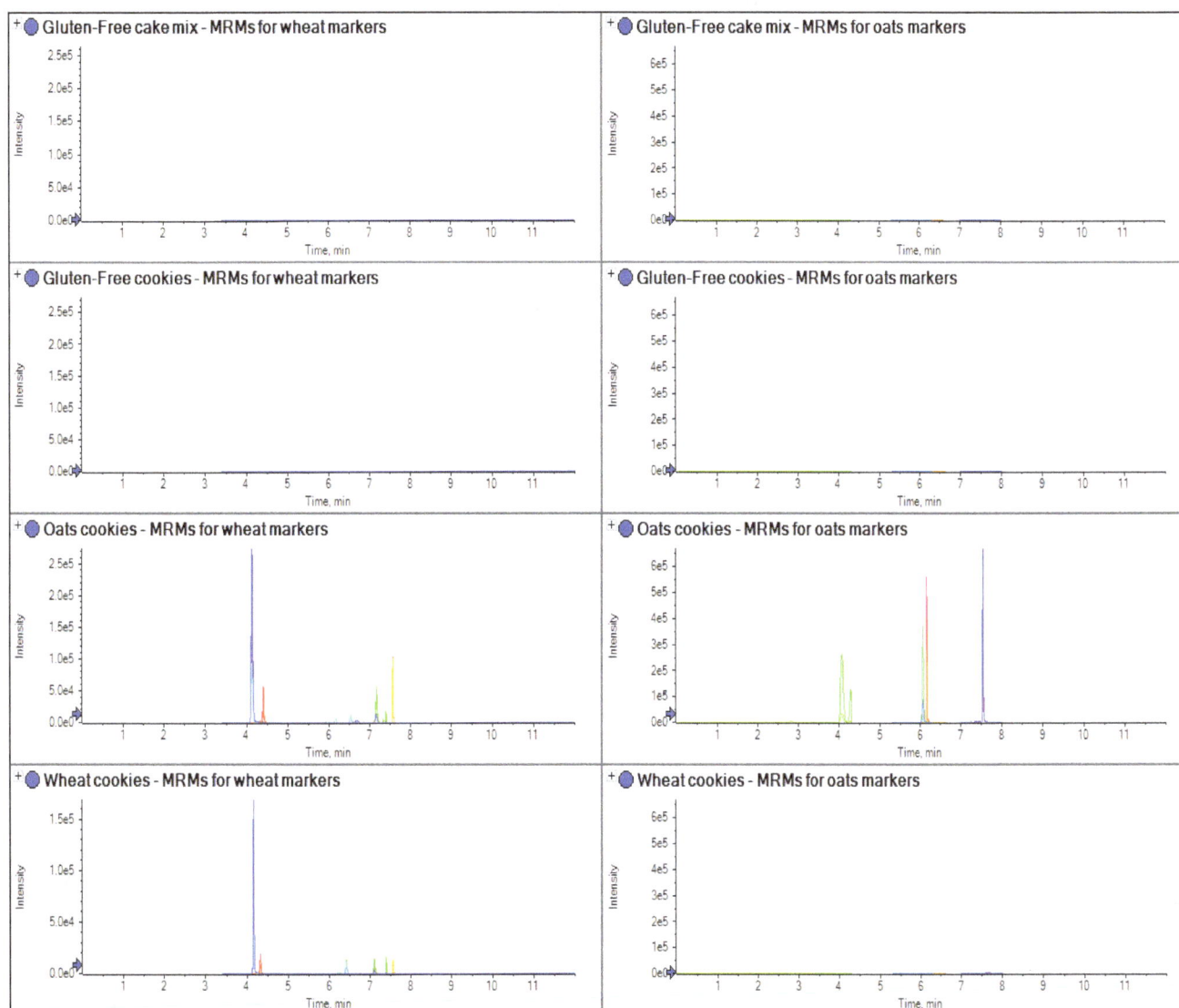

Figure 9. The calibration line obtained from the spiking of gliadin into gluten-free wheat from the range of 5–200 ppm for wheat peptide 3. Inlayed in the calibration line is the chromatogram for the 10 ppm spike of gliadin into gluten-free flour. The calibration line was a linear $1/x$ fit with $r = 0.9944$.

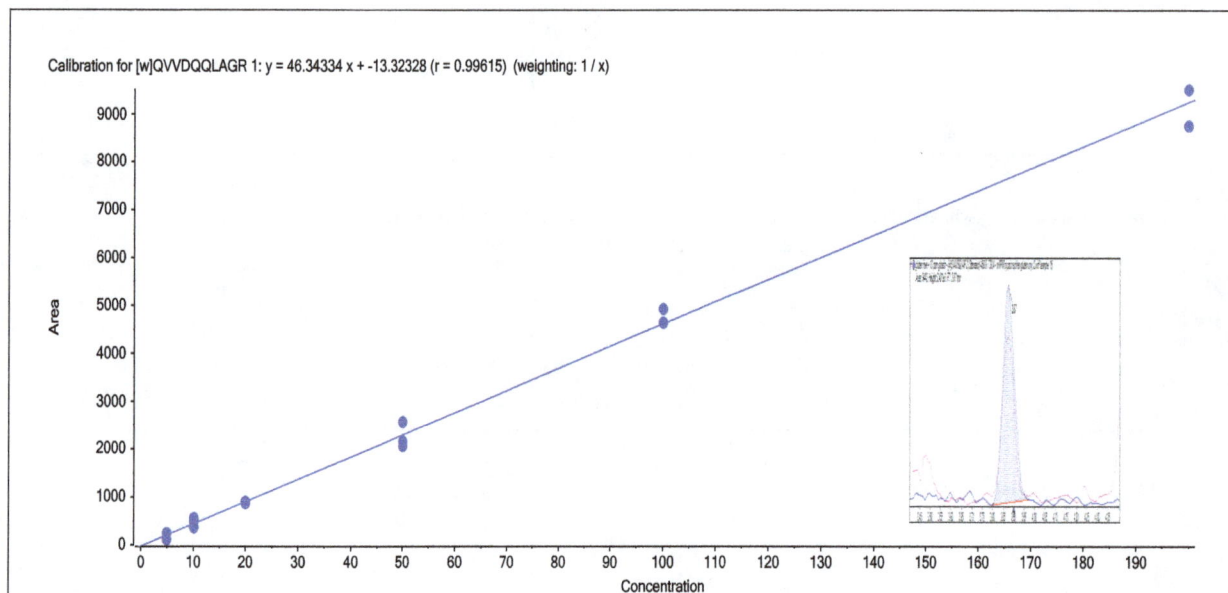

Calibration for [w]QVVDQQLAGR 1: y = 46.34334 x + -13.32328 (r = 0.99615) (weighting: 1 / x)

4. Discussion

Similar to current R5 antibody based ELISA methods, LC-MS/MS marker peptides can be found which are present in all gluten varieties, with the exception of oats as shown in Figure 2, where the use of a tryptic peptide from the protein beta amylase was present in wheat, barley and rye, but absent in oats. This marker gave a consistent MRM ratio for the three peptide MRM transitions across all three species and is a good marker to replicate the results of R5 antibody ELISA based methods which are positive to only wheat, rye and barley. However, LC-MS/MS differs to ELISA in that specific markers from the individual species of oats, rye, barley and wheat can also be developed. Figures 3–6 show how individual markers for each of these species can be used to distinguish different species by LC-MS/MS and specifically confirm whether oats had been used as a replacement to wheat. LC-MS/MS offers an advantage over ELISA based methods in that you can use multiple peptide markers with multiple MRMs for each peptide to confirm the presence of the gluten species in the sample. One question that had been asked was that if LC-MS/MS was so specific, would it be affected by the variety of the species in the sample? To test this hypothesis, several single varieties of wheat grain were obtained from a grain supplier. These were then ground and the resulting flour extracted to test the effect of variety of wheat on the peptides detected. In Figure 7 the comparison of four peptide markers for wheat, across different samples of flour, were compared. What is immediately apparent is that all the peptides are seen in all the samples with the exception of the gluten-free flour. These wheat varieties included commercial self-raising flour obtained from a supermarket, as well as hard and soft wheat varieties. From the peptide responses, you can also see that the relative responses are the same for three out of the four, with only one marker significantly higher in variety 3, but this may have been as a result of the change in matrix interference for this particular sample. This clearly indicates that the majority of LC-MS/MS markers which have been found are independent of variety used to produce the wheat flour and are just species specific.

One of the important tests for the feasibility of the use of LC-MS/MS was its ability to detect gluten in processed food. In processed food, ELISA kits have been shown to fail to pick up allergens due to processing changes in the protein structure which then prevent the antibody binding and this leads to false negatives [15]. Also, due to the fact that ELISA methodology just relies on one protein region, unspecific binding has also been shown which has led to false positives in some instances [16]. In this work, the LC-MS/MS method was applied to some cookies as well as gluten-free flour to determine its ability to detect the markers in processed food. In Figure 8 it can clearly be seen that LC-MS/MS can detect the markers in the processed food and distinguish between varieties. In the case of the oats cookie, wheat and oats had been used in its manufacture and markers for both varieties were detected (barley and rye were not present in any sample, although not shown in this figure). However, in the wheat cookie, only wheat was used and this is the only species detected. In the gluten-free products, no gluten markers were detected.

A final test was linearity of response and the sensitivity of the method. To test both of these, gluten-free flour was spiked with gliadin (wheat protein obtained commercially) from a range of 5–200 ppm. A calibration line, shown in Figure 9, for one of the peptides clearly shows that the LC-MS/MS response obtained for this wheat marker was linear—this was typical of the other markers used for wheat—and for this marker a 10 ppm spike could be easily detected. Marker peptides therefore could be detected at 5–10 ppm levels in the spiked sample of gluten-free flour even though the current sample preparation used an 80-fold dilution of the original sample.

5. Conclusions

This work has demonstrated that LC-MS/MS can be used to detect gluten in processed food and food ingredients. The work demonstrated that markers can be obtained which are specific for each individual species of gluten. The presence of these multiple markers for individual species were not variety dependent, as shown in a test of several single varieties of wheat flour (where all the same markers were detected), but some were species-dependent. As well as species dependent markers, markers for proteins that are present in rye, wheat and barley, but absent in oats, can also be added to the method to mimic the behavior of the R5 antibody based ELISA method to generally pick up the species that are high in gluten and affect people who suffer from Celiac disease. The method has been shown to detect levels of 5–10 ppm gluten proteins in gluten-free flour and offers an extended linear response which is envisaged to be a lot larger than that normally obtained for ELISA assays. Further to this, as the current method actually involves an 80-fold dilution of the sample, before injecting onto the LC-MSMS system, it offers the potential of detecting low ppm (0.5–5 ppm) when an SPE protocol is used to collect concentrate and purify the peptide markers.

The presence of multiple markers for each gluten variety and the potential of acquiring MRM triggered product ion scans [12], offers multiple points for confirmation of gluten contamination and provide confidence in the results, and reduces the risk of false positives and false negatives which can occur in ELISA assays.

Conflicts of Interest

The authors declare no conflict of interest.

References

1. Heick, J.; Fischer, M.; Pöpping, B. First screening method for the simultaneous detection of seven allergens by liquid chromatography mass spectrometry. *J. Chromatogr. A* **2011**, *1218*, 938–943.

2. Lock, S. Allergen Screening in Food by LC/MS/MS. In Proceedings of the 126th Annual Meeting AOAC International, Las Vegas, NV, USA, 30 September–4 October 2012.

3. Shefcheck, K.J.; Musser, S.M. Confirmation of the allergenic peanut protein, Ara h1, in a model food matrix using liquid chromatography/tandem mass spectrometry. *J. Agric. Food Chem.* **2004**, *52*, 2785–2790.

4. Hernando, A.; Mujico, J.R.; Mena, M.C.; Lombardía, M.; Méndez, E. Measurement of wheat gluten and barley hordeins in contaminated oats from Europe, the United States and Canada by Sandwich R5 ELISA. *Eur. J. Gastroenterol. Hepatol.* **2008**, *20*, 545–554.

5. Lock, S.; Lane, C.; Jackson, P.A.; Serna, A. The Detection of Allergens in Bread and Pasta by Liquid Chromatography Tandem Mass Spectrometry. Available online: http://www.absciex.com/Documents/Downloads/Literature/Allergens-4000-QTRAP_AB%20SCIEX_1830610–01.pdf (accessed on 16 September 2013).

6. Food Standards Agency. Labelling of 'Gluten-Free' Foods. Available online: http://www.food.gov.uk/business-industry/guidancenotes/allergy-guide/gluten/#.UiRckH9m7ps (accessed on 1 January 2012).

7. Commission regulation (EC) No 41/2009 of 20 January 2009 Concerning the Composition and Labeling of Foodstuffs Suitable for People Intolerant to Gluten (Text with EEA Relevance). Available online: http://eur-lex.europa.eu/LexUriServ/LexUriServ.do?uri=CELEX:32009R0041:en:NOT (accessed on 12 September 2013)

8. U.S. Food and Drug Administration. What Is Gluten-Free? FDA Has an Answer. Available online: http://www.fda.gov/ForConsumers/ConsumerUpdates/ucm363069.htm (accessed on 2 August 2013).

9. Lock, S. Gluten Detection in Food by LC/MS/MS. In Proceedings of the 126th Annual Meeting AOAC International, Las Vegas, NV, USA, 30 September–4 October 2012.

10. Rombouts, I.; Lagrain, B.; Brunnbauer, M.; Delcour, J.A.; Koehler, P. Improved Identification of Wheat Gluten Proteins through Alkylation of Cysteine Residues and Peptide-Based Mass Spectrometry. Available online: http://www.nature.com/srep/2013/130724/srep02279/full/srep02279.html (accessed on 24 July 2013).

11. Lock, S. Allergen Detection in Wine by LC/MS/MS. In Proceedings of the 127th Annual Meeting AOAC International, Chicago, IL, USA, 25–28 August 2013.

12. Champion, M.; Duchoslav, E.; Hunter, C. *Targeted, Hypothesis-Driven Mass Spectrometry: MRM Initiated Detection and Sequencing using the MIDAS™ Workflow for Faster, More Intelligent and Sensitive Protein Discovery and Characterization*; AB SCIEX: Warrington, UK, 2007.

13. Mriziq, K.; Hobbs, S.; Settineri, T.; Neyer, D. Higher Sensitivity and Improved Resolution Microflow UHPLC with Small Diameter Turbo V™ Source Electrodes and Hardware for Use with the Eksigent expressHT™ Ultra System. Available online: http://www.eksigent.com/ Documents/Downloads/Literature/Higher_Sens_Res_LC_Small_Diam_Electrodes_4590211–01.pdf (accessed on 16 September 2013).

14. UniProt. Available online: http://www.uniprot.org/ (accessed on 16 September 2013).

15. Pöpping, B. No Hiding for Allergens Appropriate Extraction Methodologies. In Proceedings of the 127th Annual Meeting AOAC International, Chicago, IL, USA, 25–28 August 2013.

16. Stelk, T.; Niemann, L.; Lambrecht, D.M.; Baumert, J.L.; Taylor, S.L. An innovative sandwich ELISA system based on an antibody cocktail for gluten analysis. *J. Food Sci.* **2013**, *78*, 1091–1093.

Development of Next Generation Stevia Sweetener: Rebaudioside M

Indra Prakash [1,*], Avetik Markosyan [2] and Cynthia Bunders [1]

[1] The Coca-Cola Company, Atlanta, GA 30313, USA; E-Mail: cbunders@coca-cola.com

[2] PureCircle Limited, Lengkuk Teknologi, 71760 Bandar Enstek, Negeri Sembilan, Malaysia;
    E-Mail: avetik@purecircle.com

* Author to whom correspondence should be addressed; E-Mail: iprakash@coca-cola.com

Abstract: This work aims to review and showcase the unique properties of rebaudioside M as a natural non-caloric potential sweetener in food and beverage products. To determine the potential of rebaudioside M, isolated from *Stevia rebaudiana* Bertoni, as a high potency sweetener, we examined it with the Beidler Model. This model estimated that rebaudioside M is 200–350 times more potent than sucrose. Numerous sensory evaluations of rebaudioside M's taste attributes illustrated that this steviol glycoside possesses a clean, sweet taste with a slightly bitter or licorice aftertaste. The major reaction pathways in aqueous solutions (pH 2–8) for rebaudioside M are similar to rebaudioside A. Herein we demonstrate that rebaudioside M could be of great interest to the global food industry because it is well-suited for blending and is functional in a wide variety of food and beverage products.

Keywords: specifications; rebaudioside M; rebaudioside A; purification; stability; food application; sweetener blends

1. Introduction

High-purity rebaudioside M (also known as rebaudioside X), is a natural non-calorie sweetener being commercialized jointly by PureCircle Limited and The Coca-Cola Company for food and beverage use. Rebaudioside M is one of the minor sweet components of *Stevia rebaudiana* Bertoni, a South American plant. It is a glycoside of the *ent*-kaurene diterpenoid aglycone known as steviol, and is found in nature accompanied by at least ten other sweet-tasting steviol glycosides [1–3]. Stevia sweeteners have been approved for use as a sweetener in a number of countries, including US, EU, Japan, China, Brazil and other countries [4].

Discovery of Rebaudioside M

The leaves of *Stevia rebaudiana* Bertoni have been used by the natives of Paraguay to sweeten beverages for centuries [5]. The plant is the source of a number of sweet *ent*-kaurene diterpenoid glycosides (Figure 1, Table 1), but the major sweet constituents are rebaudioside A (**1**) and stevioside (**8**).

Figure 1. Backbone figure of *Stevia* sweeteners.

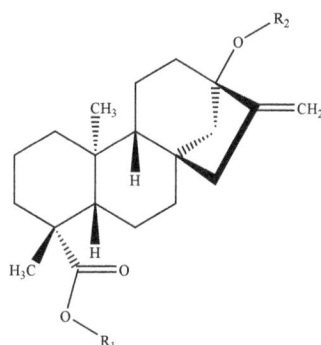

Table 1. R-groups, molecular formulas, molecular weights and potencies of the *Stevia* sweeteners.

Sweetener	Reference Number in Text	R-Groups in Backbone Figure Above		Formula	Molecular Weight (g/mol)	Potency *
		R_1	R_2			
Rebaudioside A	**1**	β-glc-	(β-glc)$_2$-β-glc-	$C_{44}H_{70}O_{23}$	967.01	200
Rebaudioside B	**2**	H	(β-glc)$_2$-β-glc-	$C_{38}H_{60}O_{18}$	804.88	150
Rebaudioside C	**3**	β-glc-	(β-glc, α-rha-)-β-glc-	$C_{44}H_{70}O_{22}$	951.01	30
Rebaudioside D	**4**	β-glc-β-glc-	(β-glc)$_2$-β-glc-	$C_{50}H_{80}O_{28}$	1129.15	221
Rebaudioside E	**5**	β-glc-β-glc-	β-glc-β-glc-	$C_{44}H_{70}O_{23}$	967.01	174
Rebaudioside F	**6**	β-glc-	(β-glc, β-xyl)-β-glc-	$C_{43}H_{68}O_{22}$	936.99	200
Rebaudioside M	**7**	(β-glc)$_2$-β-glc-	(β-glc)$_2$-β-glc-	$C_{56}H_{90}O_{33}$	1291.3	250
Stevioside	**8**	β-glc-	β-glc-β-glc-	$C_{38}H_{60}O_{18}$	804.88	210
Steviolbioside	**9**	H	β-glc-β-glc-	$C_{32}H_{50}O_{13}$	642.73	90
Rubusoside	**10**	β-glc-	β-glc-	$C_{32}H_{50}O_{13}$	642.73	114
Dulcoside A	**11**	β-glc-	α-rha-β-glc-	$C_{38}H_{60}O_{17}$	788.87	30

glc = glucose; rha = rhamnose; xyl = xylose; * Potency from [1,6,7].

In our continuing research to discover a natural non-caloric sweetener, we have focused on minor steviol glycosides and found a novel minor steviol glycoside, rebaudioside M (**7**) that is more potent, has higher sweetness intensity, and very slight licorice or bitter aftertaste than other steviol glycosides.

Work to elucidate the chemical structures of *S. rebaudiana* sweeteners began in the early twentieth century, but proceeded slowly. The structures **1** and **8** (Figure 2) were not fully determined until 1970 [8–10]. During the 1970s, additional sweet components, including rebaudiosides A–E, were isolated from *S. rebaudiana* leaves and characterized by Osamu Tanaka and co-workers at Hiroshima University in Japan [11]. Several novel steviol glycosides have been reported from the commercial extracts of the leaves of *S. rebaudiana* in the last few years [6,12–19]. Recently we reported the structure elucidation and isolation of rebaudioside M from *S. rebaudiana* Bertoni (**7**) [20,21]. In addition, rebaudioside M received a Letter of No Objection concerning its Generally Recognized as Safe (GRAS) status from US FDA [22].

Figure 2. Comparison of the chemical structures of rebaudioside A (**1**), stevioside (**8**) and rebaudioside M (**7**).

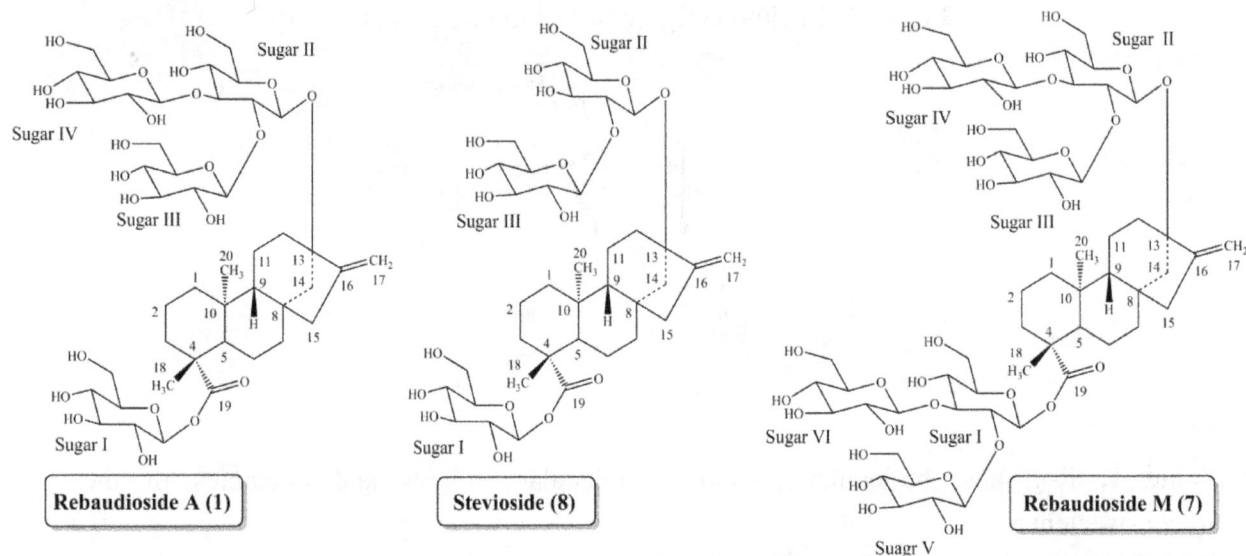

2. Experimental Section

2.1. Purification of Rebaudioside M

Rebaudioside M (**7**) was obtained from *Stevia rebaudiana* Bertoni cultivar AKH L1 leaves [23]. Two kilograms of *S. rebaudiana* leaves were extracted by 40 L water at 40 °C for two hours. The filtrate was separated and treated with a flocculant (calcium oxide) to remove the mechanical particles, proteins, polysaccharides and coloring agents. The resulting precipitate was separated by filtration and the filtrate was deionized by Amberlite FCP22 (H$^+$) and Amberlite FPA53 (OH$^-$) ion-exchange resins. The deionized filtrate was fractionated by seven columns, packed with Diaion HP20 (Mitsubishi Chemical, Japan). The columns with higher rebaudioside M content were eluted using aqueous ethanol. The obtained glycoside eluate was treated with activated carbon, deionized, evaporated and dried *in vacuo*. The obtained material was recrystallized twice from aqueous methanol to afford about 1.1 g of rebaudioside M with >98% purity by HPLC [21]. The highly purified stevia extract meets the international Joint Expert Committee on Food Additives (JECFA) purity requirement for steviol glycosides, which requires ≥95% total steviol glycosides [7].

2.2. Properties of Rebaudioside M

Pure rebaudioside M (**7**) (>95%) crystallizes out as a crystalline form which is slightly soluble in water (0.1 g/100 mL at 25 °C in 5 min) and in ethanol. Its amorphous material has a solubility of 1.1%–1.3% in water at 25 °C. The thermodynamic equilibrium solubility in water is 0.26% at 25 °C [24].

2.3. Stability of Rebaudioside M

As a dry powder, rebaudioside M (**7**) is stable for at least one year at ambient temperature and under controlled humidity conditions. In solution, it is most stable in pH 4–8 and noticeably less stable below pH 2. As expected, the stability decreases with increasing temperature. Its stability is very similar to rebaudioside A [25].

In aqueous solutions (pH 2–8), the major reaction pathways leading to loss of rebaudioside M are as follows (Figure 3): isomerization of the C-16 olefin to form the C-15 isomer (**12**), hydration of the C-16 olefin to yield compound **13**, hydrolysis of the glycosidyl ester at C-19 to form compound **2**, and isomerization of the C-16 olefin to form the C-15 isomer in **2** to form compound **16**. All of these compounds (**2**, **12**, **13** and **16**) are sweet. Compound **12** is sweeter than compounds **2**, **13** and **16**, and has similar taste properties as that of compound **7**. Upon metabolism *in vitro*, rebaudioside M (**7**), just like rebaudioside A (**1**), is primarily converted to its aglycone steviol (**14**) and to the glucuronic acid of steviol, generally known as steviol glucuronide (**15**) [26].

Rebaudioside M (**7**) has similar stability as that of rebaudioside A (**1**) in both low and high pH applications. In heat-processed beverages, such as flavored ice-tea, juices, sport drinks, flavored milk, drinking yogurt and non-acidified teas, the sweetener shows good stability during High Temperature-Short Time heat processing and on subsequent product storage [27].

2.4. Sensory Panel for Rebaudioside M and Other Sweeteners

To determine the sensory attributes of rebaudioside M (**7**) we compared it to rebaudioside A (**1**), which possesses well defined characteristics. In addition, blending techniques with rebaudioside M and other sweetener were examined to demonstration the potential of this sweetener. Herein we describe our sensory panel testing. Concentration-response (C/R) function work was done with 60 in-house trained panelists (Two-Alternative Forced Choice, 2-AFC method), with the panelists specifying which of the two samples was perceptually sweeter. Sweetener blend work was done with 8–10 descriptive analysis (DA) panelists that have been highly trained. DA panelists go through 10–12 weeks of extensive descriptive training. The panel was calibrated at the beginning of each test session. In addition the panel also received anchor references representing early (sucrose), middle (aspartame) and late (thaumatin) AT every three test samples. Samples were given to the panel members sequentially and coded with triple digit numbers. The order of sample presentation was randomized to avoid the order of presentation bias. Water and unsalted crackers were provided in order to cleanse the palate. The panel members were asked to rate different attributes including sweetness onset, total sweetness, rounded sweetness, bitterness, acidity, leafy note, licorice, astringency, mouthfeel, mouth coating, sweet lingering, and bitter lingering. Samples were rated on a scale of zero (0) to ten (10), with zero indicating immediate onset, no intensity, watery/low viscosity, or very sharp peak, and ten indicating very delayed onset, high intensity, thick/high viscosity, or very round peak. One-way single factor ANOVA was used to analyze sensory results, where $\alpha = 0.05$.

Figure 3. Major pathways of degradation of rebaudioside M (**7**) under hydrolytic conditions.

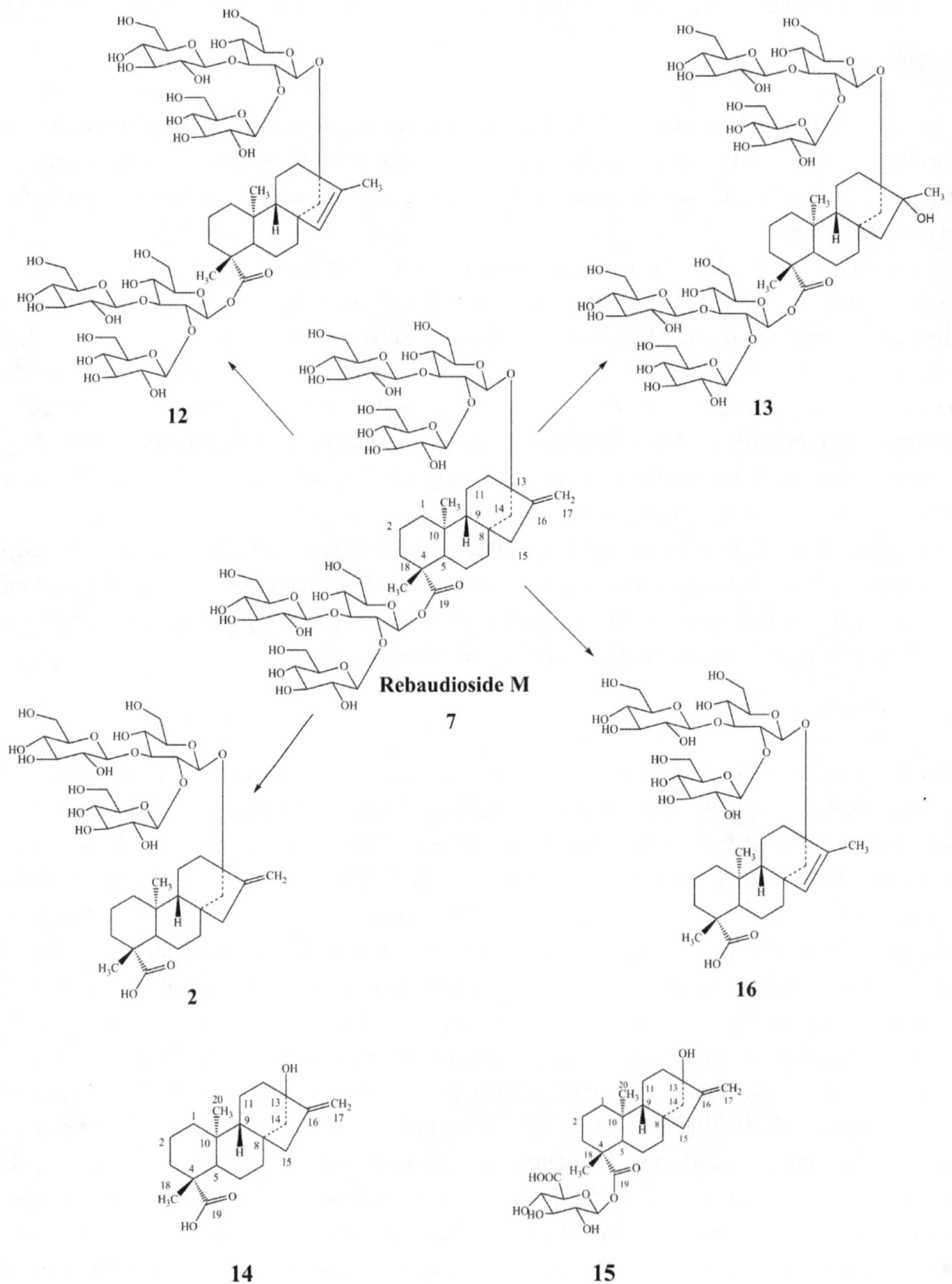

3. Results and Discussion

3.1. Rebaudioside M Sensory Attributes

To decipher the potential of rebaudioside M as a high potency sweetener, we first examined it against sucrose with the Beidler Model (Law of Mass Action) [28]. We used several reference samples of 2.5%, 5.0%, 7.5% and 10.0% sucrose solutions. Since sweetness potency is strongly dependent on sucrose equivalency level for all high potency sweeteners (HPS), it is important to state the sucrose equivalency (SE) level at which sweetness potency has been determined. Sweetness potency is also system dependent and therefore, it is important to define the medium (e.g., water, phosphoric acid at pH 2.5, *etc.*). The concentration-response (*C/R*) function in water determined for rebaudioside M is $R = 14.2 \times C/(265 + C)$ at 4 °C. Given the high R_m of 14.2, rebaudioside M can be used both in single and blended sweetener applications. From the *C/R* function the potency of water (P_w) is $P_w(5) = 347$ and $P_w(10) = 159$ (Figure 4). This model estimates that the potency of rebaudioside M is 200–350 times that of sucrose.

Figure 4. Concentration-response curves of rebaudioside M (**7**) in water at 4 °C.

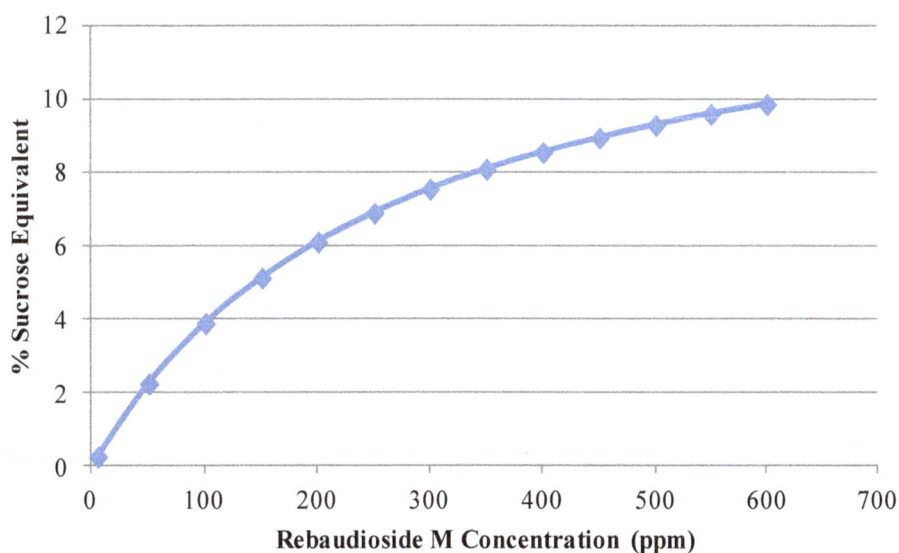

To enlist a broader picture of rebaudioside M (**7**) sensory attributes, we compared it to rebaudioside A (**1**). To compare the sensory attributes between rebaudioside M (**7**) and rebaudioside A (**1**), iso-sweet samples (8% sucrose equivalent sweetness) were made with filtered water as shown in Table 2. An 8% sugar solution in water was used as a control. The same concentrations of rebaudioside A and rebaudioside M were also used in acidified solutions (250 ppm citric acid, pH 3.2) and an 8% sugar solution acidified with citric acid was used as a standard. As described in Section 2.4, sensory panelist examined the two sweeteners and a spider plot was extracted (Figure 5).

In water solution, rebaudioside M (**7**) showed (Figure 5A) the reduced perception of bitterness, astringency, bitter lingering compared to rebaudioside A (**1**), and similar sweetness intensity. A much higher sweetness perception of rebaudioside M than rebaudioside A in acidified water was established in Figure 5B. Rebaudioside M displayed faster sweetness onset, reduced non-sweet taste (bitterness, sour, astringency) and bitterness lingering.

Table 2. Iso-sweet solutions of rebaudioside A (**1**) and rebaudioside M (**7**) in water.

Solutions	(%)	(%)
Water	99.95	99.95
Rebaudioside A 97 (dry basis)	0.0510 g	
Rebaudioside M (dry basis)		0.0423 g

Figure 5. Comparison of sensory attributes of rebaudioside M (Reb M) (**7**) and rebaudioside A (Reb A) (**1**) in water (**A**). Comparison of sensory attributes of rebaudioside M and rebaudioside A in acidified water (**B**).

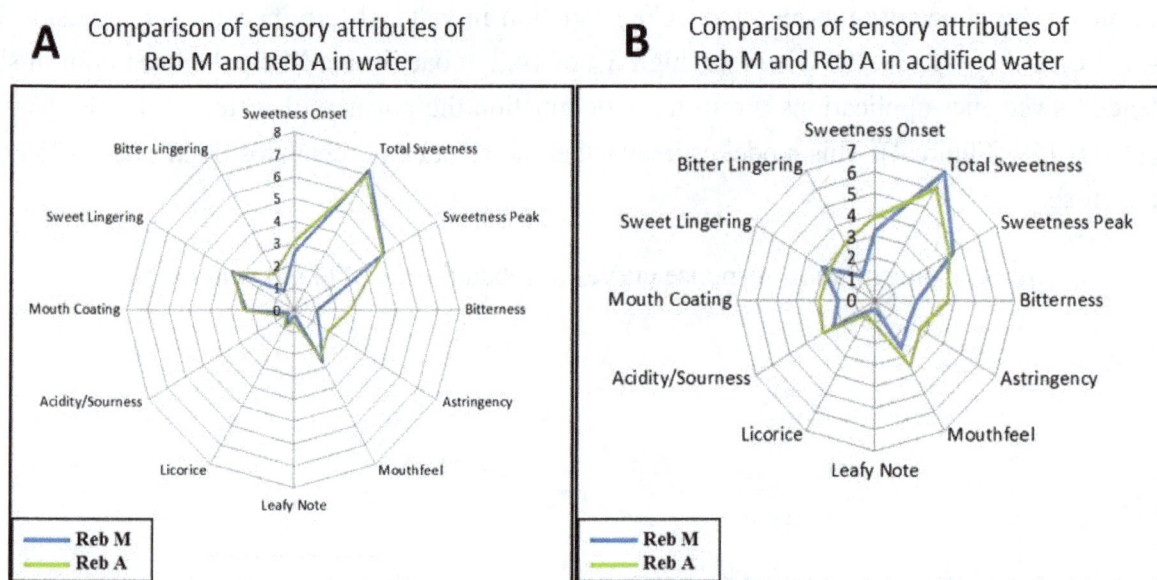

To expand the descriptive taste profile of rebaudioside M (**7**) a set of trained panelist (Section 2.4) examined 10% sucrose, aspartame at 531 mg/L, and rebaudioside M at 563 mg/L in water at 4 °C. Unlike other steviol glycosides including rebaudioside A (**1**), trained descriptive panel did not detect any significant bitter or licorice off taste when evaluated rebaudioside M in water at approximately 10% SE levels (Figure 6). Rebaudioside M had a slightly more intense sweet aftertaste then aspartame. Based on our study rebaudioside M and aspartame have similar high potency sweetener profiles.

Sweetness temporal profiles demonstrate changes in perception of sweetness over time. This property is a key to a sweetener's utility in foods and beverages, and is complementary to its flavor profile. Every sweetener exhibits a characteristic Appearance Time (AT) and Extinction Time (ET). Most high-potency sweeteners, in contrast to carbohydrate sweeteners, display prolonged ET. This can be beneficial in some products such as chewing gum, where prolonged sweetness is desirable.

With these samples in hand panelist examined the AT and ET of the three sweeteners described in Figure 6. The sweetness temporal profiles of aspartame at 531 mg/L, rebaudioside M at 563 mg/L, and sucrose at 10% in water at room temperature were compared over a period of 3 min (samples were swallowed at 5 s). The AT maximum was the shortest for sucrose, slightly longer for aspartame and longest for rebaudioside M. The ET was longest for rebaudioside M, followed by aspartame and then sucrose.

Figure 6. Descriptive taste profile of rebaudioside M (**7**) at 563 mg/L, aspartame at 531 mg/L and sucrose at 10% in water.

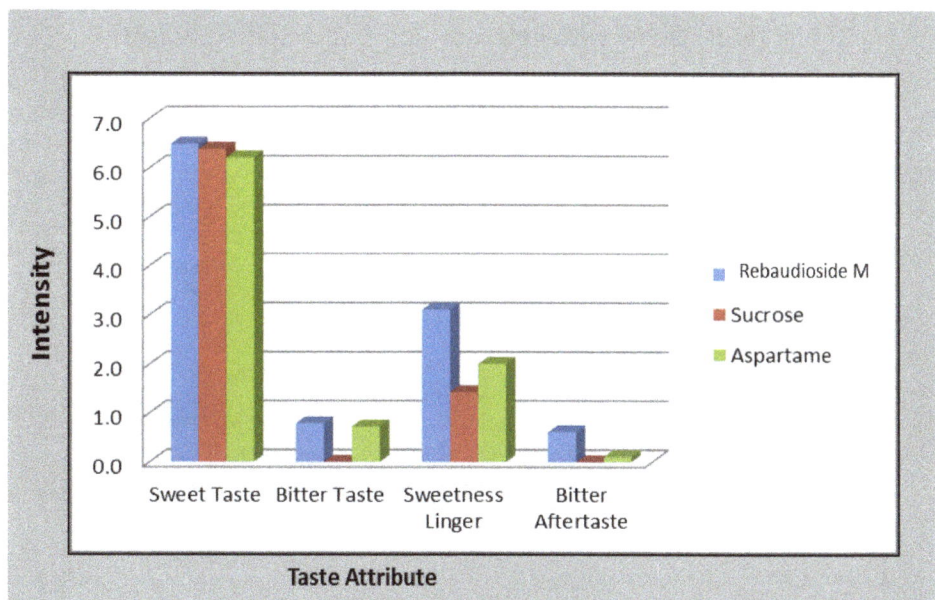

3.2. Blending

Blending of certain sweeteners (nutritive as well as non-nutritive) is often found to result in sweetness synergy. Such blends are also generally advantaged by improvements in flavor and temporal profiles as well as cost reductions in sweetener system and, often, improvement in stability [29–33].

To study the interactions between rebaudioside M (**7**) and other natural ingredients, blends were made with rebaudioside A (**1**), rebaudioside D (**4**), rebaudioside B (**2**), and erythritol at various concentrations in acidified water and sensory evaluation were performed.

Figure 7 illustrates the average response from sensory panelists that tasted the acidified solution with rebaudioside M (300 ppm) and rebaudioside A (100 ppm) or rebaudioside D (100 ppm) blends. The di-blends showed improvement in total sweetness, overall sweetness profile (peak), and leafy note. The blend with rebaudioside M and rebaudioside D showed higher improvement in sweetness intensity, overall sweetness profile, bitter lingering and sweet lingering.

The average response from sensory panelists that tasted the acidified solution with rebaudioside M and rebaudioside B blends (100 ppm or 50 ppm) is presented in Figure 8. The rebaudioside M and rebaudioside B di-blend exhibited additional rounded sweetness profile with slight improvement in sweetness intensity, onset and bitterness perception.

Figure 7. Sensory profile of di-blends of rebaudioside M (**7**) and rebaudioside A (**1**) or rebaudioside D (**4**) in acidified water.

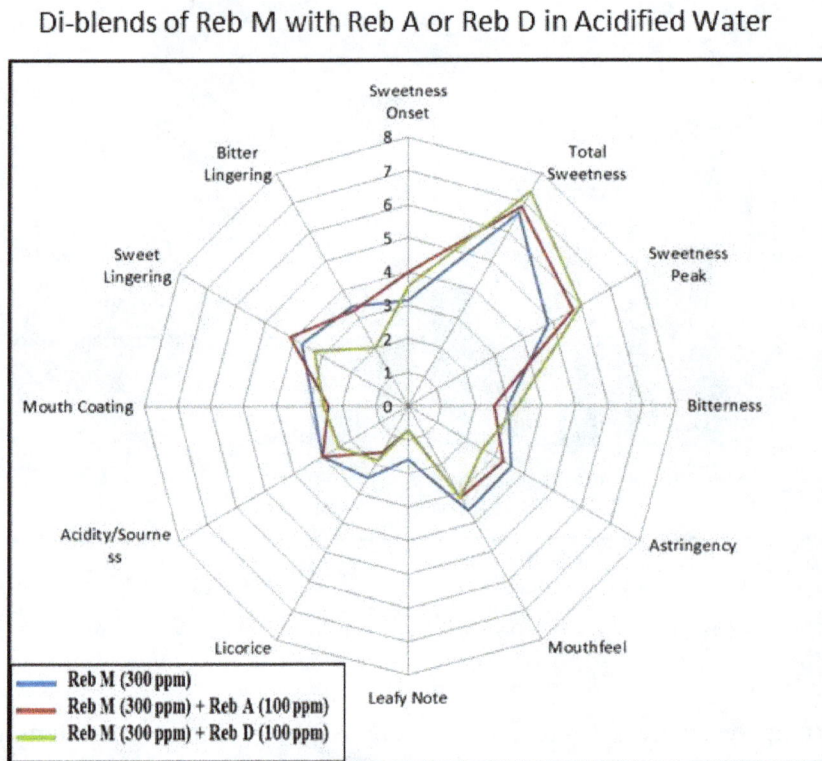

Di-blends of Reb M with Reb A or Reb D in Acidified Water

Figure 8. Sensory profile of di-blends of rebaudioside M (**7**) and rebaudioside B in acidified water.

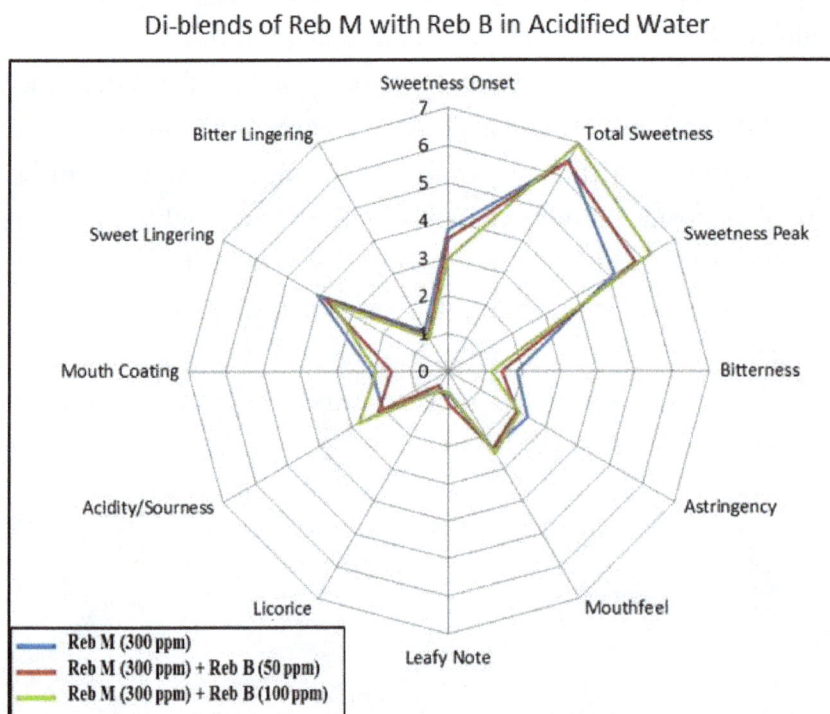

Di-blends of Reb M with Reb B in Acidified Water

The taste description of rebaudioside M and erythritol is displayed in Figure 9 based on the average response from sensory panelists that tasted the acidified solution di-blend. The blend with erythritol helps in reducing acidity, bitterness, astringency and bitter lingering. At a higher level (above 1%, 100 ppm) erythritol contributes additional sweetness, rounded sweet profile and earlier onset.

Figure 9. Sensory profile of di-blends of rebaudioside M (**7**) and erythritol in acidified water.

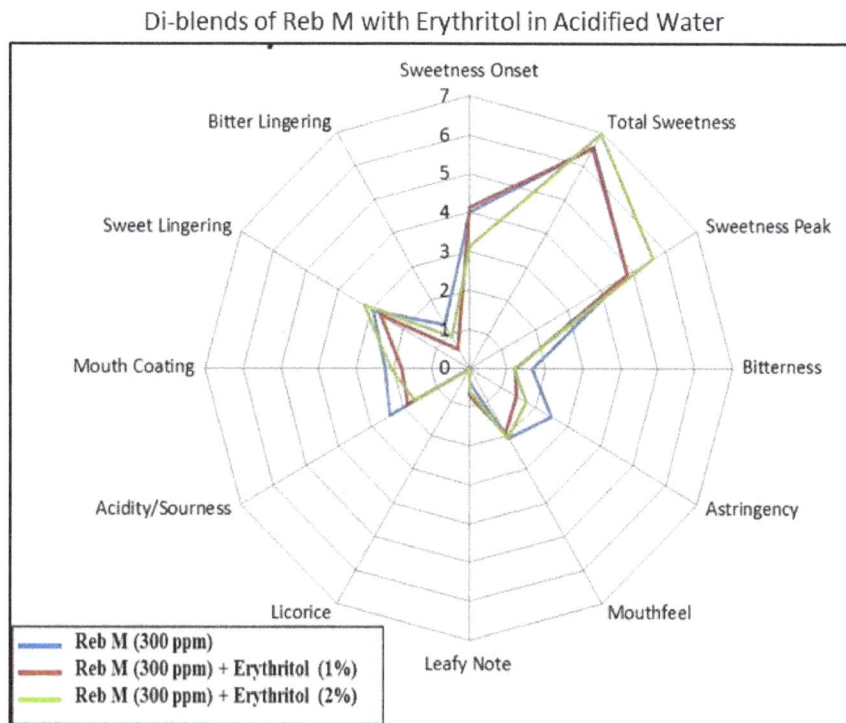

Rebaudioside M (**7**) can be used as a single sweetener system in sparkling beverages and commonly employed in blends with a number of sweeteners. A wide range of both non-caloric and caloric sweeteners are suitable partners for rebaudioside M. This work illustrates a few of the possibilities for di-blend non-caloric sweeteners and how they alter sensory attributes. In addition, a blend where rebaudioside M contributes 20%–80% of the sweetness, and sucrose provides the rest, exhibits flavor and temporal profiles very close to that of sugar (short AT). Such blends of natural sweeteners offer (i.e., rebaudioside M and sucrose) substantial energy reduction over sucrose alone. Sweet tasting amino acids such as glycine, alanine, glutamine, proline and serine as well as salts such as sodium chloride and potassium chloride also improve the taste of rebaudioside M [21]. These blends also permit the formulations of good-tasting, natural, blended sweetener systems with lower energy content than carbohydrate sweeteners alone.

3.3. Food Application

The functionality and stability of rebaudioside M (**7**) were demonstrated with a three-dimensional food matrix model representing the intended conditions of use in foods [34]. Based on experience with rebaudioside A (**1**), the key factors likely to affect rebaudioside M stability in product were considered to be product moisture, process temperature, and product pH.

Products comprised of carbonated beverages, still non-alcoholic beverages, table-top sweetener formulations, chewing gum and yogurt. All products were packed, stored (mostly at 25 °C and 60% relative humidity) and evaluated at intervals using both chemical (HPLC) and sensory analyses.

Sweetness of these products was assessed using panels consisting of 35 to 50 persons. Samples were evaluated using a five-point scale of categories ranging from 5 (much too sweet) to 1 (not at all sweet). Samples were considered satisfactory if at least 80% of the panelists rated the sweetness in category 3 (just about right) or above.

In this study key findings were determined to provide insight of the sweetness properties of rebaudioside M (7) in a wide variety of products. In the category of soft drinks (both carbonated and still) cola and lemon-lime, sweetened with rebaudioside M, remained acceptably sweet throughout 26 weeks storage. For comparison, most soft drinks are consumed within 16 weeks of production. Concerning table-top sweetener, rebaudioside M was tested in a number of formulations and all were stable for at least 52 weeks. Rebaudioside M was considered stable and functional in chewing gum for 26 weeks. Plain yogurt was evaluated, it was deemed to have no significant loss of sweetness during measured pasteurization (190 °F for 5 min) and fermentation. Rebaudioside M was stable throughout a 6 week storage period (40 °F) in plain yogurt. Table 3 provides range concentrations of rebaudioside M for sweeten of various food and beverages.

Table 3. Typical rebaudioside M (7) concentrations used to sweeten various foods and beverages.

Product	Range [a] (mg/kg or mg/L)
Carbonated soft drinks	100–600
Still beverages	50–600
Powdered soft drinks (as is)	200–2000
Tabletop (as is)	800–4000
Bakery products	200–1000
Dairy products	150–1000
Chewing gum	300–6000
Confections	100–1000
Cereals	200–1000
Edible gels	200–1000
Nutraceuticals	200–1000
Pharmaceuticals	50–1000

[a] Typical concentration when used as a single sweetener. Concentrations may vary depending upon formulation, flavor, and target consumer.

4. Conclusions

Through numerous evaluations we have determined that rebaudioside M has many beneficial properties and abundant potential as a sweetener in beverage and food products. This work illustrates that rebaudioside M has the capability to provide zero calories and has a clean sweet taste with slight bitter or licorice aftertaste. Functional in a wide array of beverages and foods, it is also well-suited for blending with other non-calorie or carbohydrate sweeteners. The molecule is stable under dry conditions and its stability is similar to rebaudioside A in aqueous food systems. High-purity zero-calorie natural stevia extract is of great interest to the global food industry because its natural source appeals to many consumers.

Conflicts of Interest

The authors declare no conflict of interest.

References

1. Kinghorn, A.D.; Kim, N.-C.; Kim, D.H.L. Terpenoid glycoside sweeteners. In *Naturally Occurring Glycosides*; Ikan, R., Ed., John Wiley & Sons: New York, NY, USA, 1999; pp. 399–429.

2. Kinghorn, A.D.; Wu, C.D.; Soejarto, D.D. Stevioside. In *Alternative Sweeteners*, 3rd ed, Revised and Expanded; O'Brien Nabors, L., Ed.; Marcel Dekker: New York, NY, USA, 2001; pp. 167–183.

3. Carakostas, M.; Prakash, I.; Kinghorn, A.D.; Wu, C.D.; Soejarto, D.D. Steviol glycosides. In *Alternative Sweeteners*, 4th ed., Revised and Expanded; O'Brien Nabors, L., Ed.; Marcel Dekker: New York, NY, USA, 2012; pp. 159–180.

4. Kinghorn, A.D.; Soejarto, D.D. Current status of stevioside as a sweetening agent for human use. In *Economics and Medicinal Plant Research*; Wagner, H., Hikino, H., Farnsworth, N.R., Eds.; Academic Press: London, UK, 1985; Volume 1, pp. 1–52.

5. Lewis, W.H. Early uses of *Stevia rebaudiana* (Asteraceae) leaves as a sweetener in Paraguay. *Econ. Bot.* **1992**, *46*, 336–337.

6. Prakash, I.; DuBois, G.E.; Clos, J.F.; Wilkens, K.L.; Fosdick, L.E. Development of Rebiana, a natural, non-caloric sweetener. *Food Chem. Toxicol.* **2008**, *46*, S75–S82.

7. Steviol Glycosides. Prepared at the 73rd JECFA (2010) and Published in FAO JECFA Monographs 10 (2010), Superseding Specifications Prepared at the 69th JECFA (2008) and Published in FAO JECFA Monographs 5 (2008). An ADI of 0–4 mg/kg bw (Expressed as Steviol) Was Established at the 69th JECFA (2008). Available online: http://www.fao.org/ag/agn/jecfa-additives/specs/monograph10/additive-442-m10.pdf (accessed on 1 November 2013).

8. Kinghorn, A.D.; Soejarto, D.D. Intensely sweet compounds of natural origin. *Med. Res. Rev.* **1989**, *9*, 91–115.

9. Kinghorn, A.D.; Soejarto, D.D. Sweetening agents of plant origin. *CRC Crit. Rev. Plant Sci.* **1986**, *4*, 79–120.

10. Kinghorn, A.D.; Compadre, C.M. Less common high-potency sweeteners. In *Alternative Sweeteners*, 2nd ed., Revised and Expanded; O'Brien Nabors, L., Gelardi, R.C., Eds.; Marcel Dekker: New York, NY, USA, 1991; pp. 197–218.

11. Kohda, H.; Kasai, R.; Yamasaki, K.; Murakami, K.; Tanaka, O. New sweet diterpene glucosides from *Stevia rebaudiana*. *Phytochemistry* **1976**, *15*, 981–983.

12. Ohta, M.; Sasa, S.; Inoue, A.; Tamai, T.; Fujita, I.; Morita, K.; Matsuura, F. Characterization of novel steviol glycosides from leaves of *Stevia rebaudiana* Morita. *J. Appl. Glycosci.* **2010**, *57*, 199–209.

13. Chaturvedula, V.S.P.; Prakash, I. A new diterpene glycoside from *Stevia rebaudiana*. *Molecules* **2011**, *15*, 2937–2943.

14. Chaturvedula, V.S.P.; Prakash, I. Structures of the novel diterpene glycosides from *Stevia rebaudiana*. *Carbohydr. Res.* **2011**, *34*, 1057–1060.

15. Chaturvedula, V.S.P.; Rhea, J.; Milanowski, D.; Mocek, U.; Prakash, I. Two minor diterpenoid glycosides from the leaves of *Stevia rebaudiana*. *Nat. Prod. Commun.* **2011**, *6*, 175–178.

16. Chaturvedula, V.S.P.; Clos, J.F.; Rhea, J.; Milanowski, D.; Mocek, U.; DuBois, G.E.; Prakash, I. Minor diterpene glycosides from the leaves of *Stevia rebaudiana*. *Phytochem. Lett.* **2011**, *4*, 209–212.

17. Chaturvedula, V.S.P.; Mani, U.; Prakash, I. Structures of the novel α-glucosyl linked diterpene glycosides from *Stevia rebaudiana*. *Carbohydr. Res.* **2011**, *346*, 2034–2038.

18. Prakash, I.; Campbell, M.; San Miguel, R.I.; Chaturvedula, V.S.P. Synthesis and sensory evaluation of *ent*-kaurane diterpene glycosides. *Molecules* **2012**, *17*, 8908–8916.

19. Prakash, I.; Campbell, M.; San Miguel, R.I.; Chaturvedula, V.S.P. Catalytic hydrogenation of the sweet principles of *Stevia rebaudiana*, rebaudiosde B, rebaudioside C and rebaudioside D and sensory evaluation of their reduced derivatives. *Int. J. Mol. Sci.* **2012**, *13*, 15126–15136.

20. Prakash, I.; Chaturvedula, V.S.P.; Markosyan, A. Isolation, characterization and sensory evaluation of a hexa β-D-glucopyranosyl diterpene from *Stevia rebaudiana*. *Nat. Prod. Commun.* **2013**, *8*, 1523–1526.

21. Prakash, I.; Markosyan, A.; Chaturvedula, V.S.P.; Campbell, M.; San Miguel, R.; Purkayastha, S.; Johnson, M. Methods for Purifying Steviol Glycosides and Uses of the Same. PCT Patent Application WO 2013/096420, 27 June 2013.

22. Agency Response Letter GRAS Notice No. GRN 000473. Purified Steviol Glycosides with Rebaudioside X (also Known as Rebaudioside M) as the Principal Component. Available online: http://www.fda.gov/Food/IngredientsPackagingLabeling/GRAS/NoticeInventory/ucm382202 (accessed on 1 June 2013).

23. Alvarez, E.R.B. Stevia Plant Named 'AKH L1'. Patent Application Number 20120090062, 12 April 2012.

24. Prakash, I.; Markosyan, A.; Chaturvedula, V.S.P.; Ma, G. The stability of rebaudioside M. 2013, Unpublished work.

25. Purkayastha, S.; Pugh, G.; Lynch, B.; Roberts, A.; Kwok, D.; Tarka, S.M. *In virto* metabolism of rebaudioside B, D, and M under anaerobic conditions: Comparison with rebaudioside A. *Regul. Toxicol. Pharmacol.* **2013**, *68*, 259–268.

26. Prakash, I.; Chaturvedula, V.S.P.; Markosyan, A. Structural characterization of the degradation products of minor natural sweet ditrpene glycoside rebaudioside M under acidic conditions. *Int. J. Mol. Sci.* **2014**, *15*, 1014–1025.

27. Lavia, A.; Hill, J. Sweeteners with Masked Saccharin Aftertaste. French Patent 2,087,843, 1972.

28. DuBois, G.E.; Walters, D.E.; Schiffman, S.S.; Warwick, Z.S.; Booth, B.J.; Pecore, S.D.; Gibes, K.; Carr, B.T.; Brands, L.M. ACS Symposium Series 450. In *Sweeteners: Discovery, Molecular Design, and Chemoreception*; Walters, D.E., Orthoefer, F.T., DuBois, G.E., Eds.; American Chemical Soceity: Washington, DC, USA, 1990; pp. 261–276.

29. Paul, T. Physical chemistry of foodstuffs. V. Degree of sweetness of sugars. *Chemiker-Zeitung* **1921**, *45*, 38–39.

30. Schiffman, S.; Booth, B.; Carr, B.; Losee, M.; Sattely-Miller, E.; Graham, B. Investigation of synergism in binary mixtures of sweeteners. *Brain Res. Bull.* **1995**, *38*, 105–120.

31. Scott, D. Saccharin-Dipeptide Sweetening Compositions. British Patent 1,256,995, 1971.

32. Verdi, R.J.; Hood, L.L. Advantages of alternative sweetener blends. *Food Technol.* **1993**, *47*, 94–102.

33. Walters, E. High intensity sweetener blends. In *Food Product Design*; Weeks Publishing Co.: Northbrook, IL, USA, 1993.

34. Pariza, M.; Ponakala, S.; Gerlat, P.; Andress, S. Predicting the functionality of direct food additives. *Food Technol.* **1998**, *52*, 56–60.

3

Effect of Kimchi Fermentation on Oxalate Levels in Silver Beet (*Beta vulgaris* var. cicla)

Yukiko Wadamori [†,*]**, Leo Vanhanen** [†] **and Geoffrey P. Savage** [†]

Food Group, Department of Wine, Food and Molecular Biosciences, Lincoln University, Lincoln 7647, Canterbury, New Zealand; E-Mails: leo.vanhanen@lincoln.ac.nz (L.V.); Savage@lincoln.ac.nz (G.P.S.)

[†] These authors contributed equally to this work.

* Author to whom correspondence should be addressed; E-Mail: yukiko.wadamori@lincolnuni.ac.nz

Abstract: Total, soluble and insoluble oxalates were extracted and analyzed by high performance liquid chromatography (HPLC) following the preparation of kimchi using silver beet (*Beta vulgaris* var. cicla) stems and leaves. As silver beet contains high oxalate concentrations and consumption of high levels can cause the development of kidney stones in some people, the reduction of oxalate during preparation and fermentation of kimchi was investigated. The silver beet stems and leaves were soaked in a 10% brine solution for 11 h and then washed in cold tap water. The total, soluble and insoluble oxalate contents of the silver beet leaves were reduced by soaking in brine, from 4275.81 ± 165.48 mg/100 g to 3709.49 ± 216.51 mg/100 g fresh weight (FW). Fermenting the kimchi for 5 days at 19.3 ± 0.8 °C in 5 L ceramic jars with a water airtight seal resulted in a mean 38.50% reduction in total oxalate content and a mean 22.86% reduction in soluble oxalates. The total calcium content was essentially the same before and after the fermentation of the kimchi (mean 296.1 mg/100 g FW). The study showed that fermentation of kimchi significantly ($p < 0.05$) reduced the total oxalate concentration in the initial mix from 609.32 ± 15.69 to 374.71 ± 7.94 mg/100 g FW in the final mix which led to a 72.3% reduction in the amount of calcium bound to insoluble oxalate.

Keywords: total, soluble and insoluble oxalates; kimchi; lactobacilli; calcium

1. Introduction

Kimchi, a fermented traditional food has been made in Korea for several hundred years [1]. It is a method of preserving food for use in the winter and it is gaining popularity as a functional food because of the phytochemicals formed during the fermentation process [2]. It is typically consumed raw and served as a side dish with practically every meal. The most popular kimchi in Korea, Baechu kimchi, is made principally from Chinese cabbage [3] although almost all vegetables cultivated in Korea have been used in kimchi preparation [2]. Other ingredients are commonly added and these vary between producers, for example, red pepper powder, ginger, garlic and radish [1]. Other minor ingredients that can be included are Indian mustard greens, carrots and dropwort. Seasoning ingredients, green onions, Chinese leeks, onions, pine nuts, gingko nuts and sugar, are also added in variable amounts. It is also common to add fish sauce made from shrimps, oysters, Alaskan pollack, squid, flounder or yellow corvine [3]. The beneficial effects of kimchi on human health may be derived from the nutrients in kimchi, such as vitamins, minerals and phytochemicals present, in the ingredients used to make the kimchi or the fermentation products produced by the lactic acid bacteria [4].

The main ingredient of traditional kimchi, Chinese cabbage, is cut and soaked in a 10% to 15% salt solution for several hours until the leaves and stems become soft. The salted Chinese cabbage is then washed under running tap water and thoroughly drained. The seasonings are prepared and chopped and then stuffed between the drained cabbage leaves. The stuffed Chinese cabbage is then packed into a traditional container, a ceramic jar with a lid, and stored in a cool place. Traditionally, this jar was buried in the ground for days to months depending on the environmental temperature. Kimchi is processed in a traditional ceramic jar called an onggi and it is stored in special refrigerators which can control the optimum temperature of the fermentation [4]. Jeong et al. [5] showed that kimchi fermented in an onggi had higher nutritional profiles than when it was fermented in polyethylene, polypropylene, stainless steel or glass containers [5].

The microorganisms which exist in the raw materials and are found in fermented kimchi, are mostly lactic acid producing bacteria and yeasts [3]. The most important feature of kimchi for human health is that it contains very high levels of lactic acid-producing bacteria (10^8 to 10^9 CFU/g) [6] such as *Leuconostoc mesenteroides*, *Lactobacillus brevis*, *Lactobacillus plantarum*, *Pediococcus cerevisiae*, *Streptococcus faecalis*, *Enterococcus faecalis*, *Pediococcus pentosaceous*, *Weissella koreanis* and *Lactobacilli* spp. [7–9]. *Leuconostoc mesenteroides* and *Lactobacillus plantarum* are found in the largest numbers. Total aerobic bacteria and fungi are reduced during kimchi fermentation and yeasts increase under the low temperature conditions [7]. Temperature is the most crucial factor in determining the balance of microbial populations during fermentation [10]. Studies have identified a range of important factors, such as the final acidity should fall between 0.6% and 0.7% and the pH should be between 4.2 and 4.3 for optimum consumer acceptance [7,11–13].

Oxalate is found in many kinds of edible plants with variable concentrations [14–16]. As consumption of additional oxalate in the diet can cause the development of kidney stones in susceptible people it is important to identify high oxalate containing foods and, if possible, reduce these levels by processing. As kimchi can be made from a range of vegetables and fruits, some kimchi may be produced from plants and spices which contain higher levels of oxalate. In addition, oxalates occur in two forms in plants, water soluble oxalate bound to Na^+ or K^+, or water insoluble oxalate bound to divalent ions such as Ca^{2+} and Mg^{2+}. Simpson *et al.* [17] presented a pH speciation diagram for oxalic acid which showed in an ideal solution that the two forms of oxalate are affected by the pH of the medium. At a pH lower than six, the proportion of fully deprotonated divalent oxalate ions ($C_2O_4^{2-}$) decreases markedly with a correspondingly reduced potential for binding with divalent mineral cations (especially Ca^{2+}) to form insoluble oxalates. Therefore, a reduction in the pH of kimchi during the fermentation will have an effect on the proportions of soluble and insoluble oxalates in the final kimchi mix. The overall effect will be to increase the soluble oxalate content of the fermenting mix. This would provide increased levels of soluble oxalates that could then be used by the anaerobic bacteria that form a large proportion of the fermentation biomass in the kimchi. Some lactic acid producing bacteria have been reported to break down oxalate and use it as a carbon and energy source *in vitro* [18,19]. *Lactobacillii plantarum* is the dominant strain of bacteria commonly found in kimchi fermentations and it has been shown that two strains of *Lactobacillus acidophilus* and one strain of *Lactobacillii plantarum* also have the ability to degrade oxalate [20].

The leaching of soluble oxalate during washing and soaking in brine solution may reduce the oxalate content of kimchi. In addition, small amounts of liquid removed after fermentation may contain leached oxalates and this fraction is not consumed.

The stems and leaves of silver beet are consumed widely in New Zealand and could be an inexpensive and interesting alternative to Chinese cabbage for communities in New Zealand who regularly consume kimchi. However, silver beet is a high-oxalate containing vegetable and needs to be eaten in small quantities by people predisposed to oxalate kidney stones. It is known that addition of calcium sources can reduce the amount of soluble oxalate available for absorption from the gastrointestinal tract but the effect of fermentation has not been previously studied [21,22].

The aim of this study was to investigate the use of the leaves and stems of silver beet, a non-traditional vegetable, to make kimchi. This study involved the measurement of the effect of soaking silver beet stems and leaves in brine and the effect of low pH and microbial fermentation on the oxalate content in silver beet during the production of traditional kimchi.

2. Experimental Section

2.1. Preparation of Kimchi

Ten kilograms of fresh, fully grown silver beet plants (*Beta vulgaris* var. cicla), seven onions (*Allium cepa*), three apples (*Malus pumila*), one head of garlic (*Allium sativum*) and one piece of fresh ginger (*Zingiber officinale*) were purchased from a local grower (Crazy Dave, Christchurch, Canterbury, New Zealand). Ten large white radishes (*Raphanus sativus*) and seven bunches of Chinese chives (*Allium tuberosum*), all grown in New Zealand, were purchased from a local Chinese shop (Chinese market, Christchurch, Canterbury, New Zealand). Salt (Cerebos Skellerup Ltd., East Tamaki, Auckland, New Zealand) and red chilli powder (*Capsicum annuum*, Ducsan, Jung-gu, Seoul, South Korea), fermented small shrimps (packed in Shirley Fish Market, Christchurch, New Zealand, produced in Korea), Korean fish sauce (Chung Jung One, Samseong-dong, Gangnam-gu, Seoul, Korea), sugar (Chelsea Sugar Company, Birkenhead, Auckland, New Zealand) and glutinous rice flour

(Samutsakorn, Mahachai, Samut Sakhon, Thailand) were all purchased from a local Korean shop (Kosco, Christchurch, Canterbury, New Zealand).

Ten kilograms of silver beet leaves and stems were washed under running tap water and chopped into 40 mm pieces. The prepared silver beet was then soaked in 20 L of 10% salt water and allowed to soak for 2 h at room temperature (19.3 ± 0.8 °C) then 9 h in the fridge at 5.4 °C. The salted silver beet leaves and stems were squeezed by hand to remove excess water and then washed gently under running tap water and squeezed again to remove excess water.

White radishes (3.22 kg) were cut into fine thin strips. Chinese chives (0.58 kg) were cut into 30 mm pieces. Seven onions (1228 g), three small apples (192 g), fresh ginger (65 g) and one garlic clove (15 g) were chopped into small pieces and then ground together using a mixer (Multipro, Kenwood, Japan). Glutinous rice powder (100 g) and water (1.33 kg) were mixed and heated in a saucepan until gelatinized. The chopped and processed vegetables were pre-mixed with the cooled gelatinized rice, 980 g of chilli powder, 33 g of salt, 300 g of fermented small shrimps, 730 g of Korean fish sauce and 200 g of sugar. This mixture was then placed in a 40 L plastic flexi-tub food container. The salted silver beet leaves and stems were then added and thoroughly mixed by hand. The mixed ingredients were then placed into three traditional (five liter) Korean kimchi jars and a plastic film was placed over the top of the mixture to keep it moist. A weight and a cap were put on each jar and water was poured into the upper rim in order to prevent air entering the jars during fermentation. The jars were allowed to stand at room temperature (mean 19.3 ± 0.8 °C) for 5 days.

2.2. Dry Matter and pH Determination

The dry matter (DM) content of each sample was determined by drying in an oven (Watvic, Watson Victor Ltd., New Zealand) to a constant weight at 105 °C [23]. The pH in the initial and the mix at the end of the 5 day fermentation period were measured at ten different points in the mix using a SevenEasy pH Meter (Mettler Toledo, GmbH, Schwerzenback, Switzerland).

2.3. Total and Soluble Oxalate Analysis

Six representative samples of kimchi were collected from each jar before and after fermentation for oxalate analysis. The liquid remaining at the bottom of each jar was also sampled. For the extraction of the total oxalic acid, 0.2 g of ground oven-dried sample was weighed accurately into a 100 mL flask and 40 mL of 0.2 M HCl was added and the mixture was then incubated with shaking for 20 min at 80 °C. After cooling, each mixture was transferred to a 100 mL volumetric flask and made up to volume with 0.2 M HCl. Three extractions were carried out on each sample. Fifty milliliters of the aliquot was transferred in a 50 mL centrifuge tube then centrifuged at 3500 rpm for 15 min. The liquid sample was filtered through a 0.45 μm cellulose acetate syringe filter (Sartorius A.G., Göttingen, Germany) into a HPLC vial for analysis of the oxalic acid concentration.

For the extraction of the soluble oxalic acid, the same procedure as followed as in the extraction for total oxalic acid except that nanopure water (18.2 Megaohm-cm, Arium 611 uv, Sartorius A.G., Germany) was used to extract the soluble oxalates.

The oxalic acid concentration of the extracts was determined using an HPLC method where a 20 μL of sample extract was analysed using a 300 × 7.80 mm Rezex ion exclusion column (Phenomenex Inc.,

CA, USA) using isocratic elution at 0.6 mL/min. with 0.025 M sulphuric acid (HPLC Grade, Baker Chemicals, Phillipsburg, NJ, USA) as the mobile phase [15]. The HPLC system consisted of a Spectra-Physics Isocratic pump and a Spectra-Physics UV/V detector set at 210 nm. Data capture was performed using PeakSimple version 3.59 (SRI Instruments, Torrance, CA, USA). The oxalate peak was identified by comparison of the retention time with an oxalate standard (99.999%, Sigma, St. Louis, MO, USA). Insoluble oxalate was calculated as the difference between the total oxalate (acid extract) and soluble oxalate (water extract) [24]. The oxalate results are presented as mg/100 g fresh weight (FW).

Standard curves, containing 2–20 mg oxalic acid/100 mL in water or 0.2 M HCl were prepared and used to quantify the soluble and total oxalic acid contents of the samples.

2.4. Total Calcium Analysis

Duplicate samples of oven dried kimchi were initially ground in a coffee mill (Sunbeam, model EM0400, China), then 0.3 g accurately weighed into a 75 mL Teflon PFA Kevlar-shielded digestion vessel (CEM Corporation, Matthews, NC, USA). To this, 5 mL of a 4:1 nitric acid (69%): hydrogen peroxide (Aristar grade, Merck Ltd., KGaA, Darmstadt, Germany) acid mixture was added. The digestion vessels were then capped and the oven dried kimchi samples were digested using a MARSXpress™ microwave digester (CEM Corporation, Matthews, NC, USA), programmed to ramp from ambient to 90 °C over 10 min then from 90 °C to 170 °C over 10 min where it was held for 10 min. Once cooled, 10 mL of deionized water (18.2 Megaohm-cm) was added to make a final volume of 15 mL.

Mineral analysis was carried out on a Varian Axial 720 Inductively Coupled Plasma Optical Emission Spectrophotometer (ICP-OES, Varian, Palo Alto, CA, USA) with SP3 auto-sampler. Minerals were identified and quantitated using an ICP multi-element standard solution (CertiPUR, Merck, KGaA, Darmstadt, Germany) containing 23 elements or a single element standard, as required. Data and standard curves were processed using ICP-Expert™ II (Varian, Palo Alto, CA, USA). A water test standard and the ICP multi-element standard solution containing 0.50 µg/g of each element were run in triplicate with each batch to determine the standard error. The overall standard error was ±0.013 mg/kg. Limits of quantitation (LOQ) were performed using 10 times the standard deviation of the blank (5% nitric acid) for each mineral. Individual mineral LOQ values ranged from 0.12 to 12.24 µg/L, with a mean of 1.81 µg/L. The measurement of the total calcium in the kimchi mix and the calculation of the calcium content of the insoluble oxalate allowed the calculation of the proportion of insoluble calcium to total calcium content of the kimchi mix.

2.5. Statistical Analysis

All calculations were performed using Excel 2010. GenStat Release 12.2 for Windows 7 (VSN International Ltd., Hemel Hempstead, Hertfordshire, UK) was used to determine the accumulated analysis of variance. The mean values were compared using Fishers LSD method ($p < 0.05$).

3. Results and Discussion

The mean pH of the initial kimchi mix was 5.06 ± 0.307 and this reached 4.34 ± 0.042 after 5 days fermentation at 19.3 ± 0.8 °C. This decrease in pH was due to the formation of lactic acid by the bacteria. The dry matter content of the raw and processed silver beet leaves and stems, and the kimchi mixes are shown in Table 1. This shows that soaking silver beet leaves in salt water significantly reduced the moisture content of the leaves. The average dry matter of raw and processed silver beet leaves and stems, and the kimchi mixes, ranged from 7.1% in the raw leaves and stems of silver beet to 14.8% in kimchi mix at end of experiment.

Table 1. Dry matter content of the raw and processed silver beet leaves and stems, and the kimchi mixes (%).

Food sample	Mean dry matter content (%) (±SE)
Raw leaves and stems of silver beet	7.07 ± 0.09
After soaking in salt water	14.00 ± 0.65
Kimchi mix at start of experiment	13.44 ± 0.12
Kimchi mix at end of experiment	13.99 ± 0.14

The dry matter content of the kimchi increased during fermentation as some liquid was released from the kimchi mass. The total, soluble and insoluble oxalate contents of the raw and processed silver beet leaves and stems are shown in Table 2.

Table 2. Mean total soluble and insoluble oxalate contents of the raw and processed silver beet used to make kimchi (mg/100 g fresh weight).

Silver beet	Total oxalate	Soluble oxalate (% of total oxalate)	Insoluble oxalate
Raw leaves and stems of silver beet	4275.81 ± 165.48	1880.67 ± 378.95 (44.0%)	2395.15 ± 397.71
Raw leaves and stems of silver beet after soaking in 10% salt water	3709.49 ± 216.51	1811.68 ± 209.42 (48.8%)	1897.82 ± 207.29

The oxalate values of the raw commercially produced silver beet leaves and stems used in this experiment are higher than previously reported data [15,17] but it has been previously noted that the levels of oxalates in silver beet leaves are very variable [17]. Soaking the silver beet leaves and stems in brine for 11 h reduced total oxalate content on a wet matter basis by 13.2% with only a small reduction in the soluble oxalate content (3.67%).

The oxalate content of the onions, apples, garlic, fish sauce, fermented shrimp, radishes and Chinese chives were measured but no oxalate was detected. The oxalate contents of fresh ginger and dry red chilli powder are shown in Table 3.

Table 3. Mean total soluble and insoluble oxalate contents of materials used to make kimchi. FW, fresh weight; DM, dry matter.

Food sample	Total oxalate	Soluble oxalate	Insoluble oxalate
Fresh ginger (mg/100 g FW)	1312.66 ± 30.99	1174.06 ± 65.43	138.60 ± 96.41
Red chilli powder (mg/100 g DM)	274.01 ± 9.8	277.22 ± 4.58	-

Total oxalate contents of the fresh ginger was very high and were similar to values reported in an earlier study [25] where dried ground ginger powder contained 1528 ± 92 mg/100 g of total oxalate and 1339 ± 38 mg/100 g of soluble oxalates.

Silver beet leaves and stems consisted of 56.7% of the total fresh kimchi mix and, therefore, supplied the greatest proportion of oxalates to the initial kimchi mix. Chilli powder and fresh ginger also supplied some oxalates although they only represented 5.6% and 0.4% respectively, of the total mix. The initial and final total, soluble and insoluble oxalate contents of the kimchi mixes are shown in Table 4. The total, soluble and insoluble oxalate contents of the kimchi were significantly reduced by 38.5%, 22.9% and 70.4%, respectively, following five days of lactic acid bacteria fermentation in kimchi jars at a mean temperature of 19.3 ± 0.8 °C. The proportion of soluble oxalate to total oxalate in the kimchi mix increased from 67.1% to 84.1%.

Table 4. Mean total soluble and insoluble oxalate contents of kimchi before and after fermentation (mg/100 g FW).

	Jar	Total	Soluble (% of total oxalate)	Insoluble
Before fermentation	A	661.63 ± 6.6	446.38 ± 0.6	215.25 ± 6.0
	B	615.90 ± 11.7	407.20 ± 21.0	208.70 ± 27.3
	C	550.43 ± 4.2	372.20 ± 13.2	178.30 ± 15.5
Mean		609.32 ± 15.69[a]	408.58 ± 12.16[a] (67.1%)	200.75 ± 10.23[a]
After fermentation	A	372.70 ± 17.9	299.30 ± 16.2	73.40 ± 26.5
	B	385.56 ± 9.5	323.97 ± 8.4	61.60 ± 14.6
	C	365.90 ± 13.7	322.35 ± 6.0	43.60 ± 11.3
Mean		374.71 ± 7.94[b]	315.19 ± 23.21[b] (84.1%)	59.52 ± 10.54[b]

Mean values before and after fermentation with a different letter, differ significantly (Fishers LSD, 95% level of confidence).

The mean liquid leachate remaining in each of the jars after five days of fermentation was 230 ± 17.8 mL which was 1.58% of the total kimchi mix in the three jars. As this leachate is not consumed, the removal of this liquid results in a small loss of oxalates from the kimchi available for consumption. The mean total and soluble oxalate contents of this fraction were, respectively, 353.78 ± 11.79 and 322.88 ± 8.60 mg.

It is most interesting to note that large reductions in total, soluble and insoluble oxalates occurred during the fermentation of kimchi (38.5% for total oxalate, 22.9% for soluble oxalate and 70.4% for insoluble oxalate). This is the first time this has been observed during the fermentation of kimchi and is presumably the result of lactic acid bacteria fermentation that occurred when all the oxygen was removed from the atmosphere of the jar during the initial stages of the microbial fermentation of kimchi. There are two reasons for the reduction of oxalate contents during the lactic acid fermentation in the kimchi: (1) as the pH was reduced, the form of oxalate would change from insoluble oxalate bound to calcium ions to soluble oxalate [17]; and (2) this, then, will increase the soluble oxalate content of the liquid fraction of the kimchi mix and the increased soluble oxalate content could then be used as an energy source by the oxalotrophic bacteria present in the kimchi fermentation reducing both the total and the soluble oxalate contents of the final product in the process. This supports the observations made in an earlier study where *Lactobacillii plantarum,* was reported to be found in the kimchi in large numbers [7–9] and were found to degrade oxalate in other studies [20].

The total calcium content of the kimchi mix before and after fermentation was essentially the same, mean 296.0 mg calcium/100 g FW (Table 5). If it is assumed that insoluble oxalate is predominantly calcium oxalate [17] then it is possible to calculate the amount of calcium unavailable in this molecule and compare this with the total calcium in the kimchi mix. Table 4 shows that there was a 70.4% reduction in the insoluble oxalate content when the values found in the initial and final kimchi mixes were compared. This corresponds to a reduction calcium bound in insoluble oxalate (Table 5). The pH of the kimchi mix reduced from an initial 5.06 to 4.34 when fermentation was completed and the speciation diagram of oxalate with pH [17] shows that this pH change will result in a large reduction in the potential of oxalate to bind to divalent cations (especially calcium). Overall, fermentation led to 72% reduction of calcium bound in insoluble oxalate.

Table 5. Mean total calcium, calculated calcium bound in insoluble oxalate and % bound calcium in kimchi before and after fermentation (mg/100 g FW).

Kimchi	Total calcium (mg/100 g FW)	Calcium in insoluble oxalate (mg/100 g FW)	Insoluble calcium/total calcium (%)
Before fermentation	286.94 ± 9.23	62.79 ± 3.57	21.84 ± 0.56
After fermentation	305.17 ± 7.73	18.61 ± 2.72	6.06 ± 0.74

4. Conclusions

The use of silver beet to prepare kimchi has not been considered before but this experiment has shown that considerable reductions in the oxalate content do occur during the fermentation process. This experiment confirms that lactic acid bacteria responsible for traditional kimchi fermentation have oxalotrophic activity. Although the soluble oxalate as a proportion of total oxalate increased, the absolute amount in the fermented kimchi available for absorption decreased.

Acknowledgments

The authors wish to thank Janette Busch for proofreading the text.

Conflicts of Interest

The authors declare no conflict of interest.

References

1. Kang, Y. Supplementary Food for Health Using Kimchi as Principal Raw Material and Method for Reducing The Same. U.S. Patent 20060029692 A1, 9 February 2006.

2. Park, K.Y.; Kil, J.H.; Jung, K.O.; Kong, C.S.; Lee, L.M. Functional properties of kimchi (Korean fermented vegetables). *Acta Hort. (ISHS)* **2006**, *706*, 167–172. Avalaible online: http://www.actahort.org/books/706/706_19.htm (accessed on 17 December 2013).

3. Cheigh, H.S.; Park, K.Y. Biochemical, microbiological, and nutritional aspects of kimchi (Korean fermented vegetable products). *Crit. Rev. Food Sci. Nutr.* **1994**, *34*, 175–203.

4. Lee, D.; Kim, S.; Cho, J.; Kim, J. Microbial population dynamics and temperature changes during fermentation of kimjang kimchi. *J. Microbiol.* **2008**, *46*, 590–593.

5. Jeong, J.-K.; Kim, Y.-W.; Choi, H.-S.; Lee, D.S.; Kang, S.-A.; Park, K.-Y. Increased quality and functionality of kimchi when fermented in Korean earthenware (onggi). *Int. J. Food Sci. Technol.* **2011**, *46*, 2015–2021.

6. Chun, J.K.; Kim, K.M.; Woo, D.H. Automation of kimchi fermentation based on pattern analysis. *Food Eng. Program* **1999**, *3*, 181–185.

7. Mheen, T.-I.; Kwon, T.-W. Effect of temperature and salt concentration on kimchi fermentation. *Korean J. Food Sci. Technol.* **1984**, *16*, 443–450.

8. Kim, H.S.; Chun, J.K. Studies on the dynamic changes of bacteria during the kimchi fermentation. *J. Korean Nucl. Soc.* **1966**, *6*, 112–118.

9. Han, H.U.; Lim, C.R.; Park, H.K. Determination of microbial community as an indicator of Kimchi fermentation. *Korean J. Food Sci. Technol.* **1990**, *22*, 26–32.

10. Cho, J.; Lee, D.; Yang, C.; Jeon, J.; Kim, J.; Han, H. Microbial population dynamics of kimchi, a fermented cabbage product. *FEMS Microbiol. Lett.* **2006**, *257*, 262–267.

11. Lee, J.H. Current studies on the community of lactic acid bacteria in kimchi a traditional Korean fermented food. *Milk Sci.* **2009**, *58*, 153–159.

12. Cha, Y.J.; Kim, H.; Cadwallader, K.R. Aroma-active compounds in kimchi during fermentation. *J. Agric. Food Chem.* **1998**, *46*, 1944–1953.

13. Park, K.Y.; Rhee, S.H. Functional foods from fermented vegetable products: Kimchi (Korean fermented vegetables and functionality). In *Asian Functional Foods*; Shi, J., Ho, C.-T., Shahidi, F., Eds.; CRC Press: Boca Raton, FL, USA, 2005; pp. 341–380.

14. Noonan, S.C.; Savage, G.P. Oxalate content of foods and its effect on humans. *Asia Pac. J. Clin. Nutr.* **1999**, *8*, 64–74.

15. Savage, G.P.; Vanhanen, L.; Mason, S.M.; Ross, A.B. Effect of cooking on the soluble and insoluble oxalate content of some New Zealand foods. *J. Food Comp. Anal.* **2000**, *13*, 201–206.

16. Radek, M.; Savage, G.P. Oxalates in some Indian green leafy vegetables. *Int. J. Food Sci. Nutr.* **2008**, *59*, 246–260.

17. Simpson, T.S.; Savage, G.P.; Sherlock, R.; Vanhanen, L.P. Oxalate content of silver beet leaves (*Beta vulgaris* var. cicla) at different stages of maturation and the effect of cooking with different milk sources. *J. Agric. Food Chem.* **2009**, *57*, 10804–10808.

18. Goldfarb, D.S.; Modersitzki, F.; Asplin, J.R. A randomized, controlled trial of lactic acid bacteria for idiopathic hyperoxaluria. *Clin. J. Am. Soc. Nephrol.* **2007**, *2*, 745–749.

19. Muller, H. Oxalsaure als Kohlenstoffquelle fur Mikroorganismen. *Arch. Mikrobiol.* **1950**, *15*, 137–148.

20. Weese, J.S.; Weese, H.E.; Yuricek, L.; Rousseau, J. Oxalate degradation by intestinal lactic acid bacteria in dogs and cats. *Vet. Microbiol.* **2004**, *101*, 161–166.

21. Massey, L.K. Food oxalate: Factors affecting measurement, biological variation, and bioavailability. *J. Am. Diet. Assoc.* **2007**, *107*, 1191–1194.

22. Johansson, S.; Savage, G.P. The availability of soluble oxalates in stir-fried silver beet (*Beta vulgaris* var. cicla) leaves eaten with yoghurt. *Int. J. Food Sci. Technol.* **2011**, *46*, 2232–2239.

23. AOAC International. *AOAC Official Methods of Analysis of AOAC*, 17th ed.; AOAC International: Gaithersburg, MD, USA, 2002.

24. Holloway, W.D.; Argall, M.E.; Jealous, W.T.; Lee, J.A.; Bradbury, J.H. Organic acids and calcium oxalate in tropical root crops. *J. Agric. Food Chem.* **1989**, *37*, 337–341.

25. Das, S.G.; Savage, G.P. Total and soluble oxalate content of some Indian spices. *Plant Foods Hum. Nutr.* **2012**, *67*, 186–190.

4

Characterization of Botanical and Geographical Origin of Corsican "Spring" Honeys by Melissopalynological and Volatile Analysis

Yin Yang, Marie-José Battesti, Jean Costa and Julien Paolini *

Laboratory of Natural Product Chemistry, UMR CNRS 6134, Grimaldi Campus, Corsican University, BP 52, Corte 20250, France; E-Mails: yang@univ-corse.fr (Y.Y.); mjbattesti@univ-corse.fr (M.-J.B.); costa@univ-corse.fr (J.C.)

* Author to whom correspondence should be addressed; E-Mail: paolini@univ-corse.fr

Abstract: Pollen spectrum, physicochemical parameters and volatile fraction of Corsican "spring" honeys were investigated with the aim of developing a multidisciplinary method for the qualification of honeys in which nectar resources are under-represented in the pollen spectrum. Forty-one Corsican "spring" honeys were certified by melissopalynological analysis using directory and biogeographical origin of 50 representative taxa. Two groups of honeys were distinguished according to the botanical origin of samples: "clementine" honeys characterized by the association of cultivated species from oriental plain and other "spring" honeys dominated by wild herbaceous taxa from the ruderal and/or maquis area. The main compounds of the "spring" honey volatile fraction were phenylacetaldehyde, benzaldehyde and methyl-benzene. The volatile composition of "clementine" honeys was also characterized by three lilac aldehyde isomers. Statistical analysis of melissopalynological, physicochemical and volatile data showed that the presence of *Citrus* pollen in "clementine" honeys was positively correlated with the amount of linalool derivatives and methyl anthranilate. Otherwise, the other "spring" honeys were characterized by complex nectariferous species associations and the content of phenylacetaldehyde and methyl syringate.

Keywords: honey; clementine and asphodel; melissopalynological analysis; HS-SPME; GC

1. Introduction

The specificity of Corsican honeys is linked with the environmental characteristics of the island (biodiversity of flora, bioclimatic conditions and topography), the endemic black honeybee and typical hive management. Organoleptic and melissopalynological analysis have permitted Corsican honeys to be classified into six ranges: "spring", "spring maquis", "honeydew maquis", "chestnut grove", "summer maquis" and "autumn maquis", according to the harvest season and the geographic location of the apiaries [1]. These honeys have been certified by two official designations of origin: the national Appellation d'Origine Contrôlée (AOC) and the European Protected Designation of Origin (PDO), both marketed as "Miel de Corse-Mele di Corsica" [2,3].

The organoleptic properties of the "spring" honey range are a light color (the lightest among the six ranges) associated with low-to-medium olfactory and aromatic intensities, sometimes with a slight acidity [1–3]. These honeys are described in terms such as floral, fresh fruit, or dry vegetal according to the vocabulary of odor and the aroma wheel [4]. Moreover, the physicochemical characteristics of "spring" honeys are low values of coloration and electrical conductivity. Finally, these honeys are harvested from April to May at low altitudes (below 400 m) on the coast, plains or valleys [1–3].

The Corsican "spring" honeys can be classified into two categories. First, honeys harvested in the oriental plain of the island. These cultivated zones are dominated by clementine orchards (*Citrus sinensis* × *reticulata*) associated with other *Citrus* species, *Actinidia sinensis* and various fruit trees. They are always surrounded by maquis; an evergreen scrub of vegetation from Mediterranean area. Second, honeys collected in ruderal and/or littoral maquis areas for their first flowering. Ruderal zones are characterized by herbaceous plants, especially *Asphodelus ramosus* subsp. *ramosus* (syn: *A. microcarpus* Salz et Viv.) associated with various species of *Fabaceae*, *Boraginaceae*, wild *Brassicaceae*, *Apiaceae* and *Asteraceae*. The coastal areas also showed a diversity of nectariferous and polleniferous resources [5].

Unifloral honeys from the *Citrus* genus, produced principally from oranges or lemons, are often found in the Mediterranean region (Italy, Spain, Greece, France and North Africa), but also in Israel, USA, Brazil and Mexico [6,7]. The nectar of *Asphodelus* species is frequently found in the composition of honeys from Mediterranean regions (Italy, Sicily, Corsica and Sardinia), but asphodel unifloral honey is produced mainly in Sardinia [8,9]. In Corsica, the *Asphodelus* genus is represented by three species: *A. ramosus* subsp. *ramosus*, *A. cerasiferus* and *A. fistulosus* [10]. *A. ramosus* subsp. *ramosus*, which flowers from March to May, was the more visited species.

The certification of geographical and botanical origins of Corsican honeys is conventionally based on the melissopalynological analysis of the entire pollen spectrum [5,11]. Furthermore, sensory characteristics and physicochemical parameters are also necessary to specify the botanical origin of honey [5,11,12]. However, this traditional approach is not precise enough to determine the predominant botanical origin exactly, especially when nectar resources are under-represented in the pollen spectrum. For this reason, the chemical composition of honeys has been used to complete the classical approaches of botanical origin determination. Thus, various extraction methods, such as headspace solid-phase microextraction (HS-SPME), simultaneous steam distillation-solvent extraction and ultrasound-assisted extraction associated with gas chromatography (GC) have been developed for the analysis of the volatile fraction of honeys [13]. Some volatile components, including methyl anthranilate, lilac aldehyde and *p*-menth-1-en-9-al, were therefore suggested as the chemical markers of citrus (species not specified) unifloral honey [13–15]. Moreover, Alissandrakis *et al.* [16] showed that the volatile fractions of citrus flowers (four species) and the corresponding honeys were dominated by linalool derivatives. The phenolic compound hesperetin was also proposed as a botanical indicator of Spanish citrus honeys for its high levels in nectar and honey [17]. Methyl syringate and/or phenylacetaldehyde were identified as characteristic components of nectar from *A. microcarpus* Salz et Viv. and corresponding unifloral honeys [18,19].

Several techniques (HS-SPME, infrared spectroscopy and ^{1}H-nuclear magnetic resonance spectroscopy) have been used to distinguish Corsican and non-Corsican honeys, but these studies did not provide results for the differentiation of the botanical origin of different ranges of Corsican honey [20–22].

According to the geographical and botanical origins of Corsican "spring" honeys certified by melissopalynological analysis, the chemical composition of volatile fractions of honey samples was established using HS-SPME, GC and GC/mass spectrometry (MS). The aim of the study is to establish for the first time a multidisciplinary method for the qualification of Corsican "spring" honeys, based on relationships between the pollen spectrum, volatile chemical markers and some physicochemical parameters.

2. Experimental Section

2.1. Honey and Flower Sampling

In total, 41 Corsican "spring" honeys (samples 1–41) were selected from our reference bank of honey with AOC and PDO appellations. All these samples were directly packaged in a sealed pot and stored below 14 °C according to the optimal conditions of honey conservation indicated by Gonnet *et al.* [23]. The honey samples of three years of harvest (2004–2006) collected in April to June were provided from 12 Corsican producers. The apiaries were located from littoral to 400 m (principally under 100 m) in the oriental cultivated plain or in ruderal and/or maquis zone of thermo- and meso-Mediterranean levels. Clementine (*Citrus sinensis* × *reticulate*, six samples) and Asphodel (*Asphodelus ramosus* subsp. *ramosus*, six sample locations) flower specimens were collected in March–May 2009–2012. The nectar secretion during harvest period was ensured by the observation of foraging nectar by honeybees. Flowers samples were analyzed within 48 h.

2.2. Melissopalynological Analysis

In this study, melissopalynological analysis was performed using the method described by Yang *et al.* [24]. Identification of pollen in the "spring" honey was based on the comparison with laboratory's own reference pollen-slides library and also carried out with the palynological expertise practice [5,11] developed for the characterization and the AOC and PDO control of Corsican honeys. Pollen analysis was allowed to establish a total pollen spectrum (qualitative analysis) and pollen density (quantitative analysis) for each honey sample. The identified taxa in the pollen spectrum were expressed in term of relative frequency (RF) and the pollen density was expressed as the absolute number of pollen grain in 10 g of honey (PG/10 g).

2.3. Physicochemical Analysis

According to the description of Corsican honeys [1,5], two physicochemical parameters, coloration and electrical conductivity were chosen to complete the botanical origin characterization of Corsican "spring" honey. The honey coloration was measured using a Lovibond Comparator apparatus [25]. Results were expressed as millimeters (mm) Pfund. Electrical conductivity was measured at 20 °C with a conductivity meter micro CM2210 (CRISON, Spain) following the method described by Bogdanov [26] and expressed as milliSiemens per centimeter (mS/cm).

2.4. HS-SPME Extraction

Volatile fractions of honey and flower samples were extracted by HS-SPME with a divinylbenzene/carboxen/polydimethylsiloxane (DVB/CAR/PDMS, 30 μm) fiber (Supelco Sigma Aldrich). The optimization of HS-SPME parameters was performed using two honey samples (9 and 24) and two flower samples (clementine and asphodel flowers). These samples and subsequent analyses (all honey and flower samples studies) were performed in triplicate to ensure that the coefficient of variation (CV: ratio of standard deviation to the mean) of the major compounds and the sum of the total peak areas were always <15%. The samples analyzed were placed in a 20 mL vial. The parameter optimization was based on the sum of the total peak areas measured using a gas chromatography-flame ionization detection (GC-FID) system. For each sample (both honeys and flowers): the temperatures (25 °C, 50 °C and 70 °C), the equilibration times (30, 60 and 90 min) and the extraction times (15, 30 and 45 min) were tested in various experiments. The honey concentration in distilled water was optimized after six different experiments (0.5 g/mL, 1 g/mL, 1.5 g/mL and 2 g/mL) with Na_2SO_4 addition (1 g and 2 g). The maximum sum of the total peak areas was obtained from 4 g of honey sample with 4 mL of water and 2 g of Na_2SO_4 at a temperature of 70 °C, an equilibrium time of 90 min, and an extraction time of 30 min. The flower weight was optimized after three different experiments (1 g, 3 g and 5 g). For the Asphodel flowers, the maximum sum of the total peak areas was obtained from 3 g of sample at a temperature of 70 °C, an equilibrium time of 90 min, and an extraction time of 30 min. Otherwise, the best sampling conditions of Clementine flowers were 1 g of sample at room temperature (25 °C) with an extraction time of 15 min. Before sampling, the fiber was reconditioned for 5 min in the GC injection port at 280 °C. After sampling, the SPME fiber was consecutively inserted into the GC-FID and GC-MS injection ports for 5 min for desorption of volatile components, both techniques using the splitless injection mode.

2.5. GC-FID and GC-MS Analysis

GC-FID analyses were performed using a PerkinElmer (Waltham, MA, USA) AutoSystem XL GC apparatus equipped with a FID system and a fused-silica capillary column (30 m × 0.25 mm, film thickness 1 μm) coated with Rtx-1 (PDMS). The oven temperature was programmed from 60 to 230 °C at 2 °C/min and then held isothermally at 230 °C for 35 min. The injector and detector temperatures were maintained at 280 °C. The samples were injected with an SPME inlet liner (0.75 mm i.d.; Supelco) using hydrogen as the carrier gas (1 mL/min). The retention indices of the compounds were determined relative to the retention times of a series of n-alkanes (C_5–C_{30}) with linear interpolation. The relative concentrations of components were calculated from the GC peak areas without using correction factors. Samples were also analyzed with a PerkinElmer TurboMass detector (quadrupole), coupled to a GC PerkinElmer AutoSystem XL, equipped with a fused-silica Rtx-1 capillary column. The ion source temperature was 150 °C, and the ionization energy was 70 eV. Electronic ionisation (EI) mass spectra were acquired over the mass range of 35–350 Da (scan time 1 s). Other GC conditions were the same as described for the GC-FID analysis. Identification of the components was based on: (1) the comparison of their GC retention indices (RI) on a nonpolar column, determined relative to the retention time of a series of *n*-alkanes with linear interpolation to the retention times of authentic compounds or data with the laboratory's library; (2) the comparison of the RI and spectra with commercial mass spectra libraries [27,28].

2.6. Statistical Analysis

The statistical analysis of melissopalynological data was carrying out the methodology previously described by Battesti *et al.* [11]. In the case of "spring" honey, the inclusion of *Citrus* and *Asphodelus* pollen during the nectar foraging is low or very low because of pollen maturity or floral morphology. The "under-representation" of these pollen types and entire pollen spectrum were taken into account for the characterization and comparison of pollen spectrum from "spring" honeys. Principal component analysis (PCA) was carried out using the "PCA" function and canonical correspondence analysis (CCA) was performed with "CCA" function from R software (R Foundation—Institute for Statistics and Mathematics, Austria). CCA is a multidimensional exploratory statistical method in order to demonstrate the correlation between two sets of variables obtained from the same individual.

3. Results and Discussion

3.1. Determination of Geographical and Botanical Origins of Corsican "Spring" Honeys

The analysis of 41 Corsican "spring" honeys allowed the determination of 92 taxa, including 64 nectariferous taxa and 28 only-polleniferous taxa (Table 1). A biogeographical analysis (biogeographical code: BC [5]) showed the diversity of biogeographical origins of these taxa. Mediterranean species (28 taxa, BC 1–3) associated with Eurasian and Atlantic species (13 taxa, BC 5–6) were well represented in the pollen spectrum. Additionally, cultivated species (four taxa, BC 99) were reported in more than 40% of honey samples. This distribution was consistent with the database of the characterization of the Corsican honey taxa directory [5,11].

To define the most representative taxa of Corsican "spring" honey, the presence ratio (PR) and the relative frequency (RF) distributions (mean, minimum, maximum, standard deviation and coefficient variation) of each taxon were reported. The pollen directory showed that 50 taxa (**T1–T50**) could be considered as regionally characteristic species of Corsican "spring" honey for their significant PR (>10%) and/or RF_{max} (>3%). This distribution of taxa was characterized by a wide diversity of nectariferous taxa in variable proportions associated with several only-polleniferous species. Among these taxa, two main only-polleniferous taxa, *Quercus* sp. **T1** (*Qeurcus* sp. (deciduous), *Q. ilex* and *Q. suber*) and *Cistus* sp. **T2** (*C. creticus*, *C. monspeliensis* and *C. salviifolius*), were present in all the samples analyzed, followed by *Castanea sativa* **T3** and *Fraxinus ornus* **T4** (PR > 90%). Additionally, we did not find a common predominant nectariferous taxon, unlike two previous studies [24,29]: "chestnut grove" honey predominated by *C. sativa* with PR = 100%, $FR_{max} > 80\%$ and $FR_{mean} = 92.99\%$ and "spring maquis" honey predominated by the "normal" pollen type of *Erica arborea* with PR = 100%, $FR_{max} > 45\%$ and $FR_{mean} = 47.7\%$. Quite the contrary, this directly demonstrates a diversity of nectariferous taxa with various pollen representation types: for example, "over-represented" (**T7** and **T13**), "normal" (**T5**, **T6**, **T8**, **T9** and **T14**) and "under-represented" (**T17**, **T18** and **T22**) pollen types [5].

Table 1. Statistical analysis and biogeographical characteristics of Corsican "spring" honeys' taxa.[*]

No [a]	Type [b]	Taxa	PR [c]	Relative frequency (RF) [d]					BC [g]
				Mean	Min.	Max.	SD [e]	CV [f]	
T1	P	*Quercus* sp.	100	13.2	0.8	35.7	9.7	73.8	21-35-55-58
T2	P	*Cistus* sp.	100	8.5	0.3	33.3	6.3	74.4	21-29
T3	P	*Castanea sativa* [h]	90	10.3	0.3	33.8	8.7	84.2	59
T4	P	*Fraxinus ornus*	90	3.3	0.3	22.3	5.0	153.2	58
T5	N, P	*Erica arborea*	85	7.8	0.2	35.5	8.7	112.3	21
T6	N, P	*Genista* form [i]	83	6.0	0.3	31.5	8.0	134.8	14-21-29-51-62
T7	N, P	*Lotus* sp.	76	5.3	0.3	52.8	9.5	178.3	21-51
T8	N, P	*Salix* sp.	73	6.3	0.2	29.9	7.4	117.1	51-52
T9	N, P	*Trifolium* sp.	71	14.2	0.4	53.5	16.8	117.9	21-31-51
T10	N, P	*Rubus* sp.	71	3.6	0.4	11.7	3.4	94.6	31-35
T11	N, P	*Prunus* form [j]	66	3.0	0.2	24.1	4.7	155.6	99-54
T12	P	*Eucalyptus* sp.	63	2.1	0.3	15.5	3.1	148.4	99
T13	N, P	*Echium* sp.	59	10.5	0.6	71.1	15.6	148.2	31
T14	N, P	*Apiaceae*	59	4.0	0.2	17.5	4.4	109.6	nd
T15	P	*Actinidia sinensis*	49	4.2	0.3	16.1	4.5	107.7	99
T16	N, P	*Brassicaceae* others	49	2.8	0.3	14.7	3.3	118.6	nd
T17	N, P	*Lavandula stoechas*	49	1.8	0.4	10.1	2.2	124.1	21
T18	N, P	*Citrus* sp.	44	6.1	0.2	16.1	5.2	86.6	99
T19	N, P	*Vicia* form	44	3.0	0.3	11.8	3.2	107.1	nd
T20	P	*Pistacia lentiscus*	44	3.0	0.5	9.3	2.6	88.0	29
T21	N, P	*Asteraceae Galactites* form	44	1.9	0.2	5.2	1.7	93.6	21
T22	N, P	*Asphodelus ramosus* subsp. *ramosus*	44	0.7	0.2	2.9	0.7	96.5	21
T23	P	*Scrophulariaceae* others	39	0.9	0.3	4.5	1.0	114.4	nd
T24	P	*Phillyrea* sp.	37	3.0	0.3	13.3	3.8	125.5	25
T25	P	*Olea* sp.	37	1.0	0.4	3.6	0.8	74.3	21
T26	N, P	*Viburnum tinus*	34	1.9	0.3	16.2	4.2	225.9	21
T27	N, P	*Asteraceae* (fenestrated type)	29	1.1	0.3	3.2	1.0	94.0	21-94
T28	N, P	*Rosa* sp.	27	1.2	0.3	4.5	1.3	108.1	31-51
T29	P	*Myrtus communis*	24	0.9	0.3	1.6	0.5	52.0	21
T30	N, P	*Fabaceae* others/*Dorycnopis* form	24	0.6	0.3	1.4	0.3	54.5	nd
T31	P	*Plantago* sp.	24	0.5	0.3	0.9	0.2	33.8	nd
T32	N, P	*Asteraceae Achillea* form	22	0.8	0.2	2.6	0.7	94.0	21-94
T33	P	*Poaceae*	22	0.6	0.2	1.2	0.3	54.0	nd
T34	N, P	*Crataegus monogyna*	20	2.1	0.3	7.9	2.7	130.8	51

Table 1. *Cont.*

No [a]	Type [b]	Taxa	PR [c]	Relative frequency (RF) [d]					BC [g]
				Mean	Min.	Max.	SD [e]	CV [f]	
T35	N, P	*Jasione montana*	17	2.0	0.3	10.0	3.5	176.0	54
T36	N, P	*Rosaceae* others	17	1.2	0.3	3.0	1.0	85.6	nd
T37	N, P	*Asteraceae Dittrichia* form	17	1.0	0.2	1.9	0.8	74.4	21-94
T38	N, P	*Rhamnus* sp.	15	1.0	0.3	3.3	1.1	117.6	21
T39	N, P	*Psoralea bituminosa*	15	0.7	0.3	1.6	0.5	75.6	31
T40	N, P	*Knautia* sp.	15	0.5	0.3	0.9	0.3	52.8	31
T41	N, P	*Lupinus angustifolius*	12	4.8	0.3	18.9	8.0	166.1	21
T42	P	*Cytinus hypocistis*	12	0.9	0.4	1.8	0.7	77.6	29
T43	N, P	*Hedera helix*	12	0.8	0.3	1.3	0.4	46.1	65
T44	N, P	*Liliaceae others*	12	0.4	0.3	0.6	0.2	45.8	nd
T45	N, P	*Allium* sp.	12	0.4	0.3	0.6	0.2	42.9	21-25
T46	N, P	*Acacia dealbata*	12	0.4	0.3	0.4	0	13.3	99
T47	P	*Alnus* sp.	7	2.3	0.3	6.1	3.3	146.5	51
T48	N, P	*Dorycnium* sp.	10	1.5	0.3	3.2	1.2	83.3	35
T49	N, P	*Rosmarinus officinalis*	10	1.7	0.4	3.0	1.3	77.9	21
T50	P	*Vitis vinifera*	7	1.5	0.4	3.0	1.3	89.4	99

[a] Order of taxa were classified by decreasing presence ratio (PR). [b] Type of taxa: P, polleniferous taxa; N, nectariferous taxa [5]. [c] PR: presence ratio, number of honey samples presented/41 samples, expressed as %. [d] Mean, Min., Max. values expressed as relative frequency RF (number of specify pollen counted/total pollen counted). [e] SD: standard deviation. [f] CV: coefficient variation. [g] Biogeographical Code, according to Battesti [5]: 1—*Endemic*: 14 Mediterraneo-montane origin; 2—*Steno-Mediterranean*: 21 Wider stenomedit., 25 Western stenomedit., 29 Western macaronesian stenomedit.; 3—*Eury-Mediterranean*: 31 Wider eurymedit., 35 Western eurymedit.; 5—*Eurasian*: 51 Wider eurasian, 52 Eurasian, 54 European-caucasian, 55 European, 58 South east european, 59 Southern European; 6—*Atlantic*: 62 Subatlantic, 65 Atlantic Mediterranean; 94 sub-Cosmopolitan; 99 Cultivated plants; nd: not defined. [h] *Castanea sativa*, taxa of "over-represented" type, could be considered as only-polleniferous taxon according to its RF (<40%) and lower pollen density taking into account its over-represented pollen type [6,24]. [i] *Genista* form contained essentially *Genista corsica*, and also *Cytisus villosus*, *Calicotome spinosa* and *Calicotome villosa*. [j] *Prunus* form contained *Prunus* sp. and other fruit tree. * Forty two other determined taxa (PR < 10%): *Populus* sp., *Boraginaceae* others, *Rumex* sp., *Ostrya carpinifolia*, *Ilex aquifolium*, *Platanus* sp., *Silene gallica*, *Stachys glutinosa*, *Anthyllis hermanniae*, *Papaver* sp., *Urticaceae*, *Reseda* sp., *Aesculus hippocastanum*, *Carpobrotus* sp., *Cercis siliquastrum*, *Potentilla* form, *Ranunculaceae*, *Corylus avellana*, *Asteraceae Helichrysum* form, *Arbutus unedo*, *Erica* others, *Cupressaceae*, *Sambucus ebulus*, *Anemone hortensis*, *Smilax aspera*, *Cynoglossum creticum* form, *Amaryllidaceae*, *Cyperaceae*, *Helleborus lividus* subsp. *corsicus*, *Mercurialis annua*, *Robinia pseudoacacia*, *Clematis* sp., *Chenopodiaceae*, *Caryophyllaceae* others, *Borago officinalis*, *Centaurea* sp., *Verbascum* sp., *Teucrium* sp., *Centaurium erythrae*, *Veronica* sp., *Asteraceae* others, *Buxus sempervirens* (according to decreasing PR).

According to these considerations, two groups of honeys could therefore be distinguished, based not by their FR distributions, but by characteristic associations of taxa (Table 2, Table S1-supplementary materials). The first group included 18 samples (group I: 1–18) and was characterized by the association of cultivated taxa: *Citrus sp.* **T18** and *A. sinensis* **T15** (PR 100% in 18 samples) followed by *Prunus* form **T11** and *Olea* sp. **T25**. *Citrus* sp. contained essentially *C. sinensis* × *reticulata*, which possessed an under-represented pollen type, principally due to nectar secretion of *Citrus* sp. flowers, often before the maturity of stamens. *Citrus* pollen varied between 0.2% and 16.1%, with an average of 6.1%. The second group (group II: 23 samples, 19–41) was characterized by the absence of a *Citrus* sp./*A. sinensis* association and the significant presence of *A. ramosus* **T22** associated with *Pistacia lentiscus* **T20**, *Phillyrea* sp. **T24**, *Apiaceae* **T14** and *Brassicaceae* **T16**. *A. ramosus* displayed an extreme "under-represented" pollen type due to the flower form (nectar protected by a large base of long stamens that prevented contact with pollen during bee foraging) and the large pollen size. *Asphodelus* pollen was present in two samples of group I (0.5%–1.3%) and 16 samples of group II (0.2%–2.9%).

In the case of honey with the "under-represented" pollen type, the contribution of other nectariferous species could not be discounted. It had to note that some honeys samples possessed dominant nectariferous taxa (RF > 45%): *Trifolium* sp. **T9** for sample 2 and 3, *Echium* sp. **T13** for sample 4 and *Lotus* sp. **T7** for sample 38. The nectar contribution of these taxa could not be neglected. Otherwise, several taxa might take part in the honey composition for their high RF in the pollen spectrum: *Trifolium* sp. **T9** and *E. arborea* **T5** were characteristic for both groups (RF_{max} 53.5% and 35.5% for group I and 44.1% and 29.9% for group II, respectively); *Echium* sp. **T13**, *Prunus* form **T11** and *Viburnum tinus* **T26** possessed a higher RF_{max} in group I (71.1%, 24.1% and 16.2%, respectively) than in group II (30.1%, 3.3% and 3.7%, respectively), while *Lotus* sp. **T7**, *Genista* form **T6**, *Salix* sp. **T8**, *Lupinus angustifolius* **T41** and *Apiaceae* **T14** were higher in group II (FR_{max}: 52.8%, 31.5%, 29.9%, 18.9% and 17.5%, respectively) than in group I (FR_{max}: 8.7%, 11.8%, 12.9%, 3.5% and 4.5%, respectively).

A quantitative analysis showed that 32 samples possessed a pollen density between 20 and 100×10^3 PG/10 g, eight samples were between 100 and 300×10^3 PG/10 g and one sample (23) could be distinguished by high pollen density (600×10^3 PG/10 g). Compared with the previous studies of Corsican "chestnut grove" and "*Erica arborea* spring maquis" honey (636.6×10^3 PG/10 g and 177×10^3 PG/10 g, respectively), the "spring" honey displayed a lower pollen density (90×10^3 PG/10 g) [24,29]. Excluding sample 23, the average pollen density of "clementine" honeys (68×10^3 PG/10 g) was slightly lower than that of other Corsican "spring" honeys (84×10^3 PG/10 g) (Table 2). The decreasing pollen richness was in accordance with the pollen representation type in the spectrum of the predominant nectariferous taxa: "over-represented" (*C. sativa*), "normal" (*E. arborea*) and "under-represented" (*Citrus* sp.) types.

Table 2. Melissopalynological and physico-chemical characteristics of Corsican "clementine" honeys and other "spring" honeys.

Melissopalynological Data			Group I—"Clementine" Honeys 18 Samples (1–18) RF [c]						Group II—Other "Spring" Honeys 23 Samples (19–41) RF [c]					
No. [a]	Type [b]	Taxa	PR	Mean	Min.	Max.	SD	CV	PR	Mean	Min.	Max.	SD	CV
Main nectariferous taxa														
T18	N, P	*Citrus* sp.	100	6.1	0.2	16.1	5.2	86.6	-	-	-	-	-	-
T5	N, P	*Erica arborea*	89	7.5	0.3	35.5	10.2	135.4	83	8.0	0.2	29.7	7.6	94.9
T8	N, P	*Salix* sp.	78	5.8	0.6	12.9	4.5	77.7	70	6.7	0.2	29.9	9.3	139.4
T11	N, P	*Prunus* form	78	4.8	0.2	24.1	6.1	128.4	57	1.2	0.3	3.3	0.9	75.5
T7	N, P	*Lotus* sp.	72	2.7	0.4	8.7	2.5	91.9	78	7.2	0.3	52.8	12.1	167.5
T10	N, P	*Rubus* sp.	67	2.3	0.4	6.5	1.8	77.9	74	4.5	0.4	11.7	4.0	88.4
T9	N, P	*Trifolium* sp.	67	12.9	0.5	53.5	19.4	150.7	74	15.2	0.4	44.1	15.2	100.2
T6	N, P	*Genista* form	67	3.1	0.3	11.8	3.4	107.1	96	7.5	0.6	31.5	9.4	125.3
T13	N, P	*Echium* sp.	44	19.3	1.3	71.1	23.3	121.0	70	6.1	0.6	30.1	7.5	123.2
T26	N, P	*Viburnum tinus*	22	4.5	0.3	16.2	7.8	174.5	43	0.8	0.3	3.7	1.1	127.7
T14	N, P	*Apiaceae*	28	2.2	0.2	4.5	2.0	88.8	83	4.5	0.3	17.5	4.8	106.0
T22	N, P	*Asphodelus ramosus* subsp. *ramosus*	11	0.9	0.5	1.3	0.5	57.3	70	0.7	0.2	2.9	0.7	104.1
T16	N, P	*Brassicaceae* others	33	1.2	0.3	2.7	1.0	77.9	61	3.5	0.4	14.7	3.8	108.2
T17	N, P	*Lavandula stoechas*	39	1.7	0.5	4.4	1.4	81.9	57	1.8	0.4	10.1	2.6	142.4
T19	N, P	*Vicia* form	28	2.3	0.7	6.0	2.3	102.4	57	3.3	0.3	11.8	3.6	107.6
T41	N, P	*Lupinus angustifolius*	17	1.5	0.3	3.5	1.8	119.1	9	9.8	0.7	18.9	12.9	131.6
		Other nectariferous taxa	100	3.3	0.3	11.0	3.0	89.9	100	5.2	1.0	10.0	2.8	53.5
Main only-polleniferous taxa														
T15	P	*Actinidia sinensis*	100	4.6	0.3	16.1	4.6	99.7	9	0.5	0.3	0.7	0.3	55.3
T2	P	*Cistus* sp.	100	6.9	0.3	20.4	5.2	76.0	100	9.8	0.3	33.3	6.9	70.9
T1	P	*Quercus* sp.	100	16.5	0.8	35.7	10.7	64.9	100	10.6	1.3	29.0	8.1	77.2
T3	P	*Castanea sativa*	89	7.9	0.3	25.0	7.3	91.7	91	12.1	0.3	33.8	9.4	77.4
T4	P	*Fraxinus ornus*	89	6.1	0.5	22.3	6.6	108.4	91	1.1	0.3	3.3	1.0	87.6
T12	P	*Eucalyptus* sp.	72	3.3	0.4	15.5	4.1	125.9	57	1.0	0.3	3.4	1.0	99.3
T25	P	*Olea* sp.	50	0.8	0.4	1.3	0.3	38.9	26	1.3	0.6	3.6	1.1	84.3
T20	P	*Pistacia lentiscus*	11	0.6	0.6	0.6	0.0	8.2	70	3.3	0.5	9.3	2.7	80.4
T24	P	*Phillyrea* sp.	17	4.7	0.3	13.3	7.4	157.4	52	2.6	0.4	9.9	2.7	103.5
		Other only-polleniferous taxa	100	1.9	0.3	8.9	2.1	110.0	78	1.7	0.3	8.5	2.0	117.8
		Pollen density (10^3 PG/10 g) [d]		68	20	202	52	77		107	22	603	126	118
Physico-chemical data [e]														
		Color		26.4	11.0	55.0	13.6	51.5		33.3	18.0	71.0	16.3	48.7
		Electrical conductivity		0.25	0.15	0.42	0.07	27.72		0.24	0.13	0.45	0.09	36.96

[a] Taxa number is given in Table 1. [b] Type of taxa: P, polleniferous taxa; N, nectariferous taxa [5]. [c] Mean, Min., Max. values expressed as relative frequency RF (number of specify pollen counted/total pollen counted). [d] Pollen density expressed as the absolute number of pollen grains in 10 g of honey (10^3 PG/10 g). [e] Unity of parameters: colour (mm Pfund); electrical conductivity (mS/cm).

3.2. Physicochemical Characteristics of Corsican "Spring" Honeys

Corsican "spring" honeys possessed light to very light colors. The mean value of coloration was 30.0 ± 15.4 mm Pfund, with great variation between 11.0 and 71.0 mm Pfund (Table 2). The two groups exhibited quite similar coloration values: 26.4 ± 13.6 mm Pfund for "clementine" honeys and 33.3 ± 16.3 mm Pfund for the other "spring" honeys. For each group, nine samples possessed a very light coloration value (<20.0 mm Pfund). Only one sample (17) of "clementine" honeys had a coloration value >50.0 mm Pfund while five samples (23, 35, 38, 39 and 41) of other "spring" honeys possessed coloration values between 50.0 and 71.0 mm Pfund.

The average electrical conductivity value of the honey samples was 0.25 ± 0.08 mS/cm with a variation of 0.13–0.45 mS/cm (Table 2). The electrical conductivity of the two groups was also quite similar: 0.25 ± 0.07 mS/cm for "clementine" honeys (range: 0.15–0.42 mS/cm) and 0.24 ± 0.09 mS/cm for other "spring" honeys (range: 0.13–0.45 mS/cm). Only three samples (17, 34 and 41) of these honeys had medium electrical conductivity (>0.4 mS/cm).

The coloration and electrical conductivity values of Corsican "spring" honeys were lower than those of "chestnut grove" and "*Erica arborea* spring maquis" honey ranges [5,24,29].

3.3. Chemical Variability of Corsican "Spring" Honeys

GC and GC/MS analysis of the headspaces of Corsican "spring" honeys allowed the identification of 43 compounds that accounted for 71.5%–96.8% of the total volatile composition (Table 3, Table S2-supplementary materials). It should be noted that the volatile fraction of "spring" honeys is rich in aldehyde (22.1%–63.1%) and alcohol (2.8%–40.2%) components.

To synthesize the chemical data, PCA was used to examine the relative distribution of the matrix of "spring" honey samples according to their volatile chemical compositions. The analyses included 17 compounds: two hydrocarbons (**C2** and **C12**), eight aldehydes (**C5**, **C9**, **C14**, **C25**, **C27**, **C28**, **C31** and **C32**), two ketones (**C23** and **C24**), two esters (**C38** and **C41**), two oxides (**C39** and **C42**) and one alcohol (**C37**). As shown in Figure 1a, the principal factorial plane (axes 1 and 2) accounted for 58.91% of the entire variability of the honey samples. Dimension 1 (42.24%) correlated negatively **C39** and negatively with other compounds. Dimension 2 (16.67%) correlated negatively with two hydrocarbons (**C2** and **C12**), two aldehydes (**C5** and **C14**) and one oxide **C42** and positively with with two ketones (**C23** and **C24**), three aldehydes (**C25**, **C27** and **C28**), one ester **C38** and one oxide other compounds.

The plot established according to the first two principal components suggested the existence of two main groups (Figure 1b). Group I contained 17 samples (1–17), which corresponded to the group of "clementine" honeys (except sample 18). This group I was characterized by the presence of lilac aldehyde isomers (**C25**, **C27** and **C28**), *p*-menth-1-en-9-al isomers (**C31** and **C32**) and methyl anthranilate **C38**, which were absent in the other honey samples (group II). It was rich in furan compounds (group I: 26.2% *versus* group II: 7.6%), but not in phenolic components (group I: 29.4% *versus* group II: 40.0%). Aldehyde components were also higher in group I (49.1%) than in group II (39.7%). Group II could be divided into subgroups IIa (five samples: 20, 33, 36, 38 and 39) and IIb (18 samples: 18, 19, 21–32, 34, 37, 40 and 41). These two subgroups were characterized by a greater abundance of phenolic compounds (group IIa: 39.9% and group IIb: 43.0%), but group IIa displayed a higher value for linear compounds (group IIa: 26.2% and group IIb: 19.2%). Additionally, subgroup IIa had a higher amount of aldehyde (group IIa: 40.7% and group IIb: 12.8%) and alcohol (group IIa: 35.6% and group IIb: 9.2%) compounds than subgroup IIb. This latter group displayed a greater abundance of ketones (group IIa: 4.3% and group IIb: 22.5%). Finally, sample 35 was characterized by 32.8% of hydrocarbons, whereas the abundance of hydrocarbons was not >25% in the other honey samples.

Table 3. Chemical composition of volatile fraction of Corsican "spring" honeys.

No [a]	Components	RI [b]	Group I "Clementine" Honeys [c]			Group II "Not-Clementine" Honeys [c]						Sample 35
						IIa			IIb			
			Mean ± SD [d]	Min.	Max.	Mean ± SD [d]	Min.	Max.	Mean ± SD [d]	Min.	Max.	
C1	3-Methyl-3-buten-1-ol	704	2.9 ± 2.76	0.3	11.3	1.8 ± 1.19	0.7	3.7	1.8 ± 1.44	0.4	6.0	2.8
C2	Methyl-benzene	741	6.5 ± 4.45	1.5	15.6	4.1 ± 2.08	2.4	7.1	6.2 ± 4.08	1.5	17.3	10.4
C3	Hexanal	773	1.1 ± 0.46	0.5	2.0	1.6 ± 2.62	0.1	6.3	1.6 ± 1.13	0.3	4.5	1.9
C4	Octane	790	1.4 ± 1.08	0.3	4.7	0.9 ± 0.77	0.3	2.2	2.6 ± 1.63	0.7	5.6	1.4
C5	3-Furaldehyde	800	2.8 ± 1.56	0.6	5.9	3.5 ± 1.37	2.5	5.9	3.5 ± 1.78	1.9	8.3	18.5
C6	2-Methyl butanoic acid	858	0.8 ± 0.91	0.1	3.3	4.6 ± 3.21	1.1	7.6	2.9 ± 4.59	0.1	20.4	2.4
C7	2-Methyl octane	873	0.4 ± 0.28	0.1	0.9	0.5 ± 0.44	0.1	1.2	0.7 ± 0.38	0.1	1.7	1.6
C8	Nonane	893	0.7 ± 0.65	0.2	2.5	1.2 ± 0.35	0.8	1.7	1.5 ± 0.92	0.2	3.5	3.8
C9	Benzaldehyde	924	5.5 ± 3.56	2.4	17.9	10.4 ± 3.85	5.4	14.8	8.8 ± 4.89	2.5	18.4	3.0
C10	Hexanoic acid	969	0.7 ± 0.25	0.4	1.4	1.7 ± 1.68	0.3	3.9	1.2 ± 0.75	0.4	3.3	-
C11	Octanal	982	1.0 ± 0.47	0.2	2.1	0.6 ± 0.12	0.5	0.7	1.6 ± 1.45	0.5	6.6	1.0
C12	2,2,4,6,6-Pentamethylheptane	992	1.1 ± 0.60	0.4	2.4	0.6 ± 0.53	0.1	1.3	2.3 ± 1.83	0.2	5.2	15.6
C13	*p*-Methylanisol	995	0.9 ± 1.13	0.1	4.9	1.1 ± 1.10	0.3	3.0	0.9 ± 1.48	0.1	6.1	-
C14	Phenylacetaldehyde	1006	10.1 ± 10.68	0.8	39.1	16.5 ± 6.93	7.2	25.7	21.2 ± 7.83	3.7	36.2	13.0
C15	*p*-Cymene	1008	0.7 ± 0.29	0.1	1.0	0.9 ± 0.42	0.6	1.2	-	-	-	-
C16	Acetophenone	1037	0.2 ± 0.08	0.1	0.4	0.4 ± 0.21	0.2	0.5	0.3 ± 0.10	0.2	0.4	-
C17	*trans*-Furanoid-linaloxide	1049	1.5 ± 1.00	0.8	4.0	1.3 ± 1.25	0.5	3.5	2.4 ± 1.20	1.0	6.3	-
C18	*cis*-Furanoid-linaloxide	1064	1.1 ± 0.40	0.7	2.0	1.0 ± 0.29	0.7	1.4	1.1 ± 0.26	0.5	1.6	-
C19	β-Phenylethanol	1077	4.2 ± 1.54	2.2	5.8	1.6 ± 0.00	1.6	1.6	3.3 ± 1.67	2.1	5.8	-
C20	Nonanal	1079	2.7 ± 1.73	0.9	7.6	1.8 ± 1.28	0.4	3.5	3.0 ± 2.29	0.5	7.2	2.9
C21	Linalol	1084	2.4 ± 1.82	0.2	6.6	1.3 ± 1.62	0.3	3.2	12.5 ± 10.42	2.1	32.3	tr
C22	Hotrienol	1085	4.1 ± 4.39	0.7	10.5	-	-	-	9.7 ± 0.00	9.7	9.7	-
C23	Isophorone	1087	2.8 ± 1.54	0.2	4.9	18.2 ± 8.02	8.8	29.3	3.3 ± 3.5	0.1	9.6	-
C24	4-Oxoisophorone	1102	0.9 ± 0.33	0.3	1.4	4.2 ± 1.87	2.3	6.4	1.5 ± 0.99	0.3	5.0	-
C25	(2S,2'S,5'S)-Lilac aldehyde	1112	5.4 ± 2.36	1.3	8.9	-	-	-	1.5 ± 0.00	1.5	1.5	-
C26	Dihydrolinalool	1116	1.2 ± 0.66	0.5	3.0	-	-	-	1.1 ± 0.00	1.1	1.1	-
C27	(2R,2'S,5'S)-Lilac aldehyde	1121	10.5 ± 3.66	4.9	16.5	-	-	-	2.4 ± 0.00	2.4	2.4	-
C28	(2R,2'R,5'S)-Lilac aldehyde	1134	4.8 ± 1.85	2.2	8.1	-	-	-	1.1 ± 0.00	1.1	1.1	-
C29	Octanoic acid	1167	1.7 ± 1.55	0.3	6.0	0.9 ± 0.38	0.3	1.3	1.6 ± 0.78	0.7	3.8	4.7
C30	Decanal	1174	1.2 ± 0.67	0.2	2.8	0.6 ± 0.25	0.2	0.8	1.5 ± 0.54	0.6	2.3	-
C31	*p*-Menth-1-en-9-al (isomer 1)	1184	1.9 ± 0.41	1.2	2.7	-	-	-	-	-	-	-
C32	*p*-Menth-1-en-9-al (isomer 2)	1186	1.7 ± 0.46	0.5	2.5	-	-	-	-	-	-	-
C33	*p*-Anisaldehyde	1208	0.7 ± 1.19	0.1	4.6	0.9 ± 0.37	0.3	1.1	0.4 ± 0.21	0.2	0.8	-
C34	2,3,5-Trimethylphenol	1248	0.4 ± 0.30	0.1	1.1	1.0 ± 0.69	0.4	2.0	0.8 ± 0.68	0.1	2.0	-

Table 3. *Cont.*

No [a]	Components	RI [b]	Group I "Clementine" Honeys [c]			Group II "Not-Clementine" Honeys [c]						Sample 35
						IIa			IIb			
			Mean ± SD [d]	Min	Max	Mean ± SD [d]	Min	Max	Mean ± SD [d]	Min	Max	
C35	4-*n*-Propylanisol	1264	1.6 ± 1.78	0.2	5.7	2.4 ± 1.29	0.8	4.3	3.8 ± 2.45	1.4	6.3	-
C36	Nonanoic acid	1271	2.7 ± 1.39	0.5	4.9	2.6 ± 0.94	1.4	3.7	3.1 ± 1.24	1.3	6.4	-
C37	3,4,5-Trimethylphenol	1290	0.5 ± 0.32	0.2	1.4	5.4 ± 2.71	2.9	9.4	0.5 ± 0.67	0.1	2.0	-
C38	Methyl anthranilate	1300	1.4 ± 0.96	0.2	3.5	-	-	-	-	-	-	-
C39	*cis-p*-Mentha-1(7),8-dien-1-hydroperoxide	1348	0.4 ± 0.14	0.2	0.7	-	-	-	-	-	-	-
C40	Decanoic acid	1362	1.2 ± 0.45	0.6	2.1	1.3 ± 0.75	0.1	1.9	1.7 ± 1.50	0.6	6.8	3.8
C41	Methyl 3,5-dimethoxybenzoate	1494	-	-	-	0.4 ± 0.26	0.2	0.7	0.5 ± 0.19	0.3	0.8	-
C42	Methyl syringate	1722	-	-	-	0.5 ± 0.50	0.1	1.4	0.9 ± 1.17	0.1	4.1	-
C43	Tricosane	2305	0.3 ± 0.17	0.1	0.5	0.5 ± 0.00	0.5	0.5	0.5 ± 0.22	0.2	0.7	-
	Total identification (%)		84.2 ± 6.95	71.5	94.5	91.2 ± 5.43	84.2	96.8	86.3 ± 5.49	78.8	96.7	86.8
	Total peak area(10⁶) [e]		3.8 ± 1.96	1.3	7.4	2.9 ± 1.27	1.6	4.5	2.4 ± 1.08	0.8	4.4	0.3
	Hydrocarbons		10.6 ± 5.23	4.7	20.8	7.7 ± 3.59	4.8	13.5	13.3 ± 5.88	5.0	23.9	32.8
	Oxygenated compounds		73.6 ± 7.23	58.2	82.8	83.6 ± 4.11	79.2	90.3	73.7 ± 7.56	58.1	81.7	54.0
	Phenolic compounds		29.4 ± 12.3	12.6	59.3	43.0 ± 7.39	34.8	53.0	39.9 ± 11.22	23.2	60.4	26.4
	Furan compounds		26.2 ± 7.85	12.2	38.6	5.8 ± 2.83	3.9	10.8	7.5 ± 1.99	4.3	11.0	18.5
	Linear compounds		21.0 ± 7.08	11.3	36.6	19.2 ± 3.88	14.8	23.4	26.2 ± 9.63	11.3	53.4	41.9
	Terpenic compounds		31.0 ± 9.88	15.4	52.2	3.5 ± 2.64	1.7	8.1	13.6 ± 13.05	1.5	45.8	0
	Ketones		2.5 ± 2.14	0	6.0	22.5 ± 9.67	11.6	35.7	4.3 ± 3.77	0.8	11.6	0
	Aldehydes		49.1 ± 8.03	34.4	63.1	35.6 ± 9.73	26.1	47.7	40.7 ± 9.78	22.1	52.3	40.3
	Esters		1.4 ± 0.96	0.2	3.5	0.2 ± 0.29	0	0.7	0.3 ± 0.27	0	0.8	0
	Alcohols		9.7 ± 6.17	3.3	27.8	9.2 ± 3.3	4.9	12.7	12.8 ± 10.40	3.0	40.2	2.8
	Acids		6.8 ± 3.31	0.4	15.4	9.7 ± 5.23	3.0	15.1	10.2 ± 5.28	5.6	27.0	10.9
	Oxides		5.2 ± 2.76	2.5	12.6	6.3 ± 2.45	3.0	9.3	5.5 ± 2.64	3.3	14.3	0

[a] Order of elution is given on apolar coloumn (Rtx-1). [b] Retention indice on the Rtx-1 apolar column. [c] Group number was given in "Chemical variability of Corsican "spring" honeys". [d] Means ± SD, Min. and Max. values expressed as percentages. [e] Total peak area was expressed in arbitrary units.

Figure 1. Principal component analysis (PCA) of Corsican "spring" honey volatile data.

(a) PCA distribution of variable

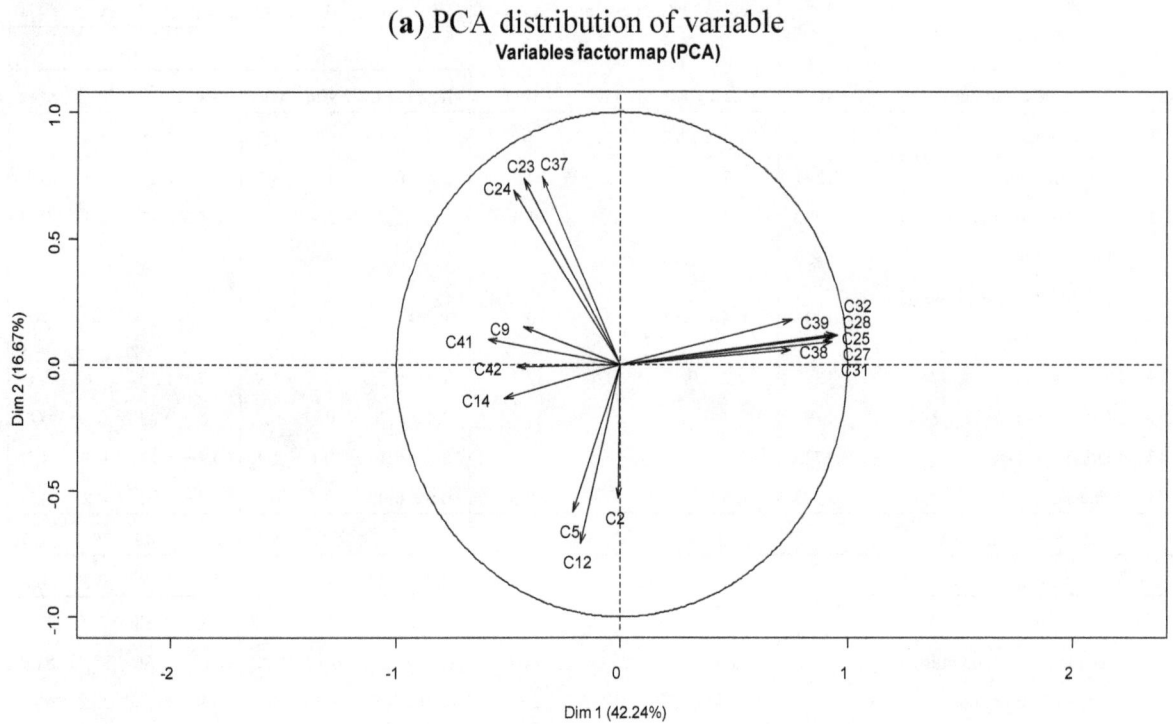

Volatile compounds number corresponding to those of Table 3.

(b) PCA distribution of honey samples (1–41)

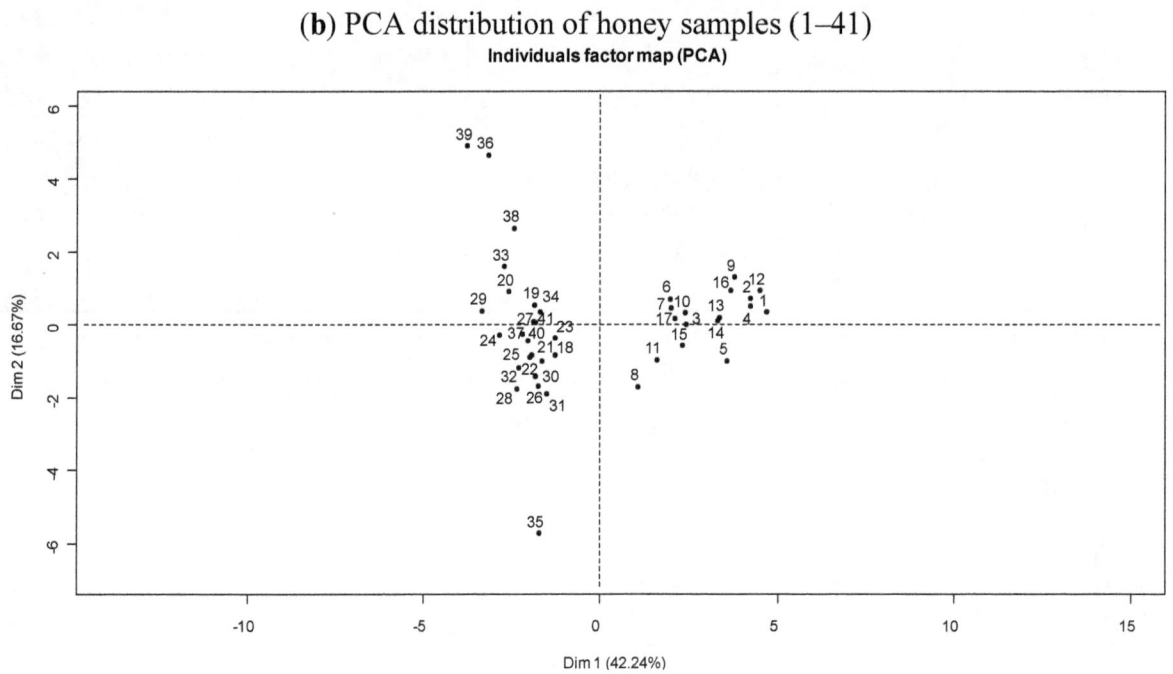

3.4. Botanical Origin and Volatile Composition of Corsican "Spring" Honeys

The 17 samples of "clementine" honeys (group I: 1–17) could be distinguished from other "spring" honeys (group II) by the presence of three lilac aldehydes (**C25**, **C27** and **C28**) and two *p*-menth-1-en-9-al isomers (**C31** and **C32**) (Table 3, Table S2-supplementary materials). These honey samples were dominated by phenolic compounds (12.6%–59.3%), followed by furan compounds (12.2%–38.6%) and linear compounds (11.3%–36.6%). The main components were phenylacetaldehyde **C14** (0.8%–39.1%), methyl-benzene **C2** (1.5%–15.6%), (2*R*,2′*S*,5′*S*)-lilac aldehyde **C27** (4.9%–16.5%), (2*S*,2′*S*,5′*S*)-lilac aldehyde **C25** (1.3%–8.9%), benzaldehyde **C9** (2.4%–17.9%) and (2*R*,2′*R*,5′*S*)-lilac aldehyde **C28** (2.2%–8.1%). Low amounts of methyl anthranilate **C38** (0.2%–3.5%) were found in the volatile fraction of the "clementine" honeys analyzed. This component is a known chemical marker of *Citrus* (species not specified) unifloral honey [13]. Additionally, various linalool derivatives, such as linalool oxides, lilac aldehydes and/or *p*-menth-1-en-9-al isomers, have also been reported as characteristic compounds of citrus unifloral honeys from Spain and Greece [15,16,30–33]. These compounds were also identified in the volatile components of Corsican "clementine" honeys. Conversely, some other linalool derivatives, such as lilac alcohol isomers (previously reported in the Spanish and Greek citrus honeys), were not detected in our honey samples. Alissandrakis *et al.* [15] showed that methyl anthranilate and lilac aldehydes could be found in honeys of mixed botanical origin with the presence of citrus nectar. These volatile compounds were also detected in the honey samples 2–4 in which were found the RF$_{max}$ of *Trifolium* sp. **T9** and *Echium* sp. **T13** taxa. Chemical investigation showed that these honey samples displayed the similar volatile composition of "clementine" honey. As these two taxa could provide great quantity of nectar and pollen [34], it appeared that they played only a polleniferous role in these honey samples.

The volatile composition of *C. sinensis* × *reticulata* flowers has not been reported previously. The HS-SPME fraction of clementine flowers is characterized by 29 compounds, which accounted for 75.5%–87.0% of the volatile composition (Table 4). Linalool (9.6%–22.6%), sabinene (13.4%–19.6%), dihydrolinalool (8.5%–14.8%) and myrcene (5.6%–6.5%) were identified as the main compounds. Linalool and dihydrolinalool were also found in low concentrations in the volatile fraction from "spring clementine" honey samples. Methyl anthranilate was detected in the volatile fraction of Corsican clementine flowers (0.1%–0.3%) and corresponding honeys.

The decrease in linalool amount and the occurrence of other linalool derivates (hotrienol, linalool oxides, lilac aldehyde isomers and *p*-menth-1-en-9-al isomers) in honey samples could be explained by the enzymatic degradation of linalool by some pathways [15]: (1) linalool can be transformed to 8-hydroxylinalool isomers by enzymatic hydroxylation at the C8 position, and then hotrienol; (2) 8-hydroxylinalool can be transformed to lilac aldehyde via (*E*)-8-oxolinalool and lilac alcohols, or *p*-menth-1-en-9-al via 8-hydroxygeraniol and (3) linalool can also be transformed via 6,7-hydroxylinalool into furanoid linalool oxide isomers under acidic conditions or by heating. These results were in accordance with those previously reported on the volatile fraction of citrus flowers and corresponding honeys [16]. It demonstrated that the flowers from *Citrus* species (orange, tangerine and sour orange) had high amounts of linalool (51.6%–80.6%) and that the honeys consisted of more than 80% of linalool derivatives (lilac aldehydes and lilac alcohols).

Table 4. Chemical composition of volatile fraction of clementine and asphodel flowers.

Components [a]	RI(Lit) [b]	RI [c]	Clementine Flower [d]			Asphodel Flower [e]			Identification [g]
			Mean ± SD [f]	Min.	Max.	Mean ± SD [f]	Min.	Max.	
3-Furaldehyde	799	800	-	-	-	1.0 ± 0.87	0.5	2.7	RI, MS
Furfural	831	836	-	-	-	3.5 ± 1.26	1.7	5.2	RI, MS
2-Furanmethanol	839	842	-	-	-	2.1 ± 1.48	0.8	4.7	RI, MS, Ref
Heptanal	882	876	-	-	-	5.4 ± 2.46	3.1	9.5	RI, MS
α-Thujene	924	922	1.2 ± 0.21	1.0	1.4	-	-	-	RI, MS
α-Pinene	932	931	3.6 ± 2.46	2.0	6.4	-	-	-	RI, MS
Benzaldehyde	929	933	-	-	-	2.7 ± 0.78	1.4	3.7	RI, MS
Tetrahydro-citronellene	937	935	6.8 ± 4.90	3.3	12.4	-	-	-	RI, MS, Ref
β-Citronellene	943	940	2.2 ± 0.15	2.0	2.3	-	-	-	RI, MS
Octen-3-ol	962	955	-	-	-	0.2 ± 0.05	0.1	0.2	RI, MS
Furfuryl acetate	964	959	-	-	-	0.7 ± 0.31	0.5	1.3	RI, MS, Ref
Sabinene	973	958	16.8 ± 3.14	13.4	19.6	-	-	-	RI, MS
2-Pentylfuran	973	966	-	-	-	0.8 ± 1.00	0.2	2.8	RI, MS
β-Pinene	978	972	1.5 ± 1.36	0.4	3.0	-	-	-	RI, MS
Myrcene	987	979	6.1 ± 0.45	5.6	6.5	-	-	-	RI, MS
Octanal	981	980	-	-	-	7.0 ± 3.12	3.5	12.6	RI, MS
(Z)-3-Hexenyl acetate	989	984	-	-	-	21.6 ± 14.27	5.2	41.8	RI, MS
(E)-3-Hexenyl acetate	1002	994	-	-	-	0.8 ± 0.54	0.1	1.5	RI, MS
α-Phellandrene	1002	995	1.5 ± 0.23	1.4	1.8	0.3 ± 0.12	0.1	0.4	RI, MS
α-Terpinene	1013	1008	0.6 ± 0.44	0.3	1.1	-	-	-	RI, MS
Phenylacetaldehyde	1012	1009	-	-	-	0.9 ± 0.67	0.2	2.1	RI, MS
p-Cymene	1015	1011	0.6 ± 0.10	0.5	0.7	-	-	-	RI, MS
p-Menth-1-ene	1017	1018	0.5 ± 0.15	0.4	0.7	-	-	-	RI, MS
Limonene	1025	1020	1.5 ± 0.70	0.8	2.2	-	-	-	RI, MS
(Z)-β-Ocimene	1029	1024	0.1 ± 0.06	0.1	0.2	-	-	-	RI, MS
(E)-2-Octenal	1034	1034	-	-	-	0.4 ± 0.29	0.1	0.8	RI, MS
(E)-β-Ocimene	1041	1036	2.6 ± 2.21	0.9	5.1	-	-	-	RI, MS
γ-Terpinene	1051	1047	1.1 ± 0.51	0.5	1.5	-	-	-	RI, MS
trans-Sabinene hydrate	1053	1050	1.0 ± 0.36	0.7	1.4	-	-	-	RI, MS
1-Octanol	1063	1057	-	-	-	6.0 ± 1.93	2.8	8.8	RI, MS
Terpinolene	1082	1078	0.1 ± 0.06	0.1	0.2	-	-	-	RI, MS
Nonanal	1076	1081	-	-	-	25.8 ± 10.1	16.5	38.2	RI, MS
Linalool	1086	1086	17.8 ± 7.14	9.6	22.6	1.7 ± 0.21	1.5	1.8	RI, MS
Tetrahydrolinalool	1099	1095	4.1 ± 3.07	0.7	6.7	-	-	-	RI, MS, Ref
Dihydrolinalool	1118	1114	10.8 ± 3.50	8.5	14.8	-	-	-	RI, MS, Ref
(E)-2-Nonen-1-ol	1149	1153	-	-	-	2.2 ± 1.65	0.6	4.6	RI, MS
1-Phenylethyl acetate	1166	1163	-	-	-	0.1 ± 0.05	0.1	0.2	RI, MS
Terpinen-4-ol	1164	1164	0.3 ± 0.20	0.1	0.5	-	-	-	RI, MS
α-Terpineol	1176	1173	tr	tr	tr	-	-	-	RI, MS
Decanal	1180	1182	-	-	-	1.6 ± 0.74	0.7	2.5	RI, MS
Undecanal	1285	1285	-	-	-	1.0 ± 1.03	0.2	2.8	RI, MS
Methyl anthranilate	1308	1302	0.2 ± 0.10	0.1	0.3	-	-	-	RI, MS
(E)-Jasmone	1356	1360	tr	tr	tr	-	-	-	RI, MS
Isocaryophyllene	1409	1405	tr	tr	tr	-	-	-	RI, MS

Table 4. *Cont.*

Components [a]	RI(Lit) [b]	RI [c]	Clementine flower [d]			Asphodel Flower [e]			Identification [g]
			Mean ± SD [f]	Min.	Max.	Mean ± SD [f]	Min.	Max.	
(*E*)-β-Farnesene	1446	1442	tr	tr	tr	-	-	-	RI, MS
(*E,E*)-α-Farnesene	1498	1492	0.1 ± 0.00	0.1	0.1	-	-	-	RI, MS
(*E*)-Nerolidol	1553	1548	tr	tr	tr	-	-	-	RI, MS
Heptadecane	1700	1698	0.2 ± 0.10	0.1	0.3	-	-	-	RI, MS
Total identification (%)			81.2 ± 5.75	75.5	87.0	85.9 ± 2.66	82.4	90.1	
Hydrocarbons			48.0 ± 8.16	39.7	56.0	-	-	-	
Oxygenated compounds			33.1 ± 11.88	19.5	41.3	85.9 ± 2.66	82.4	90.1	
Phenolic compounds			0.2 ± 0.1	0.1	0.3	3.7 ± 0.99	1.8	4.5	
Furan compounds			-	-	-	8.2 ± 4.42	4.3	16.7	
Linear compounds			0.2 ± 0.1	0.1	0.3	73.9 ± 5.41	65.0	79.3	
Terpenic compounds			80.8 ± 5.75	75.1	86.6	0.8 ± 0.85	0.1	1.9	
Ketones			tr	-	tr	-	-	-	
Aldehydes			-	-	-	43.8 ± 13.47	31.2	62.8	
Esters			0.2 ± 0.1	0.1	0.3	23.2 ± 14.58	5.8	43.1	
Alcohols			32.9 ± 11.8	19.4	41.1	11 ± 4.13	4.3	17	

[a] Order of elution is given on apolar coloumn (Rtx-1). [b] Retention indice of literature on the apolar column reported from references [27,28]. [c] Retention indice on the Rtx-1 apolar column. [d] Six clementine flower specimens were collected from Corsica oriental plain. [e] Six Asphodele flower specimens were collected from six localities of Corsica. [f] Means ± SD, Min. and Max. values expressed as percentages; tr trace (< 0.05%), [g] RI, Retention indice; MS, mass spectra in electronic impact mode. Ref., compounds identified from commercial data libraries: Konig *et al.* [27] (Samples 8, 34 and 35) and NIST [28] (Samples 3 and 11).

The 23 "not-clementine" honey samples (group II) were dominated by phenolic compounds (23.2%–60.4%) followed by linear compounds (11.3%–53.4%). The main compounds were phenylacetaldehyde **C14** (3.7%–36.2%), benzaldehyde **C9** (2.5%–18.4%) and methyl-benzene **C2** (1.5%–17.3%). Furanic compounds (average: 7.5%) were less abundant than in "clementine" honeys (average: 26.2%), and acid components (average: 10.3%) were more abundant than in the "clementine" honeys (average: 6.8%). To our knowledge, only one previous report focused on the volatile fraction of asphodel unifloral honeys from Sardinia [18]. Methyl syringate was detected in asphodel nectar in high concentrations and was therefore considered a marker of asphodel honeys [19]. A low content of this component (**C42**: 0.1%–4.1%) was reported in the volatile fraction of "spring" honey samples (18–21, 24–30, 32, 33 and 36–41). Additionally, the amount of methyl syringate was unrelated to the presence of *Asphodelus* pollen in the pollen spectrum. This result could be explained by the extreme "under-represented" type of *Asphodelus* pollen in Corsican "spring" honeys and/or by other nectar contributions in these honeys. The sample 18 exhibited the association of *Citrus* sp. and *A. sinensis*; it was grouped with the "not-clementine" honey. In this sample, the citrus nectar contribution was less important than in "clementine" honeys in accordance with the lower concentrations of lilac aldehyde isomers.

To our knowledge, the volatile composition of *A. ramosus* subsp. *ramosus* flowers is reported here for the first time (Table 4). The HS-SPME volatile fraction of asphodel flowers was dominated by oxygenated compounds, especially linear compounds. Nonanal (16.5%–38.2%), (*Z*)-3-hexenyl acetate (5.2%–41.8%), octanal (3.5%–12.6%), 1-octanol (5.7%–8.8%) and heptanal (3.1%–9.5%) were identified as major compounds. The two main components of the honey volatile fraction (phenylacetaldehyde and benzaldehyde) were detected in low concentrations in the flowers. Moreover, methyl syringate (a marker of asphodel honey) was not detected in the flowers analyzed. This result showed that a direct relationship between the volatile fractions of asphodel flowers and the corresponding "spring" honeys could not be established using HS-SPME analysis.

Finally, the characteristic compounds of the volatile fraction of Corsican "chestnut grove" (acetophenone and 2-aminoacetophenone) [24] and *Erica arborea* spring maquis" (*p*-anisaldehyde and 4-propylanisol) honeys [29] were found in low concentrations or not detected in the "spring" honeys studied.

3.5. Correlation of Melissopalynological and Chemical Data

To identify relationships between the melissopalynological analysis and volatile composition data of honey samples, CCA was applied on the matrix linked the relative amounts of the 17 volatile compounds (previously used in section "Chemical variability of Corsican "spring" honeys") and the relative frequency (explanatory variables) of eight nectariferous taxa (**T7–T9**, **T11**, **T13**, **T14**, **T18** and **T22**).

The correlations between the volatile composition and melissopalynolgical data were show in Figure 2. The first CCA axis was negatively related *Trifolium* sp. **T9**, *Prunus* form **T11**, *Echium* sp. **T13** and *Citrus* sp. **T18** to methyl-benzene **C2**, 2,2,4,6,6-pentametylheptane **C12**, three lilac aldehydes (**C25**, **C27** and **C28**), two *p*-menth-1-en-9-al isomers (**C31** and **C32**), methyl antranilate **C38** and *cis-p*-mentha-1(7),8-dien-1-hydroperoxide **C39**. The second axis negatively related *Trifolium* sp. **T9** and *Asphodelus* **T22** to methyl-benzene **C2**, 3-furaldehyde **C5**, Benzaldehyde **C9**, 2,2,4,6,6-pentametylheptane **C12**, phenlyacetaldehyde **C14** and methyl syringate **C42**.

The sample distribution showed the occurrence of two main groups, group I (17 samples: 1–17) and group II (24 samples: 18–41), which correspond to the groups defined in "Determination of geographical and botanical origins of Corsican "spring" honeys" Group I was characterized not only by the significant presence of lilac aldehyde isomers (**C25**, **C27** and **C28**), *p*-menth-1-en-9-al isomers (**C31** and **C32**) and methyl anthranilate **C38**, but also by the high abundance of taxa: *Citrus* sp. **T18**, *Echium* sp. **T13** and *Prunus* form **T11** (group I: 6.3%, 9.5% and 3.8% *versus* group II: 0.1%, 4.9% and 0.8%, respectively). According to the literature [15,16,30–33], all these compounds had been considered as characteristic components of citrus honey. From these results, it appeared that the other nectariferous taxa *Echium* sp. and *Prunus* form displayed a polleniferous role in these honey samples.

Group II included 24 samples that had great diversity. According to the sample distribution, we could distinguish 20 honey samples (18–34, 37, 40 and 41), which had higher values of phenylacetaldehyde **C14** and methyl syringate **C38** (21.4% and 0.7%, respectively). These honeys were also characterized by numerous herbaceous taxa with potential for nectar contribution, such as *Lotus* sp. **T7**, *Salix* sp. **T8**, *Apiaceae* **T14** and *Asphodelus* **T22** (3.5%, 4.4%, 3.2% and 0.6% *versus* I: 2.0%, 4.6%, 0.4% and 0.03%, respectively). As previously reported in literature data [19], the nectar contribution of *Asphodelus* **T22** in these honey samples was characterized by the presence of methyl syringate **C38**. In the same way, phenylacetaldehyde **C14** was reported as main volatile compound of

Salix honeys [35] and *Asphodelus* honey [18]. For the other nectariferous species *Lotus* sp. **T7** and *Apiaceae* **T14**, no chemical markers of nectar contribution was reported in previous studies.

Figure 2. Correlation between melissopalynological and volatile data of "spring" honey by canonical correspondence analysis (CCA).

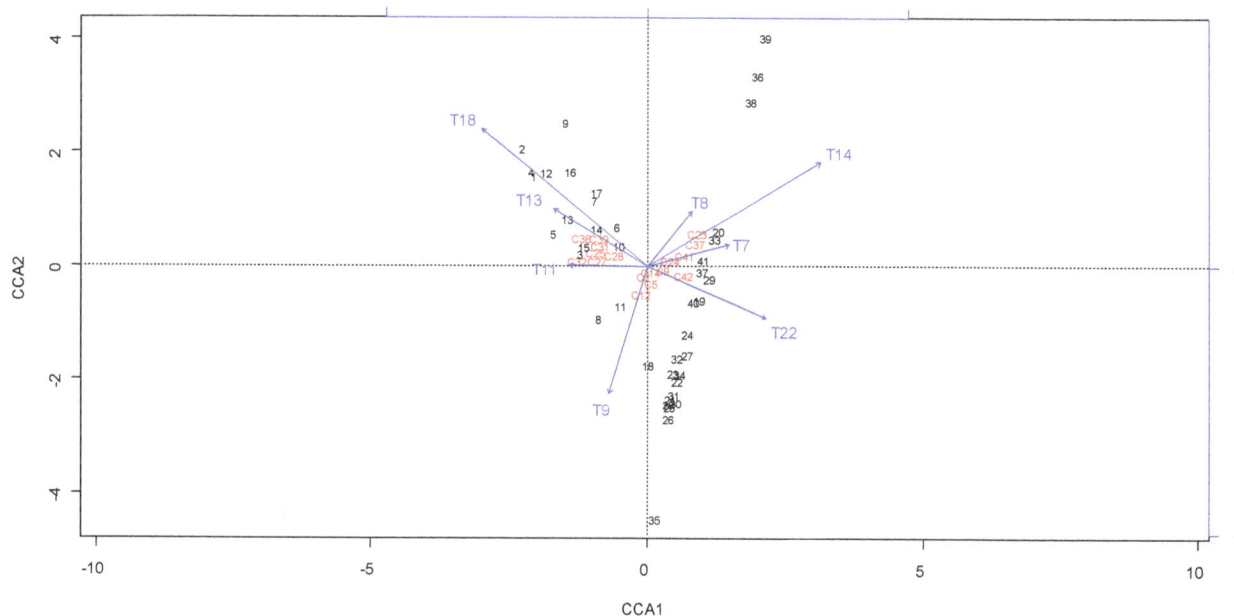

Variables: taxa number corresponding to those of Tables 2 and 3; volatile components number corresponding to those of Table 3. In the CCA plot, location of each sample indicated its compositional similarity to each other; volatile components locations indicated the similarity of their distribution to each other; length of taxa indicated the importance to the ordination, and the direction of taxa vector indicated its correlation with each axes. The perpendiculars drawn from volatile components to taxa give approximate ranking of volatile components response to the taxa variables.

4. Conclusions

Corsican "spring" honeys can be classified into two categories according to melissopalynological analysis: (1) honeys characterized by the association of cultivated plants, especially *C. sinensis* × *reticulata* with other *Citrus* species, *A. sinensis* and other fruit trees; (2) honeys without cultivated taxa, but with herbaceous species (*A. ramosus* subsp. *ramosus*, *Trifolium* sp., *Echium* sp., *Apiaceae*, *Brassicaceae*, *Lotus* sp., *etc.*), low shrub species (*Rubus* sp. and *Lavandula stoechas*) and some polleniferous taxa with precocious flowering (*P. lentiscus* and *Phillyrea* sp.).

Analysis of the volatile fraction of "spring" honeys also demonstrated the existence of two main groups in this range. The volatile fractions were often characterized by high amounts of phenylacetaldehyde, benzaldehyde and methyl-benzene. However, the chemical composition of "clementine" honeys was dominated by three lilac aldehyde isomers that were absent in the "not-clementine" honeys. The statistical analysis showed clearly that the "clementine" honeys were characterized by high volatile content (total peak area), methyl anthranilate, lilac aldehydes, *p*-menth-1-en-9-al isomers and some cultivated taxa, while the "not-clementine" honeys were characterized by phenylacetaldehyde, methyl syringate and complex taxa associations. The richness of linalool derivatives in the volatile fraction of clementine flowers suggested biochemical transformation occurring during honeybee activity or honey conservation in the hive.

Finally, it appeared that melissopalynological analysis was necessary for the certification of geographical origin and was useful for the determination of botanical origin. Moreover, analysis of the volatile composition could be used to specify the characteristics of volatile compounds in relation to the predominance and/or complexity of botanical origins of the product, especially when nectariferous species have an "under-represented" pollen type in the pollen spectrum, such as *Citrus* sp. or *Asphodelus* sp.

Acknowledgments

The authors are indebted to the Délégation Régionale à la Recherche et à la Technologie de Corse (DRRT), the Collectivité Territoriale de Corse (CTC) and European Community for partial financial support.

Conflicts of Interest

The authors declare no conflict of interest.

References

1. Battesti, M.J.; Gamisans, J.; Piana, L. *Définition du Périmètre de Production—Rapport des Experts en Vue de la Mise à l'Enquête. Demande de Reconnaissance en A.O.C. <Miel de Corse-Mele di Corsica>*; Institut National des Appelations d'Origine (INAO): Corte, France, 1997.

2. Décret n° 2010–1045 du 31 Août 2010 Relatif à l'Appellation d'Origine Contrôlée <Miel de Corse-Mele di Corsica>. Avaiable online: http://www.legifrance.gouv.fr/affichTexte.do?cidTexte=JORFTEXT000022783277 (accessed on 1 September 2013).

3. Council Regulation (EC) No 510/2006 on the Protection of Geographical Indications and Designations of Origin for Agricultural Products and Foodstuffs. "Miel de Corse/Mele di Corsica". EC No: FR-PDO-0105-0066-20.04.2011. Available online: http://eur-lex.europa.eu/LexUriServ/LexUriServ.do?uri=OJ:C:2013:134:0039:0048:EN:PDF (accessed on 1 September 2013).

4. Piana, L.; Persano Oddo, L.; Bentabol, A.; Bruneau, E.; Bogdanov, S.; Guyot Declerck, C. Sensory analysis applied to honey: State of the art. *Apidologie* **2004**, *35*, 26–27.

5. Battesti, M.J. Contribution à la Melissopalynologie Méditerranéenne: Les Miels Corses. Ph.D. Thesis, University of Marseille St. Jérôme, Marseille, France, 1990.

6. Persano Oddo, L.; Piana, L.; Bogdanov, S.; Bentabol, A.; Gotsiou, P.; Kerkvliet, J. Botanical species giving unifloral honey in Europe. *Apidologie* **2004**, *26*, 82–93.

7. Persano Oddo, L.; Piro, R. Main European unifloral honeys: Descriptive sheets. *Apidologie* **2004**, *35*, 38–81.

8. Persano Oddo, L.; Sabatini, A.G.; Accorti, M.; Colombo, R.; Marcazzan, G.L.; Piana, L.; Piazza, M.G.; Pulcini, P. *I Mieli Uniflorali Italiani—Nuove Schede di Caratterizzazione*; Ministero delle Politiche Agricole e Forestali: Gradoli, Italy, 2000.

9. Floris, I.; Palmieri, N.; Satta, A. Caratteristiche Melissopalinologiche dei Mieli di Sardegna. In *I Mieli Regionali Italiani—Caratterizzazion Melissopalinologica*; Ministero delle Politiche Agricole Alimentari e Forestali & C.R.A. Istituto Sperimentale per la Zoologia Agraria, Sezione di Apicoltura: Roma, Italy, 2007.

10. Jeanmonod, D.; Gamisans, J. *Flora Corsica*; Edisud: Aix-en-Provence, France, 2007.

11. Battesti, M.J.; Goeury, C. Efficacité de l'analyse mélitopalynologique quantitative pour la certification des origines géographique et botanique des miels: Le modèle des miels corses. *Rev. Palaeobot. Palynol.* **1992**, *75*, 77–102.

12. Von Der Ohe, W.; Persano Oddo, L.; Piana, M.L.; Morlot, M.; Martin, P. Harmonized methods of melissopalynology. *Apidologie* **2004**, *35*, 18–23.

13. Cuevas-Glory, L.F.; Pino, J.A.; Santiago, L.S.; Sauri-Duch, E. A review of volatile analytical methods for determining the botanical origin of honey. *Food Chem.* **2007**, *103*, 1032–1043.

14. Sesta, G.; Piana, M.L.; Persano Oddo, L.; Lusco, L.; Belligoli, P. Methyl anthranilate in *Citrus* honey. Analytical methode and suitability as a chemical marker. *Apidologie* **2008**, *39*, 334–342.

15. Alissandrakis, E.; Tarantilis, P.A.; Harizanis, P.C.; Polissiou, M. Aroma investigation of unifloral Greek citrus honey using solid-phase microextraction coupled go gas chromatographic-mass spectrometric analysis. *Food Chem.* **2007**, *100*, 396–404.

16. Alissandrakis, E.; Daferera, D.; Tarantilis, P.A.; Polissiou, M.; Harizanis, P.C. Ultrasound-assisted extraction of volatile compounds from citrus flowers and citrus honey. *Food Chem.* **2003**, *82*, 575–582.

17. Ferreres F.; Giner, J.M., Tomas-Barberan, F.A. A comparative study of hesperetin and methyl anthranilate as markers of the floral origin of citrus honey. *J. Sci. Food Agric.* **1994**, *65*, 371–372.

18. Jerkovic, I.; Tuberoso, C.I.G.; Kasum, A.; Marijanovic, Z. Volatile compounds of *Asphodelus microcarpus* Salzm. et Viv. Honey obtained by HS-SPME and USE analyzed by GC/MS. *Chem. Biodivers.* **2011**, *8*, 587–598.

19. Tuberoso, C.I.G.; Bifulco, E.; Jerkovic, I.; Caboni, P.; Cabras, P.; Floris, I. Methyl syringate: A chemical marker of asphodel (*Asphodelus microcarpus* Salzm. et Viv.) monofloral honey. *J. Agric. Food Chem.* **2009**, *57*, 3895–3900.

20. Donarski, J.A.; Jones, S.A.; Charlton, A.J. Application of cryoprobe [1]H Nuclear Magnetic Resonance spectroscopy and multivariate analysis for the verification of Corsican honey. *J. Agric. Food Chem.* **2008**, *56*, 5451–5456.

21. Woodcock, T.; Downey, G.; O'Donnell, C. Near infrared spectral fingerprinting for confirmation of claimed PDO provenance of honey. *Food Chem.* **2009**, *114*, 742–746.

22. Stanimirova, I.; Üstün, B.; Cajka, T.; Riddelova, K.; Hajslova, J.; Buydens, L.M.C. Tracing the geographical origin of honeys based on volatile compounds profiles assessment using pattern recognition techniques. *Food Chem.* **2010**, *118*, 171–176.

23. Gonnet, M., Vache, G. *Le Goût du Miel*; U.N.A.F.: Paris, France, 1985.

24. Yang, Y.; Battesti, M.J.; Djabou, N.; Muselli, A.; Paolini, J.; Tomi, P.; Costa, J. Melissopalynological origin determination and volatile composition analysis of Corsican "chestnut grove" honeys. *Food Chem.* **2012**, *132*, 2144–2154.

25. Aubert, S.; Gonnet, M. Mesure de la couleur des miels. *Apidologie* **1983**, *14*, 105–118.

26. Bogdanov, S. Charakterisierung von Schweizer Sortenhonigen. *Agrarforschung* **1997**, *4*, 427–430.

27. Konig, W.A.; Hochmuth, D.H.; Joulain, D. *Terpenoids and Related Constituents of Essential oils*; Library of Mass Finder 2.1, Institute of Organic Chemistry: Hamburg, Germany, 2001.

28. National Institute of Standards and Technology (NIST). Spectral Database for Organic Compounds. In *NIST Chemistry WebBook*; Available online: http://webbook.nist.gov/chemistry (accessed on 1 September 2013).

29. Yang, Y.; Battesti, M.J.; Paolini, J.; Muselli, A.; Tomi, P.; Costa, J. Melissopalynological origin determination and volatile composition analysis of Corsican "*Erica arborea* spring maquis" honeys. *Food Chem.* **2012**, *134*, 37–47.

30. Perez, R.A.; Sanchez-Brunete, C.; Calvo, R.M.; Tadeo, J.L. Analysis of volatiles from Spanish honeys by solid-phase microextraction and gas chromatography-mass spectrometry. *J. Agric. Food Chem.* **2002**, *50*, 2633–2637.

31. Castro-Vazquez, L.; Diaz-Maroto, M.C.; Perez-Coello, M.S. Aroma composition and new chemical markers of Spanish citrus honeys. *Food Chem.* **2007**, *103*, 601–606.

32. Castro-Vazquez, L.; Diaz-Maroto, M.C.; Gonzalez-Vinas, M.A.; Perez-Coello, M.S. Differentiation of monofloral citrus, rosemary, eucalyptus, lavender, thyme, and heather honeys based on volatile composition and sensory descriptive analysis. *Food Chem.* **2009**, *112*, 1022–1030.

33. Aliferis, K.A.; Tarantilis, P.A.; Harizanis, P.C.; Alissandrakis E. Botanical discrimination and classification of honey samples applying gas chromatography/mass spectrometry fingerprinting of headspace volatile compounds. *Food Chem.* **2010**, *121*, 856–862.

34. Maurizio, A.; Louveaux, J. *Pollens de Plantes Mellifères d'Europe*; Union des Groupements Apicoles Français: Paris, France, 1965.

35. De la Fuente, E.; Sanz, M.L.; Martinez-Castro, I.; Sanz, J.; Ruiz-Matute, A.I. Volatile and carbohydrate composition of rare unifloral honeys from Spain. *Food Chem.* **2007**, *105*, 84–93.

5

Efficacy of Acetic Acid against *Listeria monocytogenes* Attached to Poultry Skin during Refrigerated Storage

Elena Gonzalez-Fandos * and Barbara Herrera

Food Technology Department, CIVA Research Center, University of La Rioja, Madre de Dios 51, 26006 Logroño, La Rioja, Spain; E-Mail: anabarbaraherrera@redfarma.org

* Author to whom correspondence should be addressed; E-Mail: elena.gonzalez@unirioja.es

Abstract: This work evaluates the effect of acetic acid dipping on the growth of *L. monocytogenes* on poultry legs stored at 4 °C for eight days. Fresh inoculated chicken legs were dipped into either a 1% or 2% acetic acid solution (v/v) or distilled water (control). Changes in mesophiles, psychrotrophs, Enterobacteriaceae counts and sensorial characteristics (odor, color, texture and overall appearance) were also evaluated. The shelf life of the samples washed with acetic acid was extended by at least two days over the control samples washed with distilled water. *L. monocytogenes* counts before decontamination were 5.57 log UFC/g, and after treatment with 2% acetic acid (Day 0), *L. monocytogenes* counts were 4.47 log UFC/g. Legs washed with 2% acetic acid showed a significant ($p < 0.05$) inhibitory effect on *L. monocytogenes* compared to control legs, with a decrease of about 1.31 log units after eight days of storage. Sensory quality was not adversely affected by acetic acid. This study demonstrates that while acetic acid did reduce populations of *L. monocytogenes* on meat, it did not completely inactivate the pathogen. The application of acetic acid may be used as an additional hurdle contributing to extend the shelf life of raw poultry and reducing populations of *L. monocytogenes*.

Keywords: poultry; decontamination; meat safety; carcass; pathogen reduction; organic acids; *Listeria monocytogenes*

1. Introduction

Meat and poultry products are often identified as the source of foodborne pathogens [1]. Raw poultry is a well-recognized source of *L. monocytogenes*, and many surveys have confirmed the presence of this pathogen on fresh poultry [2–4]. Some authors have associated cases of listeriosis with the consumption of undercooked chicken [5].

The contamination of raw chicken with bacterial pathogens has important implications for public health. The reduction of poultry contamination with foodborne pathogens during slaughter is particularly important. Since hygienic practices during slaughter cannot completely prevent the contamination of poultry carcasses, decontamination treatments are gaining increasing interest in the slaughter process [6–9].

Organic acids (acetic, lactic, propionic and sorbic) are increasingly used in food products as preservatives, because of their antibacterial activity, and they occur naturally in foods. Organic acids are generally recognized as safe substances (GRAS) by the FDA and are approved as food additives by European Commission, FAO/WHO and FDA [10].

High concentrations of organic acids are required to be effective as decontaminating agents, but it is important to consider the effect of high concentrations of acids on product quality, since some alterations in the visual appearance of carcasses have been reported [6,10]. Generally, treatments with organic acids at varying concentrations result in population reductions ranging from one to three log units on meat surfaces [7–9].

Acetic acid has been investigated as an antimicrobial agent for use in meat, including poultry, to extend its shelf-life and inhibit the growth of pathogens, such as *Salmonella* or *Escherichia coli* [11–15].

The effectiveness of acetic acid for controlling meat-borne pathogens varies between studies and may be attributable to differences in acid concentration, as well as methods for acid delivery, the temperature of acids, contact time, sampling techniques, tissue type or organisms [16].

The ability of acetic acid to inhibit *L. monocytogenes* has been studied in laboratory media [17–20] and in beef and sheep [21]. However, there are few studies on the effect of acetic acid on *Listeria monocytogenes* growth on poultry [6].

The aim of this work was to evaluate the effectiveness of an acetic acid dip to control the growth of *Listeria monocytogenes* on poultry stored at 4 °C. Microbiological and sensorial quality were also evaluated.

2. Experimental Section

2.1. Preparation of Bacterial Inoculum

The *Listeria monocytogenes* serotype 1/2a strain CECT 932 was grown in tryptone soya broth (Oxoid, Hampshire, UK) at 30 °C for 18 h to achieve a viable cell population of 9 log CFU/mL. The culture was then transferred to a sterile centrifuge bottle and centrifuged at 10,000× *g* for 10 min at 4 °C. The supernatant was decanted and the pellet resuspended in sterile 0.1% peptone solution (Merck, Darmstadt, Germany) (pH 6.2) by vortexing. The washing step was repeated twice. The suspension of washed cells was diluted in a sterile 0.1% peptone solution to obtain an appropriate cell concentration for inoculation of sterile distilled water.

2.2. Inoculation of Poultry and Treatment

Ninety fresh chicken legs were obtained from a poultry processing plant (La Rioja, Spain). The legs were placed on crushed ice and transported to the laboratory.

Fresh chicken legs were inoculated with *L. monocytogenes* by dipping them into a suspension of this pathogen (7 log CFU/mL) for 5 min at room temperature. After the inoculation, the legs were removed and kept for 30 min at room temperature to allow the attachment of inoculated cells to the skin.

The inoculated poultry legs were divided into three groups, each containing 30 legs. Samples of each group were dipped for 5 min into sterile distilled water (control) (group one), 1% (v/v) (group 2) or 2% (group 3) acetic acid (Scharlau, Barcelona, Spain). After these treatments, the legs were removed and drained for 5 min and stored individually in sterile bags left open at 4 °C for 8 days. All experiments were carried out in duplicate.

Samples were taken on Days 0 (after dipping treatment), 1, 3, 6 and 8. On the sampling days, six legs of each group were taken out from storage to perform microbiological, pH and sensorial analysis.

2.3. Sensorial Analysis

The samples were evaluated for overall acceptability with regard to odor, color, texture and overall appearance by a panel of 9 members who were regular consumers of poultry meat. A structured hedonic scale [22] with numerical scores ranging from 7 (I like it very much) to 1 (I dislike it very much) was used. A score of 3 was considered the borderline of acceptability [7].

2.4. Microbiological Analyses and pH Determination

Ten grams of skin were aseptically weighed and homogenized in a Stomacher (IUL, Barcelona, Spain) for 2 min with 90 mL of sterile peptone water (Oxoid). Further decimal dilutions were made with the same diluent.

Studies were carried out to relate the weight with the surface of the poultry legs. It was found that 1 g of leg skin corresponded to an average of 6.88 cm^2 of leg skin.

The total number of mesophilic microorganisms was determined on Plate Count Agar (PCA, Merck) following the pour plate method, incubating at 30 °C for 72 h [23]. Psychrotrophs were determined on Plate Count Agar (Merck) with an incubation temperature of 7 °C for 10 days, using the pour plate method [23]. Enumeration of Enterobacteriaceae was carried out on violet red bile glucose (VRBG) (Merck) following the pour plate method with an incubation temperature of 37 °C for 48 h [23]. *Listeria* spp. were determined following the surface plate method on Palcam agar with an incubation temperature of 30 °C for 48 h [24]. Ten suspected colonies grown on Palcam agar were subcultured for purity on tryptone soya agar (TSA) (Merck) and incubated for 24 h at 30 °C. The following identification tests for *L. monocytogenes* were performed: Gram stain, catalase reaction, oxidase test, tumbling motility at 20–25 °C, umbrella motility in the SIM medium (Oxoid, Hampshire,

UK) and CAMP test [25]. Five suspected isolates were also identified by using API *Listeria* strips (BioMérieux, Marcy l'Etoile, France). The percentage of colonies identified as *L. monocytogenes* was 99%.

For pH determination, 5 g of skin were blended with 10 mL of distilled water. The pH of the homogenized sample was measured with a Crison model 2002 pHmeter (Crison Instruments, Barcelona, Spain).

2.5. Statistical Analysis

For microbiological data, an analysis of variance was performed using the SYSTAT program for Windows; Statistics version 5.0 (Evanston, IL, USA). Tukey's test for the comparison of means was performed using the same program. Plate count data were converted to logarithms prior to their statistical treatment. All experiments were carried out in duplicate. The significance level was defined at $p < 0.05$.

The data obtained from sensorial evaluation on the various sampling days were compared for statistical significance using Wilcoxon's matched pair test. To compare the data obtained on the same day with different concentrations of acetic acid, a Mann-Whitney U test was used. The significance level was defined at $p < 0.05$. The tests were carried out using the Statistica 6.0 program (Statsoft, IL, USA).

3. Results and Discussion

3.1. Microbiological Quality

The effect on mesophiles and psychrotrophs of dipping the legs into different acetic acid concentrations is shown in Tables 1 and 2, respectively. Significant differences ($p < 0.05$) in mesophile counts were found between the legs treated with 1% or 2% acetic acid and the control legs. The data obtained showed that a 5-min dip in 2% v/v acetic acid reduced mesophiles counts between 1.1 and 2.66 log cycles compared to the control legs throughout storage. After treatment, mesophile counts were about 0.89 or 1.1 log lower than in control samples, depending on the acetic acid concentration. Treatment with 1% or 2% acetic acid extended the lag phase; no growth was observed on Day 1. After six days, mesophile counts on samples treated with 1% or 2% acetic acid were 1.9 and 2.34 log units lower compared to control samples, respectively. Significant differences ($p < 0.05$) were found for these bacterial counts between the samples treated with 1% acetic acid and those treated with 2% acetic acid only on Days 6 and 8, although lower counts were observed in samples treated with 2% acetic acid on the other days.

Significant differences ($p < 0.05$) in psychrotroph counts were found between the legs treated with 1% or 2% acetic acid and the control legs. The dipping with 2% acetic acid reduced psychrotroph counts between 0.49 and 2.43 log cycles compared with the control legs throughout storage.

The results obtained agree with those reported by Fabrizio *et al.* [26], who found that immersion of chicken carcass in 2% acetic acid reduced the total counts between 1.25 and 3.3 log cycles compared to the control legs throughout storage. Dickens and Whittemore [12] also observed that the dipping of poultry carcass into 1% acetic acid reduced mesophiles by 0.6 log cycles. Similar reductions on mesophiles counts were reported by Dickens and Whittemore [11] and Dickens et al. [27].

Table 1. The effect of acetic acid on mesophile counts on poultry legs (log CFU/g).

Batch	Days of Storage				
	0	1	3	6	8
Control	5.57 ± 0.13 [a]	7.13 ± 0.04 [a]	7.69 ± 0.44 [a]	9.83 ± 0.03 [a]	10.31 ± 0.01 [a]
1% Acetic acid	4.68 ± 0.01 [b]	4.68 ± 0.06 [b]	6.29 ± 0.03 [b]	7.93 ± 0.03 [b]	8.84 ± 0.03 [b]
2% Acetic acid	4.47 ± 0.01 [b]	4.47 ± 0.02 [b]	5.87 ± 0.02 [b]	7.49 ± 0.01 [c]	8.31 ± 0.01 [c]

Notes: Mean ± standard deviation. Means within columns followed by the same letter were not significantly different ($p > 0.05$).

Table 2. The effect of acetic acid on the psychrotroph counts on poultry legs (log CFU/g).

Batch	Days of Storage				
	0	1	3	6	8
Control	5.17 ± 0.03 [a]	5.62 ± 0.01 [a]	6.88 ± 0.01 [a]	9.02 ± 0.03 [a]	9.57 ± 0.01 [a]
1% Acetic acid	4.87 ± 0.01 [b]	4.91 ± 0.01 [b]	5.72 ± 0.03 [b]	6.77 ± 0.03 [b]	8.12 ± 0.04 [b]
2% Acetic acid	4.68 ± 0.02 [c]	4.69 ± 0.03 [b]	5.63 ± 0.01 [b]	6.59 ± 0.01 [c]	7.89 ± 0.01 [c]

Notes: Mean ± standard deviation. Means within columns followed by the same letter were not significantly different ($p > 0.05$).

Acetic acid has been also applied to pig, lamb and beef carcasses, being found effective in reducing microbial counts by one log cycle [28–30].

In the present study, the treatment with 1% or 2% acetic acid extended the lag phase. Furthermore, Jiménez et al. [31] observed that the immersion of chicken breast in a 1% acetic acid solution extended the lag phase of microbial growth.

According to Gill and Landers [32], decontaminating treatments must be regarded as trivial when the numbers of bacteria recovered before and after a treatment do not differ by a minimum of 0.5 log units. In consequence, in the present study, acetic acid treatments could be considered as effective.

In a previous work, it was observed that a treatment with 2% lactic acid reduced mesophile counts between 0.67 and 2.32 log cycles compared with the control legs throughout storage [33]. In the present work, a treatment with 2% acetic acid reduced mesophile counts between 1.1 and 2.66 log cycles. Thus, the antimicrobial effect of acetic acid was higher than lactic acid, if we compare the percentage added. The antimicrobial effect of citric acid was lower than acetic and lactic acids, since a treatment with 2% citric acid reduced mesophile counts between 0.45 and 1.08 log cycles [34].

Other authors have reported that the efficacy of acetic acid was lower than that reached with lactic acid. Thus, Sakhare et al. [15] reported that lactic acid was superior to acetic acid as the decontaminating agent to reduce the microbial load on poultry carcasses at different processing steps.

Table 3 shows the effect of acetic acid treatment on the growth of Enterobacteriaceae. Significant differences ($p < 0.05$) in the Enterobacteriaceae counts were observed on legs treated with 1% or 2% acetic acid compared to the control samples. After treatment, Enterobacteriaceae counts were 1.61 log cycles lower in legs treated with 2% acetic acid than in control ones. Significant differences

($p < 0.05$) were also found between the legs treated with 1% acetic acid and those treated with 2% acetic acid on Days 0 and 8.

Table 3. The effect of acetic acid on the Enterobacteria counts on poultry legs (log CFU/g).

Batch	Days of Storage				
	0	1	3	6	8
Control	3.04 ± 0.04 [a]	4.21 ± 0.06 [a]	5.09 ± 0.01 [a]	5.70 ± 0.07 [a]	6.54 ± 0.08 [a]
1% Acetic acid	2.13 ± 0.01 [b]	2.64 ± 0.01 [b]	3.95 ± 0.01 [b]	4.55 ± 0.01 [b]	5.51 ± 0.04 [b]
2% Acetic acid	1.63 ± 0.04 [c]	2.60 ± 0.01 [b]	3.84 ± 0.11 [b]	4.34 ± 0.07 [b]	5.04 ± 0.03 [c]

Notes: Mean ± standard deviation. Means within columns followed by the same letter were not significantly different ($p > 0.05$).

In the present study, Enterobacteriaceae counts of treated samples were significantly lower than in control samples. These findings agree with those reported by Dickens and Whittemore [11], who observed that the dipping of poultry carcass into a 0.6% acetic acid solution reduced Enterobacteriaceae counts by 0.71 log cycles. After six days of storage, we observed that Enterobacteriaceae counts on treated samples were lower (about 1.15–1.36 log units) than in control samples. Furthermore, Jiménez et al. [31] reported an Enterobacteriaceae reduction of about 1.5 log units in carcasses treated with acetic acid.

3.2. pH Evolution

The pH values of the legs treated with acetic acid are shown in Figure 1. Significant differences ($p < 0.05$) were found in pH values between samples treated with 1% or 2% acetic acid and control samples. The pH was lower when the acetic acid concentration was higher. These pH differences did not decrease throughout storage. Initial pH values in legs treated with 2% acetic acid (Day 0) were 4.39 ± 0.04, 1.71 units lower than in control legs.

Figure 1. The evolution of pH in chicken legs treated with acetic acid. The data are the mean values of six replicates.

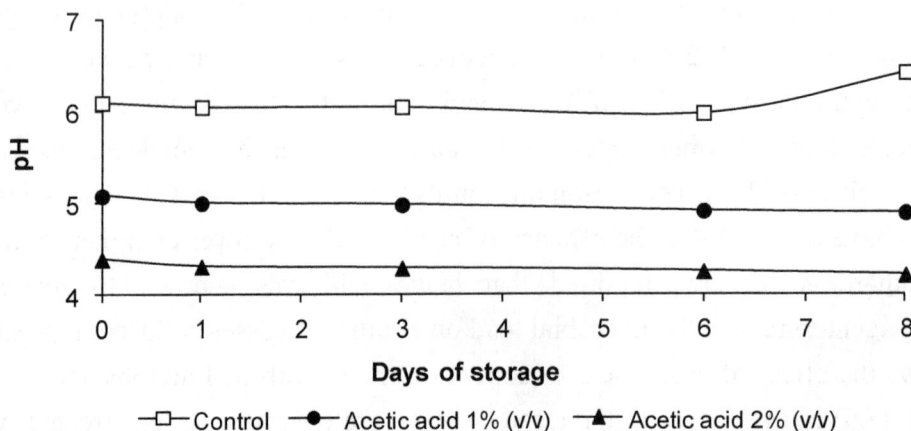

The pH data indicated that the reductions of bacterial populations may have been due to the effects of acidic pH. Thus, lower counts were observed in legs with lower pH. The antimicrobial effect of organic acids has been attributed to undissociated acid molecules that interfere with cellular metabolism or a decrease in biological activity, as a result of pH changes in the cell's environment [35,36]. In this study, the application of acetic acid reduced the surface pH immediately after treatment, thereby creating an unfavorable environment for bacterial growth. The mean skin pH value of untreated samples was 6.1. Treatment with 1% or 2% acetic acid solution resulted in a decrease in pH of about one and 1.5 units, respectively. A similar pH decrease in chicken breast after dipping with 1% acetic acid has been reported by Jimenez et al. [31].

3.3. Listeria monocytogenes

Table 4 shows the effect of acetic acid treatment on the growth of L. monocytogenes inoculated onto legs. Significant differences ($p < 0.05$) in the L. monocytogenes populations were observed on legs treated with 2% acetic acid compared to the control samples. After eight days of storage, L. monocytogenes counts were 1.31 log cycles lower in legs treated with 2% acetic acid than in control ones. Significant reductions in the L. monocytogenes populations were also observed on legs treated with 1% acetic acid on Days 1, 3, 6 and 8 of storage compared to the control samples. Significant differences were observed between legs treated with 1% and 2% acetic acid on Days 3, 6 and 8. Samples treated with 1% or 2% acetic acid displayed an extended lag phase of L. monocytogenes and lower counts throughout storage compared with control legs. While L. monocytogenes grew readily on control legs, growth was slower on acid-dipped legs, particularly those dipped in 2% acetic acid.

Table 4. The effect of acetic acid on *Listeria monocytogenes* counts on poultry legs (log CFU/g).

Batch	Days of Storage				
	0	1	3	6	8
Control	4.90 ± 0.20 [a]	5.53 ± 0.03 [a]	7.38 ± 0.03 [a]	7.74 ± 0.01 [a]	8.77 ± 0.01 [a]
1% Acetic acid	4.53 ± 0.01 [ab]	4.38 ± 0.04 [b]	6.25 ± 0.01 [b]	7.18 ± 0.06 [b]	7.88 ± 0.03 [b]
2% Acetic acid	4.25 ± 0.07 [b]	4.28 ± 0.01 [b]	5.72 ± 0.06 [c]	6.60 ± 0.01 [c]	7.46 ± 0.01 [c]

Notes: Mean ± standard deviation. Means within columns followed by the same letter were not significantly different ($p > 0.05$).

The ability of acetic acid to inhibit L. monocytogenes can be higher in laboratory media than in foods, according to the results reported by Ahamad and Marth [17]. These authors found that the presence of up 0.1% acetic acid in tryptose broth inhibited the growth of L. monocytogenes and that the degree of inhibition increased as the temperature of incubation decreased. These authors reported that L. monocytogenes growth was suppressed when acetic acid concentrations in the medium were 0.2% at all temperatures tested. According to these authors, acetic acid was the most detrimental to L. monocytogenes followed in order by lactic and citric acids. Vermeulen et al. [20] also reported that L. monocytogenes was not able to grow in nutrient broth with 0.4% acetic acid. Cunningham et al. [18] studied the response of L. monocytogenes to weak acids, including acetic acid. These authors observed that acetic acid at concentrations of 0.15% in BHI reduced L. monocytogenes counts.

George et al. [19] also found that acetic acid was more inhibitory to the growth of L. monocytogenes than lactic acid in terms of total acid added. According to Farber et al. [37], acetic acid increases the minimum pH for the growth of *L. monocytogenes* more than lactic acid. The greater effectiveness of acetic acid could be explained by its lower pKa, giving a greater proportion of acid in the undissociated form [19].

Dorsa *et al*. [13] reported that spray application of 1.5% or 3% acetic acid in beef reduced the levels of *L. innocua* after washing. After, two days of storage at 5 °C, these authors could not detect the growth of *L. innocua*.

Conflicting reports on the efficacy of acetic acid against *L. monocytogenes* may be due to variations in the media or food, pH or acid concentration [38]. The efficacy of acetic acid against *Listeria* could be higher in other types of meat, since the pH is lower. Glass and Doyle [39] reported that the *L. monocytogenes* grew well on those meat products with a pH value near or above 6.0, while on meats near or below pH 5.0, the organism grew poorly or not at all. Poultry has a higher pH than other types of meat. It should be pointed out that poultry leg muscles have a pH of 6.1, while other parts, like the breast muscles, have lower pH values (5.7–5.9) [40]. This higher pH can explain why poultry supports the growth of *L. monocytogenes* better than other meats; for that reason, decreasing the pH with acetic acid treatment could contribute to controlling the growth of *L. monocytogenes*.

Other pathogens are also inhibited by acetic acid. Fabrizio *et al*. [26] reported that immersion in a 2% solution of acetic acid reduces *Salmonella* counts by 1.41 log cycles after treatment. Jimenez *et al*. [14] observed *Salmonella* reduction of about 0.4 log cycles when 2.8% acetic acid was applied in poultry.

Moreover, Waterman and Small [41] found that *Salmonella* inoculated onto the surface of pre-acidified ground beef could not survive if the pH on the surface of the beef was 2.61 or lower, but was viable if the surface pH was 3.27. In the present study, although the pH of the acetic acid solution was low, the mean pH value on the legs was 5.08 or 4.39 after treatment, depending on the acid concentration.

It must be highlighted that Uyttendale *et al*. [4] reported that among chicken parts, *L. monocytogenes* was predominantly isolated from chicken legs and chicken wings, the parts that are still partially covered with skin. This pathogen is mainly located on the skin surface of poultry carcasses and, to a lesser, extent in the meat. On the other hand, the higher pH of leg meat may provide more favorable conditions for multiplication of *L. monocytogenes* [1].

L. monocytogenes can grow at temperatures as low as 4 °C. Thus, this bacterium is a particular foodborne hazard, because of the ability to replicate, albeit slowly, at refrigeration temperatures [1].

According to Carpenter *et al*. [6], acetic acid displays residual activity to prevent the growth of pathogens. These authors highlighted the role of organic acids in the meat industry, especially the effectiveness of organic acid washes relative to their ability to decontaminate meat tissues and subsequently inhibit the growth of pathogens; thus, organic acids contribute to a total food safety program.

Although treatments with acetic acid did reduce populations of *L. monocytogenes* on poultry meat, they were not able to reduce the pathogen to zero levels. Depending on the initial populations of the pathogen, reductions ranging from one log CFU/g may not be sufficient as the only means to improve the overall microbiological safety of poultry carcasses. These results should be considered with caution because of the high *L. monocytogenes* load of the artificially inoculated poultry (4.90 log UFC/g), and further research is needed. Overall, the data suggests that acetic acid treatments may be beneficial as part of an overall hazard analysis critical control point (HACCP) approach that can be implemented in order to enhance the microbiological safety and extend the shelf life of poultry meat.

3.4. Sensorial Quality

The changes in color, odor and overall appearance of the poultry legs are shown in Figures 2–4, respectively. Sensory quality was not adversely affected by acetic acid treatment, the scores being observed above six until Day 3. No significant differences ($p > 0.05$) in color were observed between samples treated with acetic acid and control samples until Day 3. After six days of storage, the worst score was obtained by control legs. Control legs were rejected on Day 6. When treatments were compared at Day 6 of storage, treatment with acetic acid reduced ($p < 0.05$) the presence of off-odors compared with the control. The samples treated with 1% or 2% acetic acid were not severely discolored, and unacceptable odors were not detected throughout storage. Consequently, legs receiving treatments with acetic acid remained acceptable until eight days of storage, at least two days longer than control samples.

Figure 2. The evolution of color in chicken legs treated with acetic acid. The data are the mean values of six replicates.

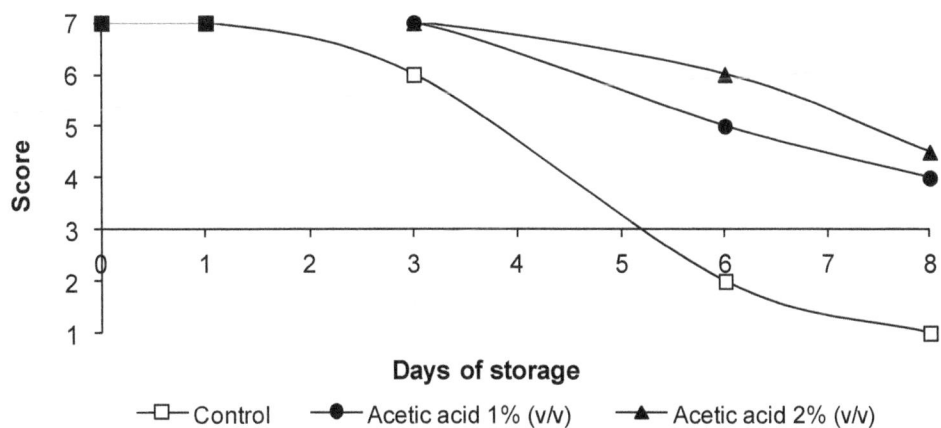

Figure 3. Evolution of odor in chicken legs treated with acetic acid. The data are the mean values of six replicates.

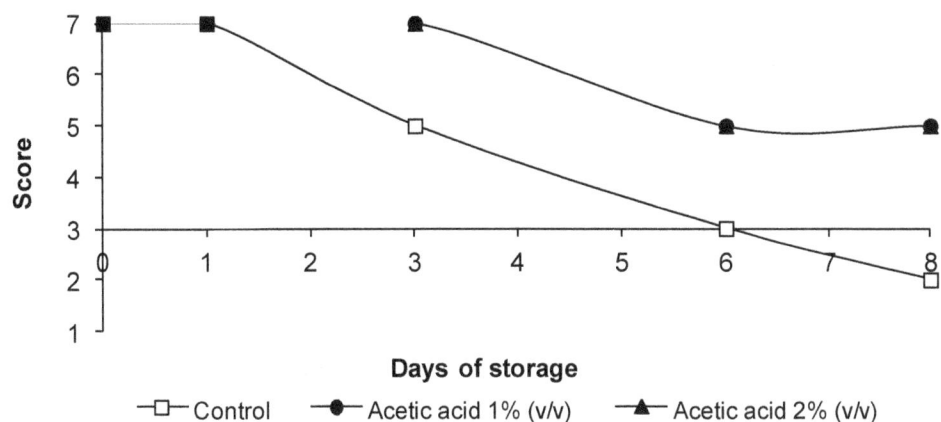

Figure 4. Overall appearance of chicken legs treated with lactic acid and acetic acid. The data are the mean values of six replicates.

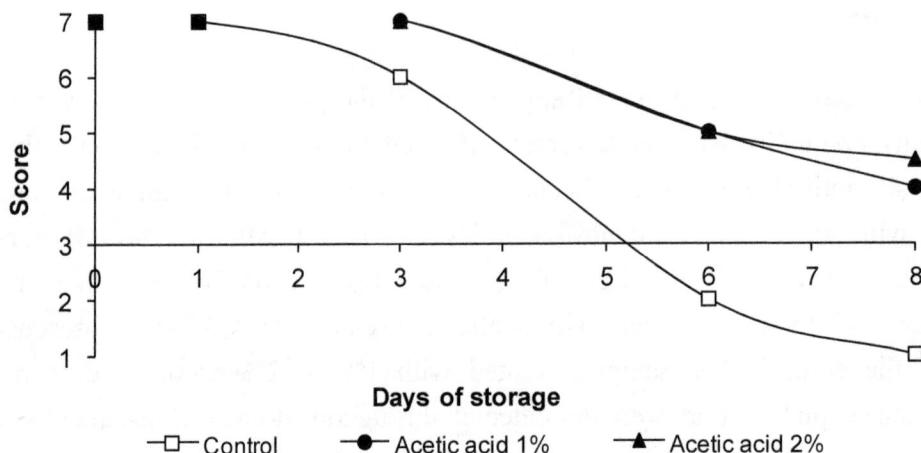

Acetic acid treatment did not have adverse effects on poultry legs quality characteristics. Other authors have also reported that solutions of organic acids (1%–3%) have no sensorial negative effects in meat when used as a decontaminant [42].

Off-odors were noticed by the panel members when the counts approached 9 log CFU/g. To compare our results with those reported by other authors, the data were transformed to log CFU/cm². It was found that 1 g of skin corresponded to an average of 6.88 cm² of skin. Thus, 9 log CFU/g corresponded to 8.16 log CFU/cm². Other authors have reported spoilage odors in poultry when counts approached 7–8 cfu/cm² [40,43,44].

After six days of storage, mesophiles and psychrotrophs reached populations above 9 log CFU/g in control legs. However, in the legs treated with 1% or 2% acetic acid, mesophile and psychrotroph counts were below 9 log CFU/g after eight days of storage at 4 °C, and signs of spoilage were not detected after eight days of storage. Sensorial scores of treated legs were above those reached by the control legs. Control legs were rejected after six days of storage.

These results agree with those reported by Dickens and Whittemore [12], who did not observe any change in skin appearance due to the 1% acetic acid treatment. Jimenez et al. [31] reported, despite the high level attained by microbial populations in poultry treated with acetic acid, that the overall aspect remained acceptable throughout the storage periods. These authors found off-odors in untreated samples, while the treated ones smelt slightly acidic and pleasant. Sakhare et al. [15] reported that acetic acid treatment at low concentrations (0.5%) after every step of poultry processing (scalding, defeathering, evisceration) did not affect the appearance of carcasses.

4. Conclusions

The shelf life of the samples washed with 1% or 2% acetic acid was extended by at least two days over the control samples washed with distilled water. Legs washed with acetic acid showed a significant ($p < 0.05$) inhibitory effect on *L. monocytogenes* compared to control legs. Sensory quality was not adversely affected by acetic acid.

This study demonstrates that while acetic acid did reduce populations of *L. monocytogenes* on meat, it did not completely inactivate the pathogen. Of the concentrations tested, treatments with 2% acetic acid were the most effective for reducing populations of *L. monocytogenes*.

The application of acetic acid cannot replace the rules of strict hygiene and good manufacturing practice, but it may be used as an additional hurdle contributing to extending the shelf life of raw poultry.

Acknowledgments

The authors thank the Regional Government of La Rioja (Spain) (Project Reference ANGI 2005/06) and the University of La Rioja (Spain) (Project Reference PROFAI 13/24) for their financial support.

Author Contributions

Elena Gonzalez-Fandos assisted with experimental procedures, statistical analysis and interpretation of data. Barbara Herrera had a major part in the experimental work. All authors contributed to revision and writing of the manuscript.

Conflicts of Interest

The authors declare no conflict of interest.

References

1. ICMSF (International Commission on Microbiological Specifications for Foods). *Microorganisms in Foods. 6. Microbial Specifications of Food Commodities*; Blackie Academic & Professional: London, UK, 1998.
2. Bailey, J.S.; Fletcher, D.L.; Cox, N.A. Recovery and serotype distribution of *Listeria monocytogenes* from broiler chickens in the southeastern United States. *J. Food Prot.* **1989**, *52*, 148–150.
3. Genigeorgis, C.A.; Dutulescu, D.; Garazabar, J.F. Prevalence of *Listeria* spp. in poultry meat at the supermarket and slaughterhouse level. *J. Food Prot.* **1989**, *52*, 618–624.
4. Uyttendale, M.R.; Neyts, K.D.; Lips, R.M.; Devebere, J.M. Incidence of *Listeria monocytogenes* in poultry products obtained from Belgian and French abbatoirs. *Food Microbiol.* **1997**, *14*, 339–345.
5. Schuchat, A.; Deaver, K.; Wenger, J.D.; Swaminathan, B.; Broome, C.V. Role of food in sporadic listeriosis: I. Case-control-study of dietary risk factors. *J. Am. Med. Assoc.* **1992**, *267*, 2041–2045.
6. Carpenter, C.E.; Smith, J.V.; Broadbent, J.R. Efficacy of washing meat surfaces with 2% levulinic, acetic, or lactic acid for pathogen decontamination and residual growth inhibition. *Meat Sci.* **2011**, *58*, 256–260.
7. Gonzalez-Fandos, E.; Dominguez, J.L. Effect of potassium sorbate on the growth of *Listeria monocytogenes* on fresh poultry. *Food Control* **2007**, *18*, 842–846.

8. Gonzalez-Fandos, E.; Herrera, B. Efficacy of malic acid against *Listeria monocytogenes* attached in poultry skin during refrigerated storage. *Poult. Sci.* **2013**, *92*, 1936–1941.

9. Gonzalez-Fandos, E.; Herrera, B. Efficacy of propionic acid against *Listeria monocytogenes* attached in poultry skin during refrigerated storage. *Food Control* **2013**, *34*, 601–606.

10. Surekha, M.; Reddy, S.M. Preservatives. Classsification and properties. In *Encyclopedia of Food Microbiology*; Robinson, R.K., Batt, C.A., Patel, C., Eds.; Academic Press: New York, NY, USA, 2000; pp. 1710–1717.

11. Dickens, J.A.; Whittemore, A.D. The effects of extended chilling times with acetic acid on the temperature and microbiological quality of processed poultry carcasses. *Poult. Sci.* **1995**, *74*, 1044–1048.

12. Dickens, J.A.; Whittemore, A.D. The effect of acetic acid and hydrogen peroxide application during defeathering on the microbiological quality of broiler carcasses. *Poult. Sci.* **1997**, *76*, 657–660.

13. Dorsa, W.J.; Cutter, C.N.; Siragusa, G. Effects of acetic acid, lactic acid and trisodium phosphate on the microflora of refrigerated beef carcass surface tissue inoculated with *Escherichia coli* O157:H7, *Listeria innocua* and *Clostridium sporogenes*. *J. Food Prot.* **1997**, *60*, 619–624.

14. Jiménez, S.M.; Caliusco, M.F.; Tiburzi, M.C.; Salsi, M.S.; Pirovani, M.E. Predictive models for reduction of *Salmonella* Hadar on chicken skin during single and double sequential spraying treatments with acetic acid. *J. Appl. Microbiol.* **2007**, *103*, 528–535.

15. Sakhare, P.Z.; Sachindra, N.M.; Yashoda, K.P.; Rao, D.N. Efficacy of intermittent decontamination treatments during processing in reducing the microbial load on broiler chicken carcass. *Food Control* **1999**, *10*, 189–194.

16. Greer, G.C.; Dilts, B. Factors affecting the susceptiblility of meatborne pathogens and spoilage bacteria to organic acids. *Food Res. Int.* **1992**, *25*, 355–364.

17. Ahamad, N.; Marth, E.H. Behaviour of *Listeria monocytogenes* at 7, 13, 21 and 35 °C in tryptose broth acidified with acetic, citric, or lactic acid. *J. Food Prot.* **1989**, *52*, 688–695.

18. Cunningham, E.; O'Byrne, C.; Oliver, J.D. Effect of weak acids on *Listeria monocytogenes* survival: Evidence for a viable but nonculturable state in response to low pH. *Food Control* **2009**, *20*, 1141–1144.

19. George, S.M.; Richardson, L.C.C.; Peck, M.W. Predictive models of the effect of temperature, pH and acetic and lactic acids on the growth of *Listeria monocytogenes*. *Food Microbiol.* **1996**, *32*, 73–90.

20. Vermeulen, A.; Gysemans, K.; Bernaerts, K.; Geeraerd, A.H.; van Impe, J.; Debevere, J.; Devlieghere, F. Influence of pH, water activity and acetic acid concentration on *Listeria monocytogenes* at 7 °C: Data collection for the development of a growth/no growth model. *Int. J. Food Microbiol.* **2007**, *114*, 332–341.

21. Dickson, J.S.; Siragusa, G.R. Survival of *Salmonella typhimurium*, *Escherichia coli* O157:H7 and *Listeria monocytogenes* during storage on beef sanitized with organic acids. *J. Food Saf.* **1994**, *14*, 313–327.

22. Anzaldúa-Morales, A. *La Evaluación Sensorial de los Alimentos en la Teoría y en la Práctica* (in Spanish); Acribia: Zaragoza, Spain, 1994.

23. ICMSF (International Commission on Microbiological Specifications for Foods). *Microorganisms in Foods. 1: Their Significance and Methods of Enumeration*, 2nd ed.; University of Toronto Press: Toronto, ON, Canada, 1978.

24. Mossel, D.A.A.; Corry, J.E.L.; Struijk, C.B.; Baird, R.M. *Essentials of the Microbiology of Foods. A Textbook for Advanced Studies*; John Wiley and Sons Ltd.: Chichester, UK, 1995.

25. Seeliger, H.P.R.; Jones, D. *Listeria*. In *Bergey's Manual of Systematic Bacteriology*; Sneath, P.H.A., Nair, N.S., Sharpe, M.E., Holt, J.G., Eds.; Williams and Wilkins: Baltimore, MD, USA, 1986; Volume 2, pp. 1235–1245.

26. Fabrizio, K.A.; Sharma, R.R.; Demirci, A.; Cutter, C.N. Comparison of electrolyzed oxidizing water with various antimicrobial interventions to reduce Salmonella species on poultry. *Poult. Sci.* **2002**, *81*, 1598–1605.

27. Dickens, J.A.; Lyon, B.G.; Whittemore, A.D.; Lyon, C.E. The effect of an acetic acid dip on carcass appearance, microbiological quality, and cooked breast meat texture and flavor. *Poult. Sci.* **1994**, *73*, 576–581.

28. Eggenberger, L.; Niebuhr, S.E.; Acuff, G.R.; Dickson, J.S. Hot water and organic acid interventions to control microbiological contamination on hog carcasses during processing. *J. Food Prot.* **2002**, *65*, 1248–1252.

29. Fu, A.H.; Sebranek, J.G.; Murano, E.A. Microbial and quality characteristics of pork cuts from carcasses treated with sanitizing sprays. *J. Food Sci.* **1994**, *59*, 306–309.

30. Hardin, M.D.; Acuff, G.R.; Lucia, L.M.; Oman, J.S.; Savell, J.W. Comparison of methods for decontamination from beef carcass surfaces. *J. Food Prot.* **1995**, *58*, 368–374.

31. Jiménez, S.M.; Salsi, M.S.; Tiburzi, M.C.; Rafaghelli, R.C.; Pirovani, M.E. Combined use of acetic acid treatment and modified atmosphere packaging for extending the shelf-life of chilled chicken breast portions. *J. Appl. Microbiol.* **1999**, *87*, 339–344.

32. Gill, C.O.; Landers, C. Microbiological effects of carcass decontaminating treatments at four beef packing plants. *Meat Sci.* **2003**, *65*, 1005–1011.

33. Gonzalez-Fandos, E.; Dominguez, J.L. Efficacy of lactic acid against *Listeria monocytogenes* attached to poultry skin during refrigerated storage. *J. Appl. Microbiol.* **2006**, *101*, 1331–1339.

34. Gonzalez-Fandos, E.; Herrera, B.; Maya, N. Efficacy of citric acid against *Listeria monocytogenes* attached to poultry skin during refrigerated storage. *Int. J. Food Sci. Technol.* **2009**, *44*, 262–268.

35. Doores, S. Organic acids. In *Antimicrobials in Foods*; Branen, A.L., Davidson, P.M., Eds.; Marcel Dekker Inc.: New York, NY, USA, 1983; pp. 75–108.

36. Cherrington, C.A.; Hinton, A.M.; Pearson, G.R.; Copra, I. Inhibition of *Escherichia coli* K12 by short-chain organic acids: Lack of evidence for induction of the SOS response. *J. Appl. Bacteriol.* **1991**, *70*, 156–160.

37. Farber, J.M.; Sanders, G.W.; Dunfield, S.; Prescott, R. The effect of various acidulants on the growth of *Listeria monocytogenes*. *Lett. Appl. Microbiol.* **1989**, *9*, 181–183.

38. Thomas, L.V. Preservatives. Sorbic acid. In *Encyclopedia of Food Microbiology*; Robinson, R.K., Batt, C.A., Patel, C., Eds.; Academic Press: New York, NY, USA, 2000; pp. 1769–1776.

39. Glass, K.P.; Doyle, M.P. *Listeria monocytogenes* in processed meat products during refrigerated storage. *Appl. Environ. Microbiol.* **1989**, *55*, 1565–1569.

40. Barnes, E.M. Microbiological problems of poultry at refrigerator temperatures. A review. *J. Sci. Food Agric.* **1976**, *27*, 777–782.

41. Waterman, S.R.; Small, P.L.C. Acid-sensitive enteric pathogens are protected from killing under extremely acidic conditions of pH 2.5 when they are inoculated onto certain solid food sources. *Appl. Environ. Microbiol.* **1998**, *64*, 3882–3886.

42. Smulders, F.J.M.; Greer, G.G. Integrating microbial decontamination with organic acids in HACCP programmes for muscle foods: Prospects and controversies. *Int. J. Food Microbiol.* **1998**, *44*, 149–169.

43. Elliot, P.H.; Tomlins, R.J.; Gray, R.J.H. Control of microbial spoilage on fresh poultry using a combiantion potassium sorbate/carbon dioxide packaging system. *J. Food Sci.* **1985**, *50*, 1360–1363.

44. Studer, P.; Schmidt, R.E.; Gallo, L.; Schmidt, W. Microbial spoilage of refrigerated fresh broilers. II. Effect of packaging on microbial association of poultry carcasses. *Lebensm. Wiss. Technol.* **1988**, *21*, 224–228.

Amino Acid Composition of an Organic Brown Rice Protein Concentrate and Isolate Compared to Soy and Whey Concentrates and Isolates

Douglas S. Kalman

Nutrition/Endocrinology Department, Miami Research Associates, 6141 Sunset Drive, Suite 301, Miami, FL 33143, USA; E-Mail: dkalman@miamiresearch.com

Abstract: A protein concentrate (Oryzatein-80™) and a protein isolate (Oryzatein-90™) from organic whole-grain brown rice were analyzed for their amino acid composition. Two samples from different batches of Oryzatein-90™ and one sample of Oryzatein-80™ were provided by Axiom Foods (Los Angeles, CA, USA). Preparation and analysis was carried out by Covance Laboratories (Madison, WI, USA). After hydrolysis in 6-N hydrochloric acid for 24 h at approximately 110 °C and further chemical stabilization, samples were analyzed by HPLC after pre-injection derivitization. Total amino acid content of both the isolate and the concentrate was approximately 78% by weight with 36% essential amino acids and 18% branched-chain amino acids. These results are similar to the profiles of raw and cooked brown rice except in the case of glutamic acid which was 3% lower in the isolate and concentrate. The amino acid content and profile of the Oryzatein-90™ isolate was similar to published values for soy protein isolate but the total, essential, and branched-chain amino acid content of whey protein isolate was 20%, 39% and 33% greater, respectively, than that of Oryzatein-90™. These results provide a valuable addition to the nutrient database of protein isolates and concentrates from cereal grains.

Keywords: brown rice; supplemental protein; essential amino acids; branched-chain amino acids; muscle protein synthesis

1. Introduction

Concentrated protein powders, commonly used as dietary supplements and in food processing, are available in a variety of flavors and forms, including ready to drink shakes, bars, bites, oats, gels and powders manufactured using soy, milk, peas, or eggs as the source of the protein. Protein concentrates are created by pushing the protein source through a very small filter that allows water, minerals, and other organic materials to pass though. The proteins, which are too big to pass through the filter, are collected, resulting in protein powder or protein concentrate. Concentrates can have substantial amounts of carbohydrate and fat. Further purification using additional filtration or a technique called ion-exchange or cross-flow microfiltration [1–3]. Results in the formation of the protein isolate. Isolates have very low levels of carbohydrates and fat and are almost exclusively pure protein.

Soy protein concentrate is made from defatted soy flour without the water-soluble carbohydrates but with the fiber. It contains approximately 70% protein. Soy protein isolate is a more purified form of soy protein which has had most of the non-protein elements removed. It is almost 90% pure protein and has a very neutral flavor. Soy protein concentrates and isolates have long been used for a variety of food processing purposes such as increasing the protein content of foods, enhancing moisture retention, and acting as emulsifiers.

Whey protein concentrate contains 70%–85% protein and up to 5% lactose. People with lactose intolerance may experience gastric discomfort when consuming large amounts of whey protein concentrates, a problem that is eliminated with whey protein isolates. Whey protein isolate is rapidly digested making it a popular supplement among athletes attempting to rebuild muscle after a strenuous workout [4,5].

The amino acid composition of protein isolates has made them popular as dietary supplements. Although the concentration of sulfur-containing amino acids is low in soy protein isolates, soy contains a high concentration of branched-chain amino acids (BCAA). BCAA are concentrated in muscle tissue and used to fuel working muscles and stimulate protein synthesis [6,7]. Whey proteins are highly bioavailable, very quickly absorbed into the body, and also have a high concentration of BCAA. Moreover, whey protein isolate is an excellent source of lysine which is often the rate-limiting amino acid in grains and legumes [8]. Like most Americans, athletes consume more than enough dietary protein, questioning the need for protein shakes. However, some evidence supports the idea that protein shakes are superior to whole foods with regard to enhancing muscle hypertrophy in the one hour window following intensive exercise [9].

Brown rice is an excellent source of protein, containing 37% of the total protein as essential amino acids and 18% as BCAA ([10], item 20036). We are not aware of a comprehensive analysis of the amino acid profiles of a brown rice protein concentrate or isolate. Thus the purpose of the present study was to compare the amino acid profile of an organic brown rice protein concentrate (Oryzatein Ultra Silk-80, Axiom Foods, Los Angeles, CA, USA) and a brown rice protein isolate (Oryzatein Ultra Silk-90, Axiom Foods, Los Angeles, CA, USA) to that of published standards for raw and cooked brown rice, soy protein isolate and concentrate, and whey protein isolate and concentrate.

2. Materials and Methods

2.1. Materials

Commercially available rice protein isolate and concentrate as prepared and sold by the manufacturer (Axiom Foods/Growing Naturals, Los Angeles, CA, USA) were analyzed for their amino acid content by Covance Laboratories (Madison, WI, USA). The samples included Organic Oryzatein Ultra 90 isolate, Batch # HZN11008-13 (received 15 February 2012) Organic Oryzatein Silk 80 concentrate, Batch # HZN11018-22 (12 March 2012) and Organic Oryzatein Silk 90 isolate, Batch # HZN12004-111 (received 15 May 2012). The accompanying paperwork to the analytical laboratory indicated that the serving size for each was ~34 g. As the products were already in powder form, no further preparation was performed prior to analysis. The products were stored at minus 20 °C until the start of analysis, which took place approximately one week after receipt of each sample.

2.2. Preparation and HPLC Analysis of Amino Acid in Rice Protein Isolate and Concentrate

Prior to analysis, all samples were mixed following Covance SOP NA-FDA-156 v.1, which outlines inspecting the sample to ensure it appears homogeneous and also mixing the sample prior to weighing. Sample sizes of ~0.1 g were hydrolyzed in 40 mL 6-N hydrochloric acid for 24 h at approximately 110 °C. An appropriate amount of phenol was added to prevent halogenation of tyrosine. Cystine and cysteine were converted to *S*-2-carboxy-ethylthiocysteine by the addition of dithiodipropionic acid and hydrolyzed for 24 h in 6-N HCl. Tryptophan was hydrolyzed from proteins (using a ~0.2 g sample size) by heating at approximately 110 °C in 4.2-N NaOH (for 20 h). Samples were analyzed by high performance liquid chromatography (HPLC) after pre-injection derivitization [3]. The primary amino acids were derivitized with *O*-phthalaldehyde (OPA) and the secondary amino acids were derivitized with fluorenylmethyl chloroformate (FMOC) before injection [1–3]. The isolate was further purified by ion-exchange or cross-flow microfiltration [3]. The result of the analysis of Oryzatein-90 represents the average of the two analyses.

The amino acid profiles of Oryzatein-80 and Oryzatein-90 were compared to the published amino acid profiles for raw ([10], item 20036) and cooked brown rice ([10], item 20037), soy protein isolate ([10], item 16122) and concentrate ([10], item 16420), and whey protein concentrate and isolate [11].

3. Results and Discussion

3.1. Total Amino Acid Contents of Brown Rice Protein Isolate and Concentration

The results of the present study indicate that an organic brown rice protein concentrate, Oryzatein-80 and its isolate, Oryzatein-90 are excellent sources of total, essential, and branched-chain amino acids and are comparable to soy and whey protein concentrates and isolates. The total amino acid (TAA) content of both the brown rice concentrate and isolate were approximately 78% by weight with 36% essential amino acids (EAA) and 18% branched-chain amino acids (BCAA) (Table 1). These profiles were very similar to those of cooked brown rice (Table 1) and raw brown rice (data not shown) except in the case of glutamic acid, which was 3% lower in the isolate and concentrate. Cooking had no effect on the amino acid profile of brown rice.

3.2. Comparison and Variation of Amino Acid Concentration between Sources of Brown Rice, Whey, and Soy

Oryzatein-90 isolate contains 79% TAA by weight, soy protein isolate is 88% TAA by weight, and whey protein isolate is almost 100% amino acids (Table 2). The amino acid profile and content of soy protein isolate was similar to that of Oryzatein-90; both contain >36% EAA and >17% BCAA (Table 2). Although Oryzatein-90 was a better source of valine and methionine, soy protein isolate was a better source of lysine. Whey protein isolate contains 46% EAA and 22% BCAA (Figure 1) and is a richer source of isoleucine, leucine, lysine and threonine but not phenylalanine compared to Oryzatein-90 (Figure 1).

Oryzatein-80 brown rice concentrate contains 77% TAA by weight, 36% EAA and 18% BCAA, similar to the isolate (Table 1). The amino acid profiles of the soy and whey protein concentrates were also similar to their isolates. Although soy protein concentrate has 63% TAA, the percent of EAA and BCAA is similar to that of Oryzatein-80 (Table 3). Whey protein concentrate, similar to the isolate, contains 99% TAA, 53% EAA, and 24% BCAA (Table 3).

The amino acid concentrations of the various isolates and concentrates all differ from the amino acid concentrations of their respective source materials, namely raw brown rice, defatted soy flour, and cow's milk. The concentrations of lysine and glutamate decrease but those of methionine, tyrosine and cystine increase in the brown rice protein isolate Oryzatein-90 compared to raw brown rice (Table 1). In contrast, there are few differences in the amino acid profiles of soy protein isolate and defatted soy flour (data not shown). Whey protein isolate has lower concentrations of phenylalanine, proline and glutamate but higher concentrations of threonine, alanine and aspartate compared to cow's milk (data not shown).

Although not addressed in this analysis and report, it appears that Oryzatein-90 may have utility over soy and whey protein isolates for several reasons: it is not genetically modified; does not contain lactose; and does not come from an animal known to be treated with growth hormones, anabolic steroids, estrogens and other hormones, antibiotics or other chemicals that may have an impact upon human health.

Moreover, a recent study comparing the effects of rice or whey protein isolate in male athletes immediately post exercise found no differences in perceived recovery or soreness between the two isolates [12].

Table 1. Amino acid composition of Oryzatein Silk 90 brown rice protein isolate, Oryzatein Silk 80 brown rice protein concentrate and cooked *Oryza sativa* L. (brown rice).

| Amino Acid | Oryzatein 90 Silk Isolate | | | | Oryzatein 80 Silk Concentrate [c] | | Oryza sativa Cooked [d] | |
| | Batch A [a] | | Batch B [b] | | | | | |
	mg/100 g	% Total AA	mg/100 g	% Total AA	mg/100 g	% Total AA	mg/100 g	% Total AA
Alanine	4469	5.8	4381	5.7	4322	5.6	151	5.8
Arginine	6321	8.2	6350	8.3	6115	7.9	196	7.6
Aspartic acid	6938	9.0	6791	8.9	6762	8.7	242	9.4
Cystine	1697	2.2	1808	2.4	1629	2.1	31	1.2
Glutamic acid	13,906	18.0	13,700	17.9	13,495	17.4	526	20.4
Glycine	3528	4.6	3410	4.5	3410	4.4	127	4.9
Histidine	1820	2.4	1670	2.2	1699	2.2	66	2.6
Isoleucine	3469	4.5	3234	4.2	3381	4.4	109	4.2
Leucine	6409	8.3	6321	8.3	6203	8.0	214	8.3
Lysine	2420	3.1	2190	2.9	2120	2.7	99	3.8
Methionine	2270	2.9	2258	3.0	2279	2.9	58	2.2
Phenylalanine	4410	5.7	4292	5.6	4116	5.3	133	5.1
Proline	2881	3.7	3675	4.8	3557	4.6	121	4.7
Serine	3910	5.1	3881	5.1	3793	4.9	134	5.2
Threonine	2919	3.8	2858	3.7	2799	3.6	95	3.7
Tryptophan	1170	1.5	1150	1.5	1120	1.4	33	1.3
Tyrosine	4263	5.5	4263	5.6	6203	8.0	97	3.8
Valine	4557	5.9	4263	5.6	4469	5.8	151	5.8
Total amino acids	77,357	100.0	76,495	100.0	77,472	100.0	2583	100.0
Total EAA/% total	29,444	38.1	66,940	36.9	28,186	36.4	958	37.1
Total BCAA/% total	14,435	18.7	13,818	18.1	14,053	18.1	474	18.4

[a] Results of Covance® Laboratories Inc. (Madison, WI, USA) Axiom Foods Organic Oryzatein Ultra 90 Batch HZN11008-13. Sample: 120021. Certificate of analysis. Report 518291-0. Received on 17 February 2012. [b] Results of Covance® Laboratories Inc. Axiom Foods Organic Oryzatein Ultra 90 Batch No. HZN12004-111. Sample 1360846. Certificate of analysis. Report 570345-0. Received 24 May 2012. [c] Results of Covance® Laboratories Inc. Axiom Foods Organic Oryzatein Ultra 80 Batch HZN11018-22. Sample: 1250431. Certificate of analysis. Report 535689-0. Received on 21 March 2012. [d] U.S. Department of Agriculture, agricultural research service. 2013. USDA National Nutrient Database for Standard Reference, Release 26 [10].

Table 2. Amino acid composition of Oryzatein brown rice protein isolate, whey protein isolate and soy protein isolate.

| Amino Acid | Oryzatein 90 Silk Isolate | | | | Soy Protein Isolate [c] | | Whey Protein Isolate [d] | |
| | Batch A [a] | | Batch B [b] | | | | | |
	mg/100 g	% Total AA	mg/100 g	% Total AA	mg/100 g	% Total AA	mg/100 g	% Total AA
Alanine	4469	5.8	4381	5.7	3589	4.1	4800	4.8
Arginine	6321	8.2	6350	8.3	6670	7.6	1779	1.8
Aspartic acid	6938	9.0	6791	8.9	10,203	11.6	10,161	10.2
Cystine	1697	2.2	1808	2.4	1046	1.2	2089	2.1
Glutamic acid	13,906	18.0	13,700	17.9	17,452	19.8	19,311	19.4
Glycine	3528	4.6	3410	4.5	3603	4.1	1421	1.4
Histidine	1820	2.4	1670	2.2	2303	2.6	1311	1.3
Isoleucine	3469	4.5	3234	4.2	4253	4.8	5600	5.6
Leucine	6409	8.3	6321	8.3	6783	7.7	10,239	10.3
Lysine	2420	3.1	2190	2.9	5327	6.0	9700	9.7
Methionine	2270	2.9	2258	3.0	1130	1.3	1689	1.7
Phenylalanine	4410	5.7	4292	5.6	4593	5.2	2579	2.6
Proline	2881	3.7	3675	4.8	4960	5.6	5739	5.8
Serine	3910	5.1	3881	5.1	4593	5.2	4921	4.9
Threonine	2919	3.8	2858	3.7	3137	3.6	7911	7.9
Tryptophan	1170	1.5	1150	1.5	1116	1.3	1889	1.9
Tyrosine	4263	5.5	4263	5.6	3222	3.7	2679	2.7
Valine	4557	5.9	4263	5.6	4098	4.7	5879	5.9
Total amino acids	77,357	100.0	76,495	100.0	88,078	100.0	99,697	100.0
Total EAA/% total	29,444	38.1	66,940	36.9	32,740	37.2	46,797	46.9
Total BCAA/% total	14,435	18.7	13,818	18.1	15,134	17.2	21,718	21.8

[a] Results of Covance® Laboratories Inc. Axiom Foods Organic Oryzatein Ultra 90 Batch HZN11008-13. Sample: 120021. Certificate of Analysis. Report 518291-0. Received on 17 February 2012. [b] Results of Covance® Laboratories Inc. Axiom Foods Organic Oryzatein Ultra 90 Batch No. HZN12004-111. Sample 1360846. Certificate of Analysis. Report 570345-0. Received 24 May 2012. [c] U.S. Department of Agriculture, Agricultural Research Service. 2013. USDA National Nutrient Database for Standard Reference, Release 26, item 16122 [10]. [d] US Dairy Export Council. 2004. Applications Monograph. Senior Nutrition [11].

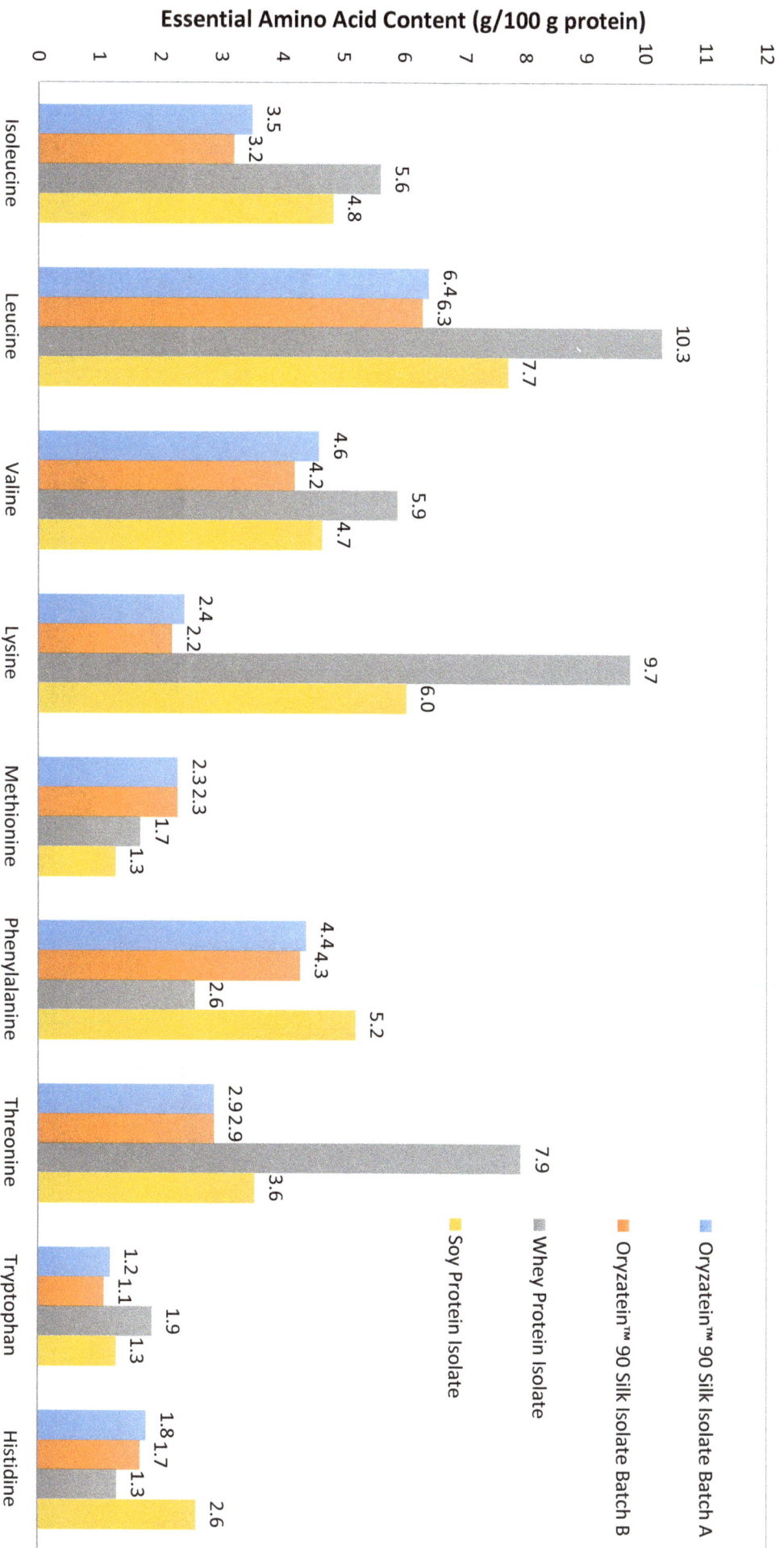

Figure 1. Essential amino acid profiles of Oryzatein-90 organic brown rice batch A and B, soy, and whey protein isolates.

Table 3. Amino Acid Composition of Oryzatein Brown Rice Protein Concentrate, Whey Protein Concentrate and Soy Protein Concentrate.

Amino Acid	Oryzatein 80 Silk Concentrate [a]		Soy Protein Concentrate [b]		Whey Protein Concentrate [c]	
	mg/100 g	% Total AA	mg/100 g	%Total AA	mg/100 g	%Total AA
Alanine	4322	5.6	2677	4.3	4900	5
Arginine	6115	7.9	4642	7.4	2100	2.1
Aspartic acid	6762	8.7	7249	11.5	10,800	10.9
Cystine	1629	2.1	886	1.4	2089	2.1
Glutamic acid	13,495	17.4	12,013	19.1	16,700	16.9
Glycine	3410	4.4	2688	4.3	1800	1.8
Histidine	1699	2.2	1578	2.5	2200	2.2
Isoleucine	3381	4.4	2942	4.7	5800	5.9
Leucine	6203	8	4917	7.8	10,239	10.4
Lysine	2120	2.7	3929	6.2	9600	9.7
Methionine	2279	2.9	814	1.3	1900	1.9
Phenylalanine	4116	5.3	3278	5.2	3300	3.3
Proline	3557	4.6	3298	5.2	5800	5.9
Serine	3793	4.9	3369	5.4	4700	4.8
Threonine	2799	3.6	2474	3.9	7200	7.3
Tryptophan	1120	1.4	835	1.3	2100	2.1
Tyrosine	6203	8	2301	3.7	1800	1.8
Valine	4469	5.8	3064	4.9	5800	5.9
Total amino acids	77,472	100	62,954	100	98,828	100
Total EAA/% total	28,186	36.4	23,832	37.9	52,490	53.1
Total BCAA/% total	14,053	18.1	10,923	17.4	24,100	24.4

[a] Results of Covance® Laboratories Inc. Axiom Foods Organic Oryzatein Ultra 80 Batch HZN11018-22. Sample: 1250431. Certificate of Analysis. Report 535689-0. Received on 21 March 2012. [b] U.S. Department of Agriculture, Agricultural Research Service. 2013. USDA National Nutrient Database for Standard Reference, Release 26, item 16420 [10]. [c] US Dairy Export Council. 2004. Applications Monograph. Senior Nutrition [11].

4. Conclusions

These results provide a valuable addition to the nutrient database of protein isolates and concentrates from cereal grains. Oryzatein-80, an organic brown rice protein concentrate or Oryzatein-90, an organic brown rice protein isolate, may be used in place of other protein isolates or concentrates without any loss of essential nutrient value.

Acknowledgments

This project was supported entirely by Miami Research Associates.

Conflicts of Interest

The author declares no conflict of interest.

References

1. Schuster, R. Determination of amino acids in biological, pharmaceutical, plant and food samples by automated precolumn derivitization and HPLC. *J. Chromatogr.* **1988**, *431*, 271–284.

2. Barkholt, V.; Jensen, A.L. Amino acid analysis: Determination of cystine plus half cystine in proteins after hydrochloric acid hydrolysis with disulfide compound as additive. *Analyt. Biochem.* **1989**, *177*, 318–322.

3. Henderson, J.W.; Ricker, R.D.; Bidlingermeyer, B.A.; Woodward, C. Rapid, accurate, sensitive, and reproducible HPLC analysis of amino acids. *Agilent Technol. Tech. Note* **2000**, *1100*, 1–10.

4. Volpi, E.; Kobayashi, H.; Sheffield-Moore, M. Essential amino acids are primarily responsible for the amino acid stimulation of muscle protein anabolism in healthy elderly adults. *Am. J. Clin. Nutr.* **2003**, *78*, 250–258.

5. Paddon-Jones, D.; Sheffield-Moore, M.; Zhang, X.J.; Volpi, E.; Wolf, S.E.; Aarsland, A.; Ferrando, A.A.; Wolf, R.R. Amino acid ingestion improves muscle protein synthesis in the young and elderly. *Am. J. Physiol. Endocrinol. Metab.* **2004**, *286*, E321–E328.

6. Blmstrand, E.; Eliasson, J.; Karlsson, H.K.R.; Köhnke, R. Branched-chain amino acids activate key enzymes in protein synthesis after physical exercise. *J. Nutr.* **2006**, *136*, 269S–273S.

7. Kimball, S.R.; Jefferson, L.S. Signaling pathways and molecular mechanisms through which branched-chain amino acids mediate translational control of protein synthesis. *J. Nutr.* **2006**, *136*, 227S–231S.

8. Ha, E.; Zemel, M. Functional properties of why, whey components, and essential amino acids: Mechanisms underlying health benefits for active people. *J. Nutr. Biochem.* **2003**, *14*, 251–258.

9. Rasmussen, B.B.; Tipton, K.D.; Miller, S.L. An oral essential amino acid-carbohydrate supplement enhances muscle protein anabolism after resistance exercise. *J. Appl. Physiol.* **2000**, *88*, 386–392.

10. USDA National Nutrient Database for Standard Reference, Release 26. Available online: http://www.ars.usda.gov/ba/bhnrc/ndl (accessed on 18 September 2013).

11. US Dairy Export Council, Applications Monograph, Senior Nutrition. Available online: http://www.usdec.org/files/PDFs/2008Monographs/SeniorNutrition_English.pdf (accessed on 18 September 2013).

12. Joy, J.M.; Lowery, R.P.; Wilson, J.M.; Purpura, M.; de Souza, E.O.; Wilson, S.M.; Kalman, D.S.; Dudeck, J.E.; Jäger, R. The effects of 8 weeks of whey or rice protein supplementation on body composition and exercise performance. *Nutr. J.* **2013**, *12*, 86–93.

7

Comparison of the Digestibility of the Major Peanut Allergens in Thermally Processed Peanuts and in Pure Form

Soheila J. Maleki, David A. Schmitt [†], Maria Galeano [‡] and Barry K. Hurlburt *

Southern Regional Research Center, Agricultural Research Service, U.S. Department of Agriculture, 1100 Robert E. Lee Blvd, New Orleans, LA 70124, USA;
E-Mails: soheila.maleki@ars.usda.gov (S.J.M.); daschmitt2004@hotmail.com (D.A.S.); mgalea-athens@hotmail.com (M.G.)

[†] Current address: Marathon Petroleum, Texas City, TX 77590, USA

[‡] Current address: Vital Source Technologies, Raleigh, NC 27601, USA

* Author to whom correspondence should be addressed; E-Mail: barry.hurlburt@ars.usda.gov

Abstract: It has been suggested that the boiling or frying of peanuts leads to less allergenic products than roasting. Here, we have compared the digestibility of the major peanut allergens in the context of peanuts subjected to boiling, frying or roasting and in purified form. The soluble peanut extracts and the purified allergens were digested with either trypsin or pepsin and analyzed by gel electrophoresis and western blot. T-cell proliferation was measured for the purified allergens. In most cases, boiled and raw peanut proteins were similarly digestible, but the Ara h 1 protein in the boiled extracts was more resistant to digestion. Most proteins from fried and roasted peanuts were more resistant to digestion than in raw and boiled samples, and more IgE binding fragments survived digestion. High-molecular-weight fragments of Ara h1 were resistant to digestion in fried and roasted samples. Ara h 1 and Ara h 2 purified from roasted peanuts were the most resistant to digestion, but differed in their ability to stimulate T-cells. The differences in digestibility and IgE binding properties of the major allergens in roasted, fried and boiled peanuts may not explain the difference between the prevalence of peanut allergy in different countries that consume peanut following these varied processing methods.

Keywords: peanut; allergy; allergen; processing; digestion

---

1. Introduction

Approximately 5% of adults and 8% of children have a food allergy, and there is evidence for an increasing prevalence. An older study indicated that peanut allergies cause the majority of the annual emergency room admissions due to food allergies and approximately 63%–67% of deaths due to anaphylaxis [1]. Most recent studies have shown an 18% increase in food allergy and a prevalence of 1.3% peanut sensitive individuals [2,3]. Despite the focus on this issue, there is still no treatment for food allergies, and the only available option is avoidance. Even with avoidance, 55% of peanut allergic individuals have at least 1–2 accidental peanut ingestions every five years [4]. With the wide number of applications for peanut and peanut products in processed foods, particularly in candy and confectionary products, and the potential for cross-contamination of peanut-free products with traces of peanuts, avoidance can be very difficult for allergic consumers. Therefore, peanut allergy is not only an increasing public health problem, but it also poses a challenge to the food industry and regulatory agencies in terms of food safety.

In order to understand the immune system-allergen recognition and response, it is important to understand the cause(s) of allergenicity in the allergenic foods and the allergenic components at a molecular level. The effects of processing on clinical symptoms caused by a food has become increasingly important, due to recent studies that demonstrate that individuals can safely become desensitized to a food by consuming that food in one processed form *versus* another [5–10]. If the processing-induced-specific changes in an allergic protein can be determined and correlated with clinical reactivity, designing safe preventative or immunotherapeutic treatments through food processing has much potential. Previously, we addressed the effects of thermal processing on some of the allergenic properties of peanut proteins [11,12]. Thermal processing, such as roasting, curing and various types of cooking, can cause multiple non-enzymatic, biochemical reactions to occur in food [13]. One of the predominant reactions that occurs during the thermal processing of foods is known as the Maillard reaction, which is important in the development of flavor and color [13]. In addition to protein cross-linking, it is known that advanced Maillard reaction products [14,15], also known as advanced glycation end-products (AGEs), could lead to the modification of amino acids, such as lysine and cysteine [16]. The majority of alterations to protein structure are due to the heat-induced interactions of sugar components with amino acids to form compounds, such as carboxymethyllysine, melanoidin and other non-cross-linking modifications to proteins that may have detrimental nutritional, physiological and toxicological consequences [13,16]. Other studies have addressed the role of food processing on the allergenic properties of ingested foods [11,12,17–24]. Some of the roasting-induced biophysical mechanisms for enhanced allergenic properties of the major peanut allergens were previously explored in a simulated roasting model [11]. Both Ara h 1 and Ara h 2 bound higher levels of IgE, and the increase in IgE binding was correlated with increased carboxymethyllysine (CML) modifications on the surface of the protein [25]. Ara h 1 was found to be inter-molecularly cross-linked to form highly stable trimers, and Ara h 2 was thought to form intra-molecular cross-links due to roasting, without forming higher orders structures. Since resistance to digestion is a classic characteristic of food allergens, we wanted to determine if different thermal processes induced different modifications of the peanut allergens, altering their stability against digestive enzymes within the context of other peanut proteins or if purified from thermally processed peanuts. The effects of different thermal processes on Ara h 1 and Ara h 2 were assessed for digestibility with trypsin and pepsin, IgE binding and stimulation of T-cells from peanut allergic individuals.

2. Experimental Section

2.1. Patient T-Cells and Sera

Sera and lymphocytes were obtained from the blood of peanut allergic individuals, which were collected after informed consent at Tulane Health Science Center (New Orleans, LA, USA) in accordance with the rules and regulations of the institutional review board. A pool of sera from previously well-characterized and -described peanut allergic patients' sera was used in this study [17].

2.2. Extract Preparation and Protein Purification

Florunner peanuts were used either raw, roasted, boiled or fried, as previously described [19]. The samples were solubilized by adding 50 mg of a defatted peanut meal to 1.8 mL of a buffer containing 60 mM Tris, 1 mM EDTA and 200 mM NaCl at pH 8.5 followed by sonication and centrifugation at $5500\times g$ for 15 min to remove insoluble material yielding CPE (crude peanut extract). Ara h 1 and Ara h 2 were purified as described [23,26].

2.3. Digestion Reactions

Trypsin digestions were set up according to Maleki *et al.* [23]. Raw, roasted, boiled or fried peanut extracts (each at a concentration of 5 mg/mL) and the purified Ara h 1 (1 mg/mL) and Ara h 2 (1 mg/mL) from raw and roasted peanuts were incubated in the presence of 1 μM trypsin (final concentration of trypsin) in PBS at pH 8.5 for various times at 37 °C. Aliquots were taken for SDS-PAGE analysis at the times indicated in each figure. Pepsin was used to make simulated gastric fluid (SGF). The peanut samples were incubated in the presence of SGF (0.5 μg/mL pepsin in phosphate buffered saline (PBS) at pH 2 at 37 °C), and aliquots were taken at the indicated time points in each figure. The digestion reaction in each time point aliquot was quenched by the addition of SDS sample buffer. Samples were then subjected to SDS-PAGE and either stained or transferred to nitrocellulose for western blot analysis. The 0 time point was taken immediately after mixing the sample with the enzymes and not before adding enzyme, so some degree of digestion may be observed.

2.4. SDS-PAGE and Western Blot Analysis

The samples from each time point of digestion were subjected to SDS-PAGE on 4%–20% Novex Tris-HCl pre-cast gels (Life Technologies, Carlsbad, CA, USA), where individual proteins were separated according to size and either stained with Gel-Code Blue (Pierce, Rockford, IL, USA) for 1 h, and digitally recorded, or transferred to PVDF membranes. The membrane was then blocked for 1 h using 5% Blotto (5% dry milk dissolved into PBS containing 0.5% Tween (PBST)). After blocking the membrane, the primary antibody was diluted in 5% Blotto, added to the membrane and incubated for 1 h. The custom-made antibodies used were chicken anti-Ara h 1 and anti-Ara h 2 (Sigma Immunosys, The Woodlands, TX, USA) at 1:5000. For IgE western blots, membranes were blocked in 2% Blotto for 15 min and incubated overnight with 1:10 dilution in PBST of patient sera from allergic individuals. After the incubation with primary antibodies, the membranes were washed 3 times with PBST and incubated with the either anti-chicken IgY at 1:100,000 or 1:10,000 anti-human IgE horseradish peroxidase (HRP)-labeled secondary antibody (Sigma Chemical Company, St. Louis, MO, USA) in 2% Blotto for 30 min. The membrane was then washed 3 times with PBST and incubated with ECL-Plus Western substrate (Amersham Bioscience Corp., Piscataway, NJ, USA). The signal was then visualized using a CCD camera system (Fuji Photo Film Co., Ltd., Duluth, GA, USA).

2.5. T-Cell Proliferation

The peripheral blood lymphocytes (PBLs) of 5 peanut allergic individuals were isolated from whole blood using standard ficoll gradient centrifugation (Sigma-Aldrich, St. Louis, MO, USA). Cells were washed and suspended in media at a concentration of 4×10^6 cells/mL. For the T-cell proliferation assays, triplicate wells of a 96 well plate at 2×10^5 PBLs/well were stimulated with media (control), CPE (50 µg/mL, data not shown), raw and light roasted Ara h 2 (10 µg/mL) and Ara h 1 (25 µg/mL) at 37 °C for 6 days. On Day 6, the cells were treated with [^{3}H]-thymidine (1 µCi/well) and re-incubated at 37 °C for 6–8 h before harvesting onto glass fiber filters (Packard, Meriden, CT, USA). T-cell proliferation was estimated by quantifying the [^{3}H]-thymidine incorporation into the DNA of proliferating cells. [^{3}H]-thymidine incorporation is reported as the stimulation index (SI), which is defined as fold stimulation above media treated (control) cells.

3. Results and Discussion

One characteristic believed to contribute to a food protein's allergenicity is resistance to digestion. Major peanut allergens, Ara h 1 and, especially, Ara h 2, are known to be resistant to degradation by digestive enzymes. In order to determine if diverse processing methods can alter the digestibility of thermally processed peanut proteins, soluble extracts were made from raw, roasted, boiled and fried peanut extracts, which were then subjected to digestion with trypsin (Figure 1) several times prior to SDS-PAGE analysis (Figure 1A).

Figure 1. SDS-PAGE and western analysis of the digestion of raw, boiled, fried and roasted peanut extracts with trypsin. Soluble protein extracts from raw, boiled, fried and roasted peanuts were digested with trypsin and subjected to SDS-PAGE (**A**), western blot analysis with anti-Ara h 1 (**B**), western blot analysis with anti-Ara h 2 antibody (**C**) and western blot analysis with pooled human IgE sera from peanut allergic individuals (**D**).

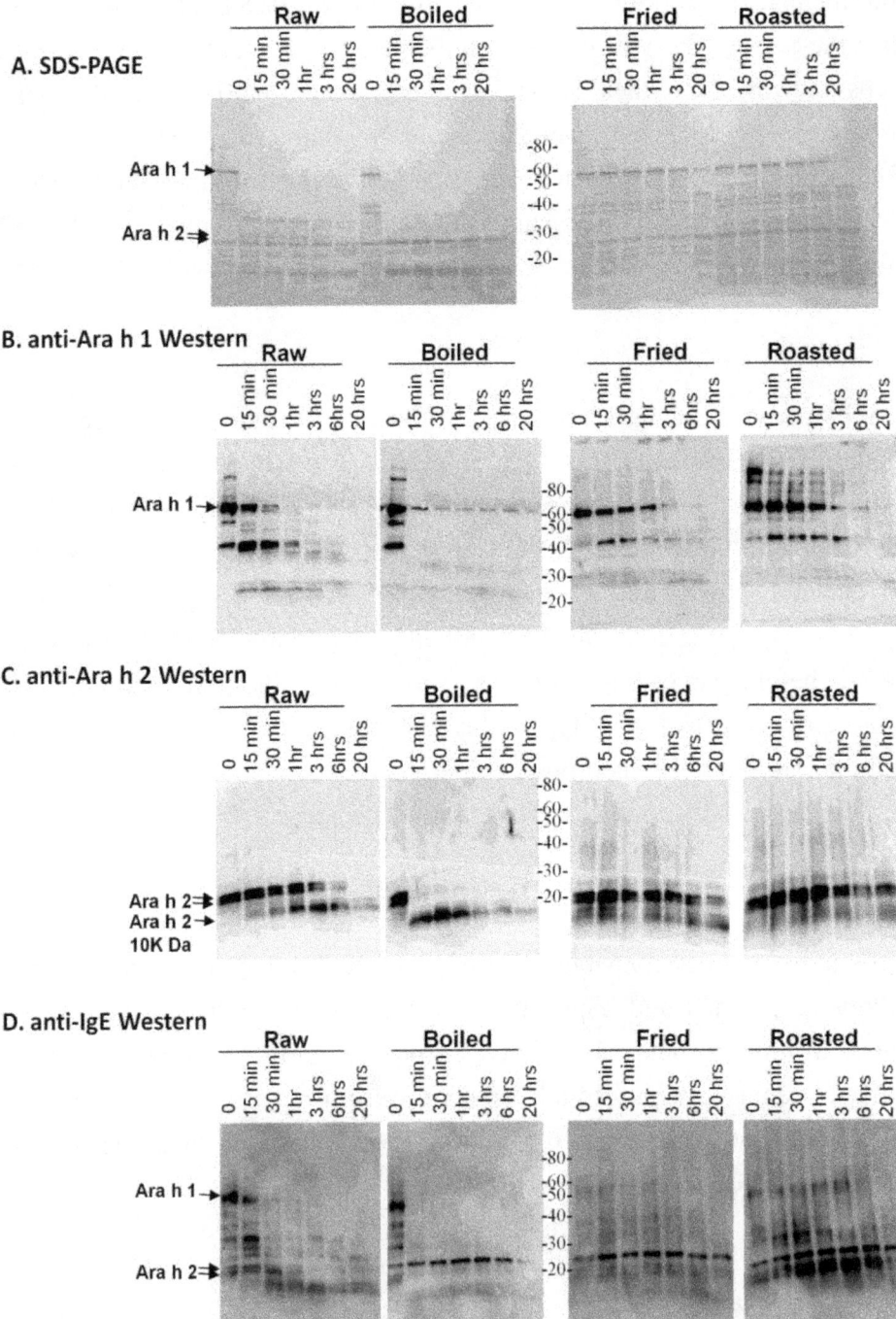

This figure demonstrates that the higher molecular weight protein bands, such as Ara h 1, are more resistant to digestion in the fried and roasted samples than in the raw and boiled samples. The digestion pattern is also different between raw and boiled samples, in which two main lower molecular weight bands, which could be fragments of other proteins, persist after 20 h of digestion with trypsin. An anti-Ara h 1 western blot on the same extracts shows that the Ara h 1 in the boiled sample is more resistant to digestion than in the raw sample (Figure 1B). Higher order structures or oligomers of Ara h 1, previously shown to exist in the simulated roasting model, are clearly recognized by the anti-Ara h 1 antibody in this western blot. Ara h 1 is more resistant to trypsin digestion in all of the thermally processed peanuts compared to the raw peanut. Digestion of Ara h 1 with trypsin *in silico* yields 84 fragments, with the largest being 3.8 kDa. Therefore, if complete digestion occurred, very few bands in the ~3–4 kDa range would be visible on the percentage of SDS-PAGE used here. It is highly likely that many trypsin digestion sites are blocked by the protein structure and or by thermal processing-induced chemical modifications An anti-Ara h 2 western blot demonstrates that the Ara h 2 in raw peanut is more stable than the boiled sample, but Ara h 2 in roasted and fried samples are more resistant to trypsin digestion than in both the raw and boiled peanuts (Figure 1C). The known Ara h 2 10 kDa digestion-resistant band [27] can be seen below the intact Ara h 2 doublet in all of the extracts. Digestion of Ara h 2 with trypsin *in silico* generates 21 fragments, the largest of which is 2 kDa. These same samples were assessed for IgE binding using western blot analysis to determine the effect of the processing on the IgE recognition pattern of allergens within the context of the extracts (Figure 1D). It appears as though the IgE binding proteins in the boiled peanut, particularly the higher molecular weight ones, are the most significantly reduced and rapidly digested with trypsin.

The raw, boiled, fried and roasted peanut extracts were subjected to digestion with pepsin (Figure 2). The SDS-PAGE indicates that most of the peanut proteins are more resistant to digestion with pepsin than to trypsin and that the proteins in the fried and roasted peanut samples seem to be minimally altered due to pepsin treatment over a 20-h period (Figure 2A). Digestion of Ara h 1 with pepsin *in silico* yields 33 fragments, the largest of which is 5 kDa. An anti-Ara h 1 western blot of the pepsin digests indicates that following 20 h of digestion, a large Ara h 1 digestion fragment, immediately below the intact Ara h 1 and several smaller Ara h 1 fragments of <40 kDa survive digestion in the roasted sample. Digestion of purified Ara h 1 with pepsin has also been reported to result in relatively large fragments capable of binding IgE [28]. The digestion pattern of Ara h 1 in the raw, boiled and fried extracts are similar, but the fragments that survive after 20 h are different in the fried peanut extracts. Chemical modification of digestion sites and surrounding amino acid residues can alter the digestion patterns by an enzyme, often by blocking the sites to be cleaved. These results indicate that the chemical or structural modifications to Ara h 1 in boiling and frying are similar in some aspects, but can also be vastly different.

Figure 2. SDS-PAGE and western analysis of the digestion of raw, boiled, fried and roasted peanut extracts with pepsin. Soluble protein extracts from raw, boiled, fried and roasted peanuts were digested with pepsin and subjected to SDS-PAGE (**A**), western blot analysis with anti-Ara h 1 antibody (**B**), western blot analysis with anti-Ara h 2 antibody (**C**) and western blot analysis pooled human IgE sera from peanut allergic individuals (**D**).

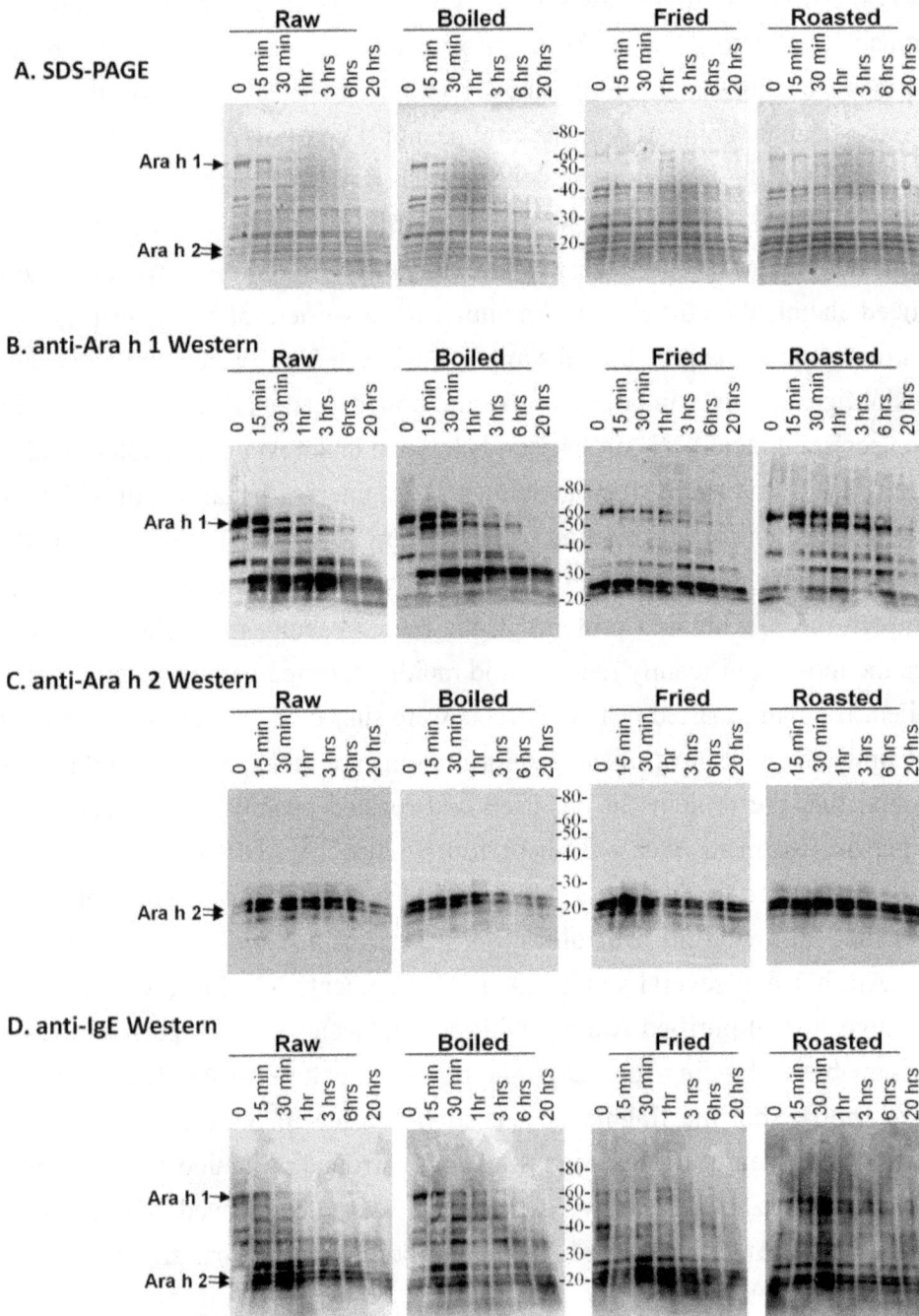

Because the most diverse digestion patterns were seen between raw and roasted peanut extracts, the major allergens, Ara h 1 and Ara h 2, were purified from raw (R), light roast (LR) and dark roast (DR) peanut and compared for their digestibility with trypsin and pepsin (data not shown for pepsin digestion). R, LR and DR Ara h 1 were subjected to trypsin digestion for the indicated times and resolved by SDS-PAGE (Figure 3, left). The IgE binding ability to the digested fragments was assessed by western blot (Figure 3, right). The R Ara h 1 is completely digested into smaller fragments following 30 min of incubation in the presence of trypsin. A strong 35-kDa band appears and survives digestion for approximately 1 h, as the intact Ara h 1 band is digested. Fragments smaller than 35 kDa survive digestion overnight, two of which are recognized by IgE, similar to what is seen in the case of the Ara h 1 digested within the context of raw peanut proteins; whereas the intact LR Ara h 1 can be seen after 3 h and DR intact Ara h 1 after overnight digestion with trypsin. The same 35-kDa fragment seen in R Arah 1 appears and survives more than 3 h in the presence of trypsin in both the LR and DR samples. While there are some higher molecular weight (>30 kDa) proteins present following 1 h of digestion in the R sample, after 20 h, some fragments of Ara h 1, approximately 25 kDa and below, remain undigested. When Ara h 1 digestion was followed within the context of raw peanut proteins, these bands are not seen after 20 h (Figure 1B). The molecular weights of the four fragments visible following overnight digestion with trypsin in SDS-PAGE, in all three digestion reactions, are ~22, 18, 13 and 13 kDa, three of which are recognized by IgE. This indicates that even though the resistance to digestion increases with the degree of roasting, the predominant trypsin digestion sites remain the same. However, the bands that are recognized by serum IgE are significantly different. The IgE binding in the DR sample is only to the higher molecular weight bands and the smaller bands in the raw extracts following 20 h of digestion. This indicates that the fragments of allergens that are recognized by serum IgE of allergic individuals change with roasting. The Ara h 1 seems to be less detectable or more digestible within the context of peanut proteins. We attribute this to both the presence of more Ara h 1 to digest and visualize in the pure samples; however, the fragments that survive digestibility within the context of R and LR peanut are similar to the purified proteins.

Figure 3. SDS-PAGE and IgE western analysis of the digestion of raw (R), light roast (LR) and dark roast (DR) Ara h 1 with trypsin. Raw Ara h 1 (*top*), light roast Ara h 1 (*middle*) and dark roast Ara h 1 (*bottom*) were digested with trypsin for the designated amounts of time. Sample taken prior to addition of trypsin is Time 0. The molecular weights of the marker proteins are indicated on the right.

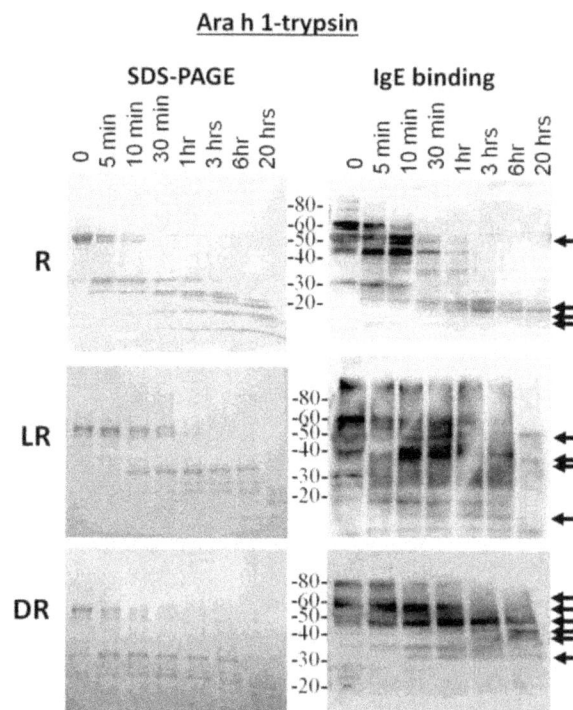

Purified samples of R, LR and DR Ara h 2 were subjected to trypsin digestion for the indicated times and resolved by SDS-PAGE (Figure 4). The intact R Ara h 2 was completely digested into smaller fragments after 1 h of incubation in the presence of trypsin. Furthermore, two fragments, a 10- and a 12-kDa band, appear as the intact Ara h 2 band disappears. These trypsin-resistant bands survive even after overnight digestion. The 10 kDa fragment becomes stronger over time, while the 12-kDa band maintains the same intensity after 30 min. In the IgE binding assay, the intact LR and DR Ara h 2 and smears thereof can both be seen after overnight digestion with trypsin. This indicates that the LR Ara h 2 and DR Ara h 2 are both more resistant to digestion with trypsin. In both LR and DR Ara h 2 samples, the 10- and 12-kDa bands, also seen in R Ara h 2, appear following the first 5–10 min of digestion, but the 12-kDa fragment appears as a smear. Interestingly, as seen in the trypsin digestion of Ara h 1, the surviving fragments are the same in all three (R, LR and DR) digestion reactions. This finding indicates that the trypsin digestion sites are not altered due to a higher degree of roasting. Interestingly, even though in SDS-PAGE, the intensity of the intact Ara h 2 is significantly decreased over time, the IgE binding remains similar, which indicates that more IgE is binding to the roasted samples. Both Ara h 1 and Ara h 2 from roasted peanuts and fragments thereof are more resistant to digestion with trypsin due to increased chemical blocking or unknown alterations of the existing digestion sites with increased time of roasting. Purified Ara h 2 digestion did not change significantly from digestion within the context of peanut proteins, which is not surprising, as the digestion sites are not altered.

Figure 4. SDS-PAGE analysis and IgE western analysis of the digestion of raw, light roast (LR) and dark roast (DR) Ara h 2 with trypsin. Raw Ara h 2 (R), light roast Ara h 2 (LR) and dark roast Ara h 2 (DR) were digested by trypsin for the designated amounts of time of 5 min (5') to overnight (O/N). Other lanes are labeled Ara h 1 and Ara h 2 control (C), undigested (U) and molecular weight marker. The molecular weights of the marker proteins are indicated on the right.

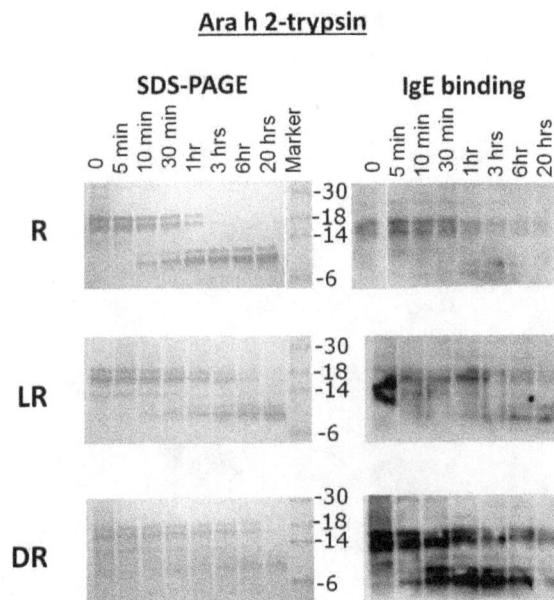

The stimulation of T-cells by purified R and LR Ara h 1 and Ara h 2 were compared (Figure 5). Interestingly, T-cell stimulation by LR Ara h 1 was significantly reduced in comparison to R Ara h 1, and the opposite was true for Ara h 2. It is known that if the IgE binding sites of an allergen are eliminated, while maintaining the T-cell proliferative characteristics of that allergen, then it can be utilized as an effective immunotherapeutic tool that alters a T-helper 2 (Th2), or an inflammatory response to a Th1 or a tolerant response [29]. In this case, the Ara h 1 allergen has a higher IgE binding and is more resistant to digestion, but T-cell proliferation is reduced. On the other hand, roasted Ara h 2 is more resistant to digestion, binds higher IgE and causes higher T-cell stimulation following roasting, indicating that Ara h 2 is more immunogenic. This is consistent with findings in the field that Ara h 2 is the most potent allergen in peanut [30].

Figure 5. T-cell response to Ara h 1 and 2 purified from raw and roasted peanut extracts. Lymphocytes from five peanut allergic individuals were stimulated with Ara h 1 (**A**) or Ara h 2 (**B**) purified from either raw or roasted peanuts. The stimulation index (SI) is shown on the *y*-axis.

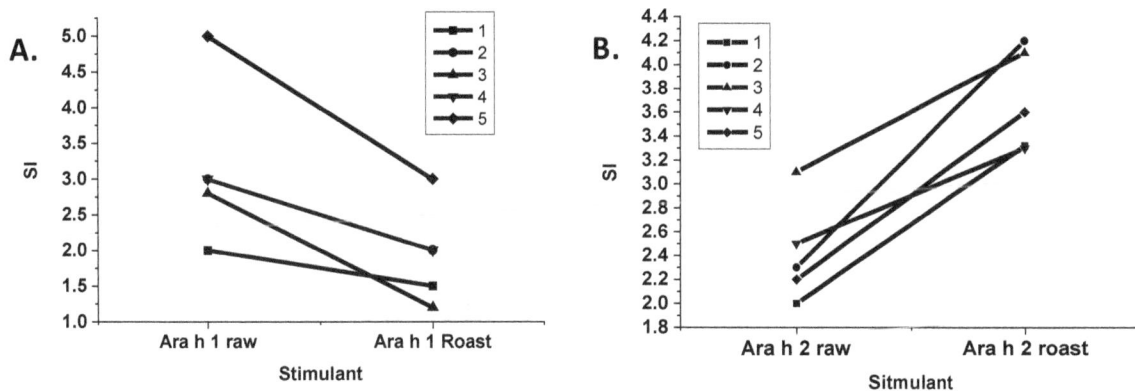

The effects of roasting on IgE binding and the allergenic potency of peanut allergens has been under debate for a long time. One study showed that in a simulated roasting model of heating crude peanut extract (CPE), Ara h 1 and Ara h 2 in the presence of reducing sugars (55 °C for 10 days), in solution, enhanced IgE binding [11]. In another similar study, the CPE and purified Ara h 1 and Ara h 2 were dried in the presence and absence of glucose (+g and −g, respectively) prior to heating at 145 °C for 20 min [31]. Following solubilization, they found that IgE binding to Ara h 1 +g was significantly reduced, but the capacity of mediator release increased. Meanwhile, both IgE binding and mediator release with Ara h 2 +g and −g was reduced. Ara h 1 purified from roasted peanuts was shown to bind higher levels of IgE than raw peanuts [17]. This observation was attributed to chemical modifications rather than major structural alterations, and the specific Maillard reaction products on the roasted Ara h 1 were identified [17,24]. Furthermore, AGE modifications were found on Ara h 1 and Ara h 3 in both raw and roasted peanut extract, but not on Ara h 2. The receptor for advanced glycation end products (RAGE) binds selectively to Ara h 1 derived from peanut extract, whereas the analysis failed to demonstrate Ara h 2 binding to RAGE. Recombinant Ara h 1 with no AGE modifications did not bind RAGE; however, after AGE modification with xylose, recombinant Ara h 1 bound to RAGE. Perhaps, the reduced AGE modifications of Ara h 2 allow more potent IgE and T-cell epitope exposure. The ability of Ara h 2 to be processed better by antigen presenting cells, due to reduced glycation, can explain the enhanced T-cell proliferation compared to the reduction seen with roasted Ara h 1. In another study, a combination of purified Ara h 2/6 from raw peanuts was heated in solution (110 °C for 15 min) and +g and −g [32]. Roasted Ara h 2/6 was also purified for comparison. They found no differences in T-cell proliferation with the raw, heat-treated and roasted Ara h 2/6, but the raw heated sample bound less IgE, due to denaturation, hydrolysis and aggregation. Discrepancies such as these can be attributed to different methods of experimentation, such as protein extraction, purification, glycation, temperatures of heating, *etc.* For example, we have found that if purified proteins are heated in the presence or absence of sugar at high temperatures for even short periods of time, then it is as if the protein or the food extract has been charred, and the subsequent findings may not apply to the actual ingested form (unpublished observation). This observation is consistent with the findings in the previous study with Ara h 2/6 that showed a significant difference between the structure and immunological properties of a purified protein glycated rapidly at high temperature and the protein purified from roasted peanuts. In the present study, we chose to assess Ara h 1 and Ara h 2 within the context of peanuts and purified from roasted peanuts in order to study the actual digestion and immunological response of these proteins in the ingested form, as opposed to using model systems.

4. Conclusions

Our results show that, while model systems are highly effective in understanding molecular events in foods, it is important to understand the effects of food processing on allergenicity and on the individual allergens in order to develop effective tools for research, diagnosis, detection and immunotherapy. While it has been shown that different processes influence the allergenic properties or immunogenicity of certain foods, it has not been shown that processing can influence sensitization or the original development of allergy to a particular food. Animal models or human studies on the ability of differently processed foods to sensitize or tolerize could be useful in assessing the influence of processing on epidemiology and the development of food allergy in various countries.

Acknowledgments

The authors would like to thank Ken Ehrlich, Si-yin Chung and Peter Bechtel for critical reading of the manuscript.

Author Contributions

S.J.M. conceived of the study, performed some of the experiments and contributed to writing the manuscript. D.A.S. and M.G. performed most of the experiments. B.K.H. contributed to the study's concept development, data analysis and to writing the manuscript.

Conflicts of Interest

The authors declare no conflicts of interest.

References

1. Bock, S.; Munoz-Furlong, A.; Sampson, H.A. Fatalities due to anaphylactic reactions to foods. *J. Allergy Clin. Immunol.* **2001**, *107*, 191–193.

2. Gupta, R.S.; Springston, E.E.; Warrier, M.R.; Smith, B.; Kumar, R.; Pongracic, J.; Holl, J.L. The prevalence, severity, and distribution of childhood food allergy in the United States. *Pediatrics* **2011**, *128*, e9–e17.

3. Liu, A.H.; Jaramillo, R.; Sicherer, S.H.; Wood, R.A.; Bock, S.A.; Burks, A.W.; Massing, M.; Cohn, R.D.; Zeldin, D.C. National prevalence and risk factors for food allergy and relationship to asthma: Results from the National Health and Nutrition Examination Survey 2005–2006. *J. Allergy Clin. Immunol.* **2010**, *126*, 798–806.

4. Bock, S.A.; Atkins, F.M. The natural history of peanut allergy. *J. Allergy Clin. Immunol.* **1989**, *83*, 900–904.

5. Netting, M.; Makrides, M.; Gold, M.; Quinn, P.; Penttila, I. Heated allergens and induction of tolerance in food allergic children. *Nutrients* **2013**, *5*, 2028–2046.

6. Tan, J.W.; Campbell, D.E.; Turner, P.J.; Kakakios, A.; Wong, M.; Mehr, S.; Joshi, P. Baked egg food challenges—Clinical utility of skin test to baked egg and ovomucoid in children with egg allergy. *Clin. Exp. Allergy* **2013**, *43*, 1189–1195.

7. Leonard, S.A.; Sampson, H.A.; Sicherer, S.H.; Noone, S.; Moshier, E.L.; Godbold, J.; Nowak-Wegrzyn, A. Dietary baked egg accelerates resolution of egg allergy in children. *J. Allergy Clin. Immunol.* **2012**, *130*, 473–480.

8. Lieberman, J.A.; Huang, F.R.; Sampson, H.A.; Nowak-Wegrzyn, A. Outcomes of 100 consecutive open, baked-egg oral food challenges in the allergy office. *J. Allergy Clin. Immunol.* **2012**, *129*, 1682–1684.

9. Turner, P.J.; Mehr, S.; Joshi, P.; Tan, J.; Wong, M.; Kakakios, A.; Campbell, D.E. Safety of food challenges to extensively heated egg in egg-allergic children: A prospective cohort study. *Pediatr. Allergy Immunol.* **2013**, *24*, 450–455.

10. Bartnikas, L.M.; Phipatanakul, W. Turning up the heat on skin testing for baked egg allergy. *Clin. Exp. Allergy* **2013**, *43*, 1095–1096.

11. Maleki, S.J.; Chung, S.Y.; Champagne, E.T.; Raufman, J.P. The effects of roasting on the allergenic properties of peanut proteins. *J. Allergy Clin. Immunol.* **2000**, *106*, 763–768.

12. Maleki, S.J. Food processing: Effects on allergenicity. *Curr. Opin. Allergy Clin. Immunol.* **2004**, *4*, 241–245.

13. Maillard, L.C.; Gautier, M.A. The reaction of amino acids with sugars: Mechanisms of melanoid formation. *CR Seances Acad. Sci.* **1912**, *III*, 66–68.

14. Chung, S.Y.; Maleki, S.J.; Champagne, E.T. Allergenic properties of roasted peanut allergens may be reduced by peroxidase. *J. Agric. Food Chem.* **2004**, *52*, 4541–4545.

15. Kitts, D.D.; Chen, X.M.; Jing, H. Demonstration of antioxidant and anti-inflammatory bioactivities from sugar-amino acid maillard reaction products. *J. Agric. Food Chem.* **2012**, *60*, 6718–6727.

16. Shahidi, F.; Ho, C.T. Process-induced chemical changes in foods. *Adv. Exp. Med. Biol.* **1998**, *434*, 1–3.

17. Nesbit, J.B.; Hurlburt, B.K.; Schein, C.H.; Cheng, H.; Wei, H.; Maleki, S.J. Ara h 1 structure is retained after roasting and is important for enhanced binding to IgE. *Mol. Nutr. Food Res.* **2012**, *56*, 1739–1747.

18. Cabanillas, B.; Pedrosa, M.M.; Rodriguez, J.; Muzquiz, M.; Maleki, S.J.; Cuadrado, C.; Burbano, C.; Crespo, J.F. Influence of enzymatic hydrolysis on the allergenicity of roasted peanut protein extract. *Int. Arch. Allergy Immunol.* **2011**, *157*, 41–50.

19. Schmitt, D.A.; Nesbit, J.B.; Hurlburt, B.K.; Cheng, H.; Maleki, S.J. Processing can alter the properties of peanut extract preparations. *J. Agric. Food Chem.* **2010**, *58*, 1138–1143.

20. Noorbakhsh, R.; Mortazavi, S.A.; Sankian, M.; Shahidi, F.; Maleki, S.J.; Nasiraii, L.R.; Falak, R.; Sima, H.R.; Varasteh, A. Influence of processing on the allergenic properties of pistachio nut assessed *in vitro*. *J. Agric. Food Chem.* **2010**, *58*, 10231–10235.

21. Liu, G.M.; Cheng, H.; Nesbit, J.B.; Su, W.J.; Cao, M.J.; Maleki, S.J. Effects of boiling on the IgE-binding properties of tropomyosin of shrimp (*Litopenaeus vannamei*). *J. Food Sci.* **2010**, *75*, T1–T5.

22. Maleki, S.J.; Hurlburt, B.K. Structural and functional alterations in major peanut allergens caused by thermal processing. *J. AOAC Int.* **2004**, *87*, 1475–1479.

23. Maleki, S.J.; Viquez, O.; Jacks, T.; Dodo, H.; Champagne, E.T.; Chung, S.Y.; Landry, S.J. The major peanut allergen, Ara h 2, functions as a trypsin inhibitor, and roasting enhances this function. *J. Allergy Clin. Immunol.* **2003**, *112*, 190–195.

24. Mueller, G.A.; Maleki, S.J.; Johnson, K.; Hurlburt, B.K.; Cheng, H.; Ruan, S.; Nesbit, J.B.; Pomes, A.; Edwards, L.L.; Schorzman, A.; *et al*. Identification of Maillard reaction products on peanut allergens that influence binding to the receptor for advanced glycation end products. *Allergy* **2013**, *68*, 1546–1554.

25. Maleki, S.; Chung, S.; Champagne, E.; Khalifah, R. Allergic and biophysical properties of peanut proteins before and after roasting. *Food Allergy Toler.* **2001**, *2*, 211–221.

26. Shin, D.S.; Compadre, C.M.; Maleki, S.J.; Kopper, R.A.; Sampson, H.; Huang, S.K.; Burks, A.W.; Bannon, G.A. Biochemical and structural analysis of the IgE binding sites on ara h1, an abundant and highly allergenic peanut protein. *J. Biol. Chem.* **1998**, *273*, 13753–13759.

27. Koppelman, S.J.; Hefle, S.L.; Taylor, S.L.; de Jong, G.A. Digestion of peanut allergens Ara h 1, Ara h 2, Ara h 3, and Ara h 6: A comparative *in vitro* study and partial characterization of digestion-resistant peptides. *Mol. Nutr. Food Res.* **2010**, *54*, 1711–1721.

28. Van Boxtel, E.L.; Koppelman, S.J.; van den Broek, L.A.; Gruppen, H. Determination of pepsin-susceptible and pepsin-resistant epitopes in native and heat-treated peanut allergen Ara h 1. *J. Agric. Food Chem.* **2008**, *56*, 2223–2230.

29. Rupa, P.; Mine, Y. Ablation of ovomucoid-induced allergic response by desensitization with recombinant ovomucoid third domain in a murine model. *Clin. Exp. Immunol.* **2006**, *145*, 493–501.

30. Chen, X.; Zhuang, Y.; Wang, Q.; Moutsoglou, D.; Ruiz, G.; Yen, S.E.; Dreskin, S.C. Analysis of the effector activity of Ara h 2 and Ara h 6 by selective depletion from a crude peanut extract. *J. Immunol. Methods* **2011**, *372*, 65–70.

31. Vissers, Y.M.; Iwan, M.; Adel-Patient, K.; Stahl Skov, P.; Rigby, N.M.; Johnson, P.E.; Mandrup Muller, P.; Przybylski-Nicaise, L.; Schaap, M.; Ruinemans-Koerts, J.; *et al*. Effect of roasting on the allergenicity of major peanut allergens Ara h 1 and Ara h 2/6: The necessity of degranulation assays. *Clin. Exp. Allergy* **2011**, *41*, 1631–1642.

32. Vissers, Y.M.; Blanc, F.; Skov, P.S.; Johnson, P.E.; Rigby, N.M.; Przybylski-Nicaise, L.; Bernard, H.; Wal, J.M.; Ballmer-Weber, B.; Zuidmeer-Jongejan, L.; *et al*. Effect of heating and glycation on the allergenicity of 2S albumins (Ara h 2/6) from peanut. *PLoS One* **2011**, *6*, e23998.

8

A Novel Proteomic Analysis of the Modifications Induced by High Hydrostatic Pressure on Hazelnut Water-Soluble Proteins

Nuria Prieto [1], Carmen Burbano [2], Elisa Iniesto [1], Julia Rodríguez [3], Beatriz Cabanillas [3], Jesus F. Crespo [3], Mercedes M. Pedrosa [2], Mercedes Muzquiz [2], Juan Carlos del Pozo [4], Rosario Linacero [1] and Carmen Cuadrado [2,*]

[1] Genetics Department, Biology Faculty, Complutense University of Madrid, Madrid 28040, Spain; E-Mails: nuph_seth@hotmail.com (N.P); e.iniesto@gmail.coM (E.I.); charolin@ucm.es (R.L.)

[2] Food Technology Department, National Institute for Agricultural and Food Research and Technology (INIA), Ctra. La Coruña Km. 7.5, Madrid 28040, Spain; E-Mails: burbano@inia.es (C.B.); mmartin@inia.es (M.M.P.); muzquiz@inia.es (M.M.)

[3] Allergy Service, Research Institute Hospital 12 de Octubre (i+12), Avenida de Córdoba s/n, Madrid 28041, Spain; E-Mails: juliarodriguez2@telefonica.net (J.R.); Beatriz.Cabanillas@ukb.uni-bonn.de (B.C.); jfcrespo@isciii.es (J.F.C.)

[4] Center for Biotechnology and Plant Genomic, Polytechnic University of Madrid-National Institute for Agricultural and Food Research and Technology (UPM-INIA), Montegancedo Campus, Boadilla del Monte, Madrid 28660, Spain; E-Mail: jc.delpozo@upm.es

* Author to whom correspondence should be addressed; E-Mail: cuadrado@inia.es

Abstract: Food allergies to hazelnut represent an important health problem in industrialized countries because of their high prevalence and severity. Food allergenicity can be changed by several processing procedures since food proteins may undergo modifications which could alter immunoreactivity. High-hydrostatic pressure (HHP) is an emerging processing technology used to develop novel and high-quality foods. The effect of HHP on allergenicity is currently being investigated through changes in protein structure. Our aim is to evaluate the effect of HHP on the protein profile of hazelnut immunoreactive extracts by comparative proteomic analysis with ProteomeLab PF-2D liquid chromatography and mass spectrometry. This protein fractionation method resolves proteins by isoelectric point and hydrophobicity in the first and second dimension, respectively. Second dimension chromatogram analyses show that some protein peaks present in unpressurized hazelnut must be unsolubilized and are not present in HHP-treated hazelnut extracts. Our results show that HHP treatment at low temperature induced marked changes on hazelnut water-soluble protein profile.

Keywords: hazelnut; high hydrostatic pressure; immunoreactivity; ProteomeLab PF-2D

---

1. Introduction

High-hydrostatic pressure (HHP) is considered an emerging processing technology used to develop novel and high-quality foods. This novel-processing technique even renders harmless foods which would be of considerable benefit to consumers. HHP treatment of foods can be used to create new products (new texture or taste) or to obtain analogue products with minimal effect on flavor, color and nutritional value and without any thermal degradation. It is well established that higher pressure has a disruptive effect on the tertiary and quaternary structure of most globular proteins, with relatively little influence on the secondary structure. Therefore, higher hydrostatic pressure can unfold proteins. The typical pressure needed for the unfolding is around 500 MPa but it varies from protein to protein, in the range from 100 MPa to 1 GPa or reaching even higher pressures in special cases. The effect of HHP on immunoreactive proteins is being currently investigated through changes in protein structure [1]. Such effects have been studied in: beef [2,3], apple [4,5], celery [4] and in nuts such as peanut [5]. However, there is scarce information on the effects of such food processing techniques on hazelnut (*Corylus avellana* L.) immunoreactive proteins.

Food allergies to hazelnut represent an important health problem in industrialized countries because of their high prevalence and severity [6]. Several hazelnut allergens are well characterized being Cor a 1 (18 KDa, Bet v 1 family) the major one. Other allergenic proteins are Cor a 2 (profilin), Cor a 8 (lipid transfer protein, LPT), Cor a 9 (11S globulin), Cor a 11 (vicilin-like protein) and Cor a 12, Cor a 13 and Cor a 14 belonging to the 2S albumins [7].Understanding how food processing affects the allergenic proteins could be important to control food allergenicity risk.

Bioinformatics tools for database searching enable the quick identification of sequences of interest. Tools for sequence comparison, motif searching or sequence profiling assist researchers to identify biologically relevant sequences similarities. The guidelines to assess potential allergenicity of proteins using bioinformatics in a step-by-step procedure are well established [8,9].

Plant comparative proteomics is becoming increasingly attractive as the rapidly expanding plant genomic and Expressed Sequence Tags (EST) databases provide new opportunities for protein identification. A partial automation of this procedure consisting of a robotic lift of protein spots embedded in the gel, followed by extraction, destaining and protein digestion, has been finalized with reasonable success to the further protein characterization and identification by mass spectrometry (MS) [10]. Proteome analyses have also been performed in a "gel free" condition by using protein fractionation procedures based entirely on liquid chromatography [11,12]. ProteomeLab PF-2D introduced a two dimensional liquid chromatography based on a high-performance chromatofocusing in the first dimension followed by high-resolution reversed-phase chromatography in the second dimension [13,14]. The ProteomeLab PF-2D has become available for sample fractionation and was more resolutive at extreme pHs, both acid and basic. It offers automation of the fractionation processes and resolves proteins by isoelectric point and hydrophobicity in the first and second dimension, respectively [15].

In this study, we have undertaken a comparative proteomic analysis of the effect of HHP at low temperature on water-soluble protein profile of hazelnut immunoreactive extracts using a novel proteomic approach that combines ProteomeLab PF-2D liquid chromatography and mass spectrometry analysis.

2. Experimental Section

2.1. Samples and High Hydrostatic Pressure (HHP) Treatments

The hazelnut (*Corylus avellana* L.) var Negreta used in this work was provided by the hazelnut collection of Institut de Recerca i Tecnología Agroalimentàries (IRTA-Mas de Bover, Tarragona, Spain) [16].

Hazelnuts were ground and defatted with *n*-hexane (34 mL/g of flour) for 4 h and air-dried after filtration of the *n*-hexane. High-pressure experiments conditions were carried out according to Omi *et al.* [17] and Kato *et al.* [18]. Hazelnut defatted flours were dissolved in distilled water (1:4 w/v) 20 h before HHP treatment. The flours were subjected to HHP, using pressures of 300, 400, 500 and 600 MPa for 15 min in a multivessel high-pressure equipment (HHP, ACB, France) at 20 °C (Figure 1).

Figure 1. Scheme of high hydrostatic pressure (HHP) treatment of hazelnut samples.

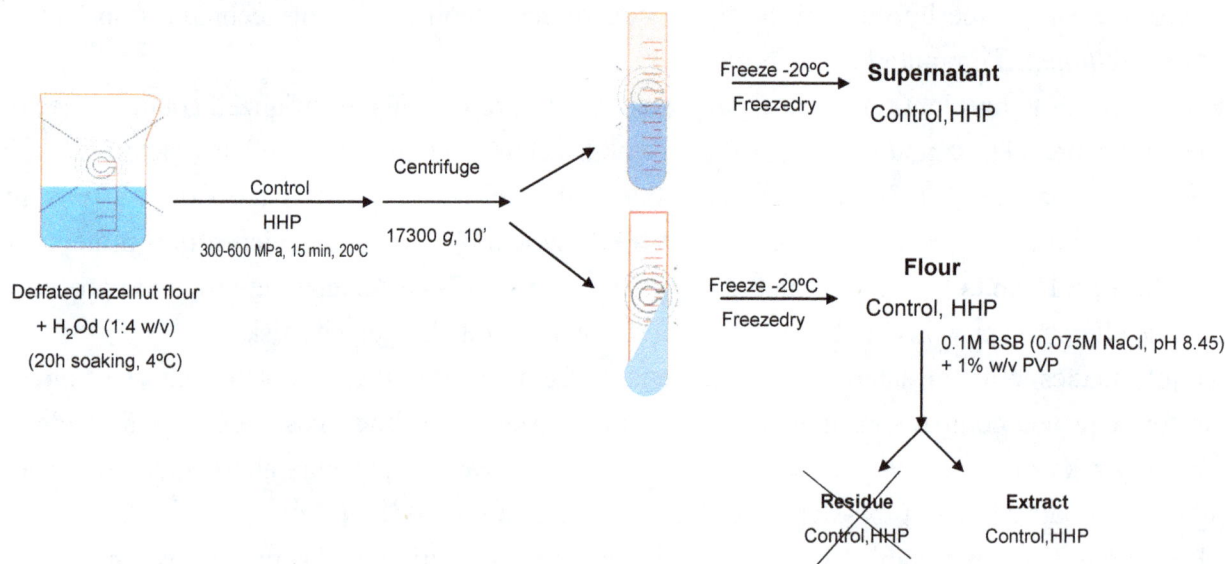

2.2. Protein Electrophoresis and Immunoblotting

1D SDS-PAGE was performed according to Laemmli [19]. Protein extracts of hazelnut unpressurized (control) and HHP-treated (300–600 MPa) of supernatant and flour were mixed with XT Sample Loading Buffer (Bio-Rad, Hercules, CA, USA) heated at 95 °C for 5 min, electrophoresed in 12% Bis-Tris precast gel (Bio-Rad, Hercules, CA, USA). Proteins were visualized with Coomassie brilliant blue R250 staining.

Western blotting was performed by electrophoretic transfer to polyvinylidene difluoride (PVDF) membranes at 250 mA for 100 min, at room temperature, essentially according to the method of Towbin *et al.* [20]. After blocking with 5% bovine serum albumin (BSA) (w/v) in phosphate buffered saline (PBS), membranes were incubated overnight with the serum pool from fifteen patients sensitized to hazelnut (serum specific IgE > 0.35 kilounits/L quantified by the CAP-FEIA assay) (1:10 dilution), washed, and then treated with mouse anti-human IgE mAb HE-2 ascitic fluid (1:3000 dilution for 2 h) [21]. After washing, a rabbit anti-mouse IgG peroxidase-conjugated antibody (1:5000 dilution for 1 h; DAKO, Glostrup, Denmark) was added. Detection of IgE-binding components was achieved by means of enhanced chemiluminescence, according to the manufacturer's instructions (Amersham Biosciences, Little Chalfont, UK).

2.3. Protein Sample Preparations and 2D-Liquid Chromatography Analysis

Total protein extracts of unpressurized and HHP-treated hazelnut previously tested for IgE-reactivity were used for comparative proteomic analysis with ProteomeLab PF-2D liquid chromatography (Beckman Coulter, Fullerton, CA, USA).

Unpressurized and HHP-treated hazelnut proteins were extracted with 0.1 M borate saline buffer (BSB, 0.075 M NaCl, pH 8.5) plus 1% (w/v) PVP at a 1:10 w/v ratio for 1 h at 4 °C by stirring. The extract was clarified by centrifugation at 27,000× g for 30 min at 4 °C, and the supernatants were dialyzed against distilled H_2O for 48 h at 4 °C using a dialysis membrane with a cut-off of 3.5 kDa and freeze-dried. The protein content of each sample was measured by the Bradford dye binding assay (Bio-Rad, Hercules, CA, USA) using bovine serum albumin (Sigma, St. Louis, MO, USA) as a standard. The nitrogen contents of the samples were determined by LECO analysis according to standard procedures based on Dumas method [22]. The total protein content was calculated as N × 5.3 [22]. The analyses were carried out in duplicates.

Protein extracts (2.5 mg of protein) from unpressurized and HHP-treated hazelnut were subjected to 2-D LC analysis using a ProteomeLab PF-2D instrument (Beckman-Coulter, Fullerton, CA, USA) and the protocol recommended by the manufacturer (ProteoSep®, Chemistry Kit, Sigma, St. Louis, MO, USA). The first-dimension separation was carried out by chromatofocusing on a High Performance ChromatoFocusing (HPCF) 1-D column (250 mm × 2.1 mm internal diameter, 300 Å pore size). The column was equilibrated at pH 8.5 with CF Start Buffer for 250 min at 0.2 mL/min. The pH gradient began after 20 min of sample injection when the CF Eluent Buffer at pH 4.0 moved through the column, gradually decreasing the pH from 8.5 to 4.0. Proteins were eluted according to their isoelectric points (pI) and, in the final step, the most acidic ones were eluted with 1 mol NaCl, 0.2% *n*-octylglucoside. All fractions were collected in 96 well plates using an automated collector.

All the different pH fractions collected from the first dimension were resolved on a reverse phase C18 column (HPRP column: 4.66 mm × 3.3 mm, 1.5 μm particle size,). Of each fraction, 200 μL was run through the column in solvent A (0.1% v/v TFA in water), and the proteins were then eluted with a linear gradient (0%–100%) of solvent B (0.08% v/v trifluoroacetic acid in acetonitrile) for 35 min. Separation was performed at 0.75 mL/min, and the temperature column was maintained at 50 °C. Eluted proteins were monitored by ultraviolet light at 214 nm of absorbance. The different fractions of the first dimension were collected in 12 plates (each 96 well) using an automated collector. All CF profiles were elaborated and compared using 32 Karat V1.01 software (Beckman-Coulter, Fullerton, CA, USA). Quantitative analysis of unpressurized and HHP-treated hazelnut protein peak areas and heights were performed using the Mapping tools software V1.0 (Beckman-Coulter, Fullerton, CA, USA).

Fractions from the second dimension were analyzed by 1D SDS-PAGE as above described and selected bands were manually excised for protein identification by mass spectrometry and database search.

2.4. Protein Identification by MS and Data Base Search

Immunoreactive hazelnut proteins modified by high hydrostatic pressure were analyzed by MS and data base search in order to determine their identification. Proteins were in-gel digested with trypsin (Sequencing Grade Modified, Promega, Madrid, Spain) in the automatic Investigator ProGest robot of Genomic Solutions. Briefly, excised gel bands were washed sequentially with 50 mM ammonium bicarbonate (NH_4HCO_3) buffer and acetonitrile. Eluted fractions were evaporated to a final volume of 10 µL. Protein digestions were carried out by incubating the samples in 50 mmol/L NH_4HCO_3 and 10 mmol/L dithiothreitol at 60 °C for 1 h. The alkylation of the reduced sulfhydryl groups was performed by adding 55 mmol/L iodoacetamide at 25 °C for 30 min in the dark. Proteins were digested by adding 1.5 µL of trypsin (125 µg/mL) and incubating at 37 °C overnight. The reaction was stopped with 1% of formic acid. Tryptic peptides were desalted and concentrated with ZipTipC18 columns according to the manufacturer's recommendation. Peptides were eluted in 0.1% trifluoroacetic acid , 50% acetonitrile for matrix-assisted laser desorption ionization time-of-flight (MALDI-TOF)-MS analysis, and with 1% formic acid, 50% methanol for electrospray MS analysis. To increase salt removal, samples were washed with 3–5 cycles of 0.1% triflouroacetic acid as wash solution. The solution was spotted directly onto a MALDI target and analyzed by MALDI-TOF/TOF off-line coupled LC/MALDI-MS/MS. MS analyses were performed automatically with a 4700 Analyzer MALDI-TOF/TOF instrument (Applied Biosystems, Carlsbad, CA, USA). First, MS spectra of all spotted fractions were acquired in the positive reflector mode for peak selection (S/N > 20, excluded precursor with 200 resolution), and further MS/MS spectra acquisition was done using the Collision Induced Dissociation of selected peaks. The search of filtered peptides was performed in batch mode using GPS Explorer V 3.5.0 software with a licensed version of MASCOT, in the Swiss-Prot Database. The MASCOT search parameters were: (1) species, *Coryllus avellana*; (2) allowed number of missed cleavages (only for trypsin digestion), 1; (3) considered modifications, cysteine as carboamidomethyl derivate and methionine as oxidized methionine; (4) peptide tolerance, ±50 ppm; (5) MS/MS tolerance, ±0.3 Da; and (6) peptide charge, +1 [23].

3. Results and Discussion

3.1. 1D Analysis and Immunoblottting

The protein migration patterns of hazelnut before (control) and after high-pressure treatments are shown by SDS-PAGE (Figure 2a). Samples were also analyzed for differences in IgE binding by immunoblot using pool sera from 15 patients sensitized to hazelnut (Figure 2b). Defatted flours that were directly solubilized in sample buffer were used for the immunoassays carried out in this study.

The results showed that electrophoretic migration patterns of high-pressure treated flour hazelnut proteins were similar to the control hazelnut.

Figure 2. (**a**) SDS-PAGE patterns of supernatant and flour protein extract from control and processed hazelnuts samples and (**b**) IgE immunoblot analysis of control and processed hazelnut samples with a serum pool from subjects with specific IgE to hazelnut (31.3 kU/L).

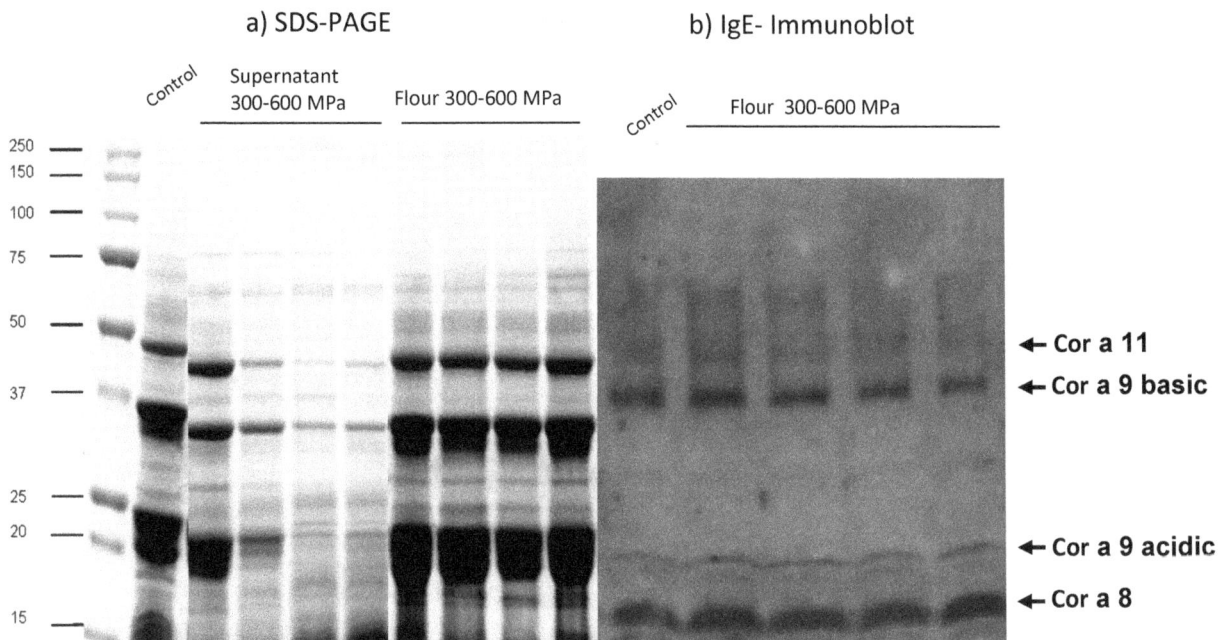

However, the band intensity of supernatants was diminishing according the pressure increase from 300 to 600 MPa (Figure 2a). The possibility that HHP resulted in variable aggregation depending on the applied pressure could explain these findings. This is in agreement with Somkuti and Smeller [1].

The IgE-immunoblot patterns are similar in control and pressure hazelnuts treated samples from 300 to 600 MPa, showing multiple immunoreactive proteins (Figure 2b). In these samples, most immunoreactive proteins fall in the central and lower area of the gel covering 15–60 kDa. The IgE immunoblotting showed that IgE of pool sera from sensitized to hazelnut recognized bands in all the samples at 50 kDa, 40 kDa, 20 kDa and 9 kDa which might correlate with Cor a 11, Cor a 9, Cor a 1 and Cor a 8 allergens [7].

3.2. Hazelnut Proteins Analyzed by Liquid Chromatography

In the ProteomeLab PF-2D system, 2.5 mg of untreated (control) and HHP (600 MPa)-treated hazelnut protein extracts were injected into the HPCF column and were recovered as follow: 40.0% in the 8.0–4.0 pH gradient, 28.6% before the initiation of the pH gradient, and the remaining 31.4% eluted as different peaks when the column was finally washed with high salinity buffer (Figure 3a). The fractions showing higher differences between untreated and HHP treated hazelnut were collected. We collected seven fractions in the control sample and 10 fractions in de HHP-treated hazelnut sample with high salinity buffer and they were subjected to a second dimension using the high performance reverse-phase chromatography column (Figure 3b).

Figure 3. Chromatographic analysis of hazelnut proteins using ProteomeLab PF-2D: (**a**) Chromatofocusing of total proteins separated by pH gradient 8.0–4.0. Fraction number 29 of control and 34 of HHP hazelnut were fractionated in the second dimension. (**b**) High performance reverse phase chromatography, fractions are separated by hydrophobicity. Proteins contained in peaks marked P1, P2 and P3 were separated by SDS-PAGE for subsequent MS analyses to determine their identity.

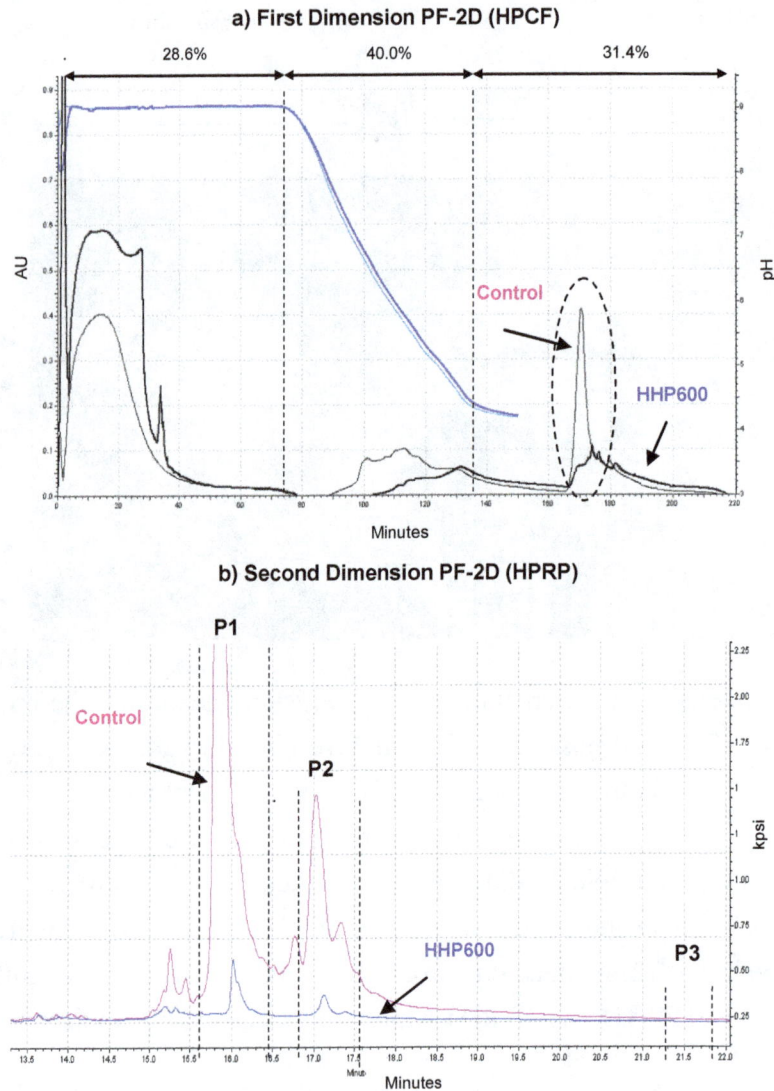

The protein profiles obtained showed important differences between the two samples (control and HHP 600 samples) with much lesser and smaller peaks in HHP 600 hazelnut. The fractions collected after the second dimension were concentrated in three ones, namely P1 (from 23 to 26), P2 (from 27 to 29) and P3 (from 34 to 35) in control as well as HHP treated hazelnut.

3.3. Proteins Identified in Raw and HHP Hazelnuts Samples

P1, P2 and P3 fractions from raw hazelnuts (control) and HHP600 hazelnut samples were further analyzed by 1D SDS-PAGE (Figure 4a,b). Control and HHP 600 hazelnut had a distinct protein band pattern, in band number as well as band intensity, in agreement with the differences showed in the second dimension absorbance profile between both hazelnut samples (Figure 3b).

Figure 4. SDS-PAGE of the different fractions collected and concentrated (P1, P2 and P3) from the second dimension chromatography column (HPRP) of control hazelnut **(a)** and HHP 600 MPa hazelnut **(b)**. Proteins were visualized by Coomassie Blue. The rectangles indicate the protein bands that were identified by MS analysis.

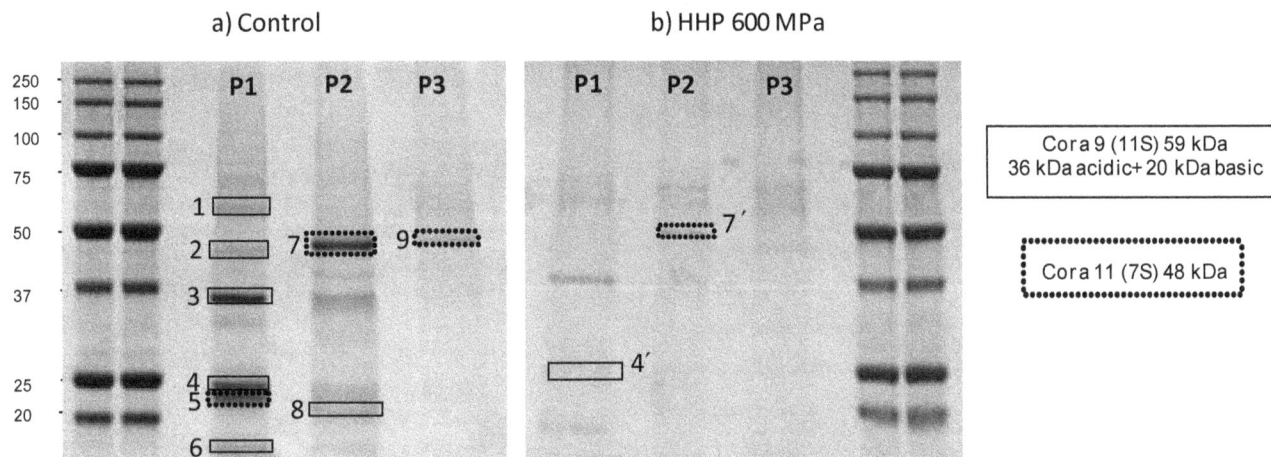

Finally, nine of these polypeptide bands of control hazelnut and two bands in HHP 600 hazelnut were manually excised for MALDI-TOF-TOF and LC-ESI-MS/MS identification. Table 1 summarizes the polypeptide identification data of raw hazelnut sample. The major bands were tryptic digested in order to carry out the analysis by MALDI-TOF/MS. A peptide mass fingerprint search allowed us to identify the bands 1, 2, 3, 4, 6 and 8 of raw hazelnut as a 11S globulin-like of *C. avellana* (59 kDa) (Table 1). The bands 5, 7 and 9 of raw hazelnut were identified as a 7S vicilin-like of *C. avellana* (48 kDa). The identification carried out in HHP600 treated sample (Table 2) showed only one band (4′) corresponding to the 11S globulin-like and another one (7′) identified as the 7S vicilin-like. According to the hazelnut allergen description [7], the bands identified as 11S globulin-like could correspond to the putative Cor a 9 allergen (59 kDa) which is composed of a 36 kDa acid subunit and a 20 kDa basic subunit. The 7S vicilin-like bands could correlate with the Cor a 11 allergen (48 kDa).

Proteome analysis is a tool that can be used both to visualize and compare complex mixtures of proteins and to gain a large amount of information about the individual proteins involved in specific biological responses. In this work, proteomic analysis by bidimensional chromatography (PF-2D) shows that the high hydrostatic pressure (HHP) induces significant changes in the proteome of hazelnut extracts in agreement with results on differential solubility after HHP process reported by other authors in several foods [1,4,18,24]. Our results show that the protein solubility became different among the fractions from 300 to 600 MPa. The identification by MALDI-TOF/MS of the proteins affected by high pressure indicated that among the proteins which are insolubilized by high pressure, there are some legumins and vicilins that could correspond to the allergen Cor 9 (11S) and Cor 11 (7S) according to data base search.

Table 1. Proteins separated by ProteomeLab PF-2D from raw hazelnuts and identified by MALDI-TOF/MS.

Band No.	No. access	Protein identification	Mascot score *	Mass (Da)	Matched peptides
1	gi|18479082	11S globulin-like protein (*C. avellana*)	73	59,605	12
2	gi|18479082	11S globulin-like protein (*C. avellana*)	161	59,605	19
3	gi|18479082	11S globulin-like protein (*C. avellana*)	136	59,605	6
4	gi|18479082	11S globulin-like protein (*C. avellana*)	187	59,605	20
5	gi|19338630	48-kDa glycoprotein precursor (*C. avellana*)	199	51,110	21
6	gi|18479082	11S globulin-like protein (*C. avellana*)	218	59,605	13
7	gi|19338630	48-kDa glycoprotein precursor (*C. avellana*)	256	51,110	24
8	gi|18479082	11S globulin-like protein (*C. avellana*)	65	59,605	11
9	gi|19338630	48-kDa glycoprotein precursor (*C. avellana*)	199	51,110	21

* Protein scores greater than 42 are significant ($p < 0.05$). Protein score is $-10 \times \mathrm{Log}\,(P)$, where P is the probability that the observed match is a random event.

Table 2. Proteins separated by ProteomeLab PF-2D from hazelnuts HHP 600 MPa and identified by MALDI-TOF/MS.

Band No.	No. access	Protein identification	Mascot score *	Mass (Da)	Matched peptides
4′	gi|18479082	11S globulin-like protein (*C. avellana*)	205	59,605	10
7′	gi|19338630	48-kDa glycoprotein precursor (*C. avellana*)	194	51,110	21

* Protein scores greater than 42 are significant ($p < 0.05$). Protein score is $-10 \times \mathrm{Log}\,(P)$, where P is the probability that the observed match is a random event.

4. Conclusions

The proteomic approach described here allows a deeper and more detailed study of the modifications induced by high pressure. The HHP treatment employed in this study is not effective to alter the immunoreactivity to hazelnut proteins but the protein solubility became different after HHP processing. Although at present there is not an accepted method to reduce the allergenicity of foods, a combination of treatments including high hydrostatic pressure with others, such as protease digestion, could be a successful strategy towards hypoallergenization. To reach this, we have to collect more information about the pressure behavior of these proteins at various environmental conditions and in presence of food additives. There is a definite need for further studies at basic scientific level, which could report the effect of pressure on the 3D structure of the allergenic proteins. The complexity of food processing demonstrates the importance of understanding its impact at the molecular level if risk assessors are to move towards knowledge-based ways of managing allergen risks.

Acknowledgments

This work was supported by BIO2008-00639, AGL2008-03453 and AGL2012-39863 research projects.

Conflicts of Interest

The authors declare no conflict of interest.

References

1. Somkuti, J.; Smeller, L. High pressure effects on allergen food proteins. *Biophys. Chem.* **2013**, *183*, 19–29.
2. Han, G.D.; Matsuno, M.; Ikeuchi, Y.; Suzuki, A. Effects of heat and high-pressure treatments on antigenicity of beef extract. *Biosci. Biotechnol. Biochem.* **2002**, *66*, 202–205.
3. Yamamoto, S.; Mikami, N.; Matsuno, M.; Hara, T.; Odani, S.; Suzuki, A.; Nishiumi, T. Effects of a high-pressure treatment on bovine gamma globulin and its reduction in allergenicity. *Biosci. Biotechnol. Biochem.* **2010**, *74*, 525–530.
4. Husband, F.A.; Aldick, T.; van der Plancken, I.; Grauwet, T.; Hendrickx, M.; Skypala, I.; Mackie, A.R. High-pressure treatment reduces the immunoreactivity of the major allergens in apple and celeriac. *Mol. Nutr. Food Res.* **2011**, *55*, 1087–1095.
5. Johnson, P.E.; van der Plancken, I.; Balasa, A.; Husband, F.A.; Grauwet, T.; Hendrickx, M.; Knorr, D.; Mills, E.N.; Mackie, A.R. High pressure, thermal and pulsed electric-field-induced structural changes in selected food allergens. *Mol. Nutr. Food Res.* **2010**, *54*, 1701–1710.
6. Piknová, L.; Pangallo, D.; Kuchta, T. A novel real-time polymerase chain reaction (PCR) method for the detection of hazelnuts in food. *Eur. Food Res. Technol.* **2008**, *226*, 1155–1158.
7. Garino, C.; Zuidmeer, L.; Marsh, J.; Lovegrove, A.; Morati, M.; Versteeg, S.; Schilte, P.; Shewry, P.; Arlorio, M.; van Ree, R. Isolation, cloning, and characterization of the 2s albumin: A new allergen from hazelnut. *Mol. Nutr. Food Res.* **2010**, *54*, 1257–1265.
8. European Food Safety Authority (EFSA). Scientific Opinion on the Assessment of Allergenicity of GM plants and microorganisms and derived food and feed. EFSA Panel on Genetically Modified Organisms (GMO Panel). Available online: http://www.efsa.europa.eu/en/efsajournal/doc/1700.pdf (accessed on 28 Jaunuary 2014).
9. Kulkarni, A.; Ananthanarayan, L.; Raman, K. Identification of putative and potential cross-reactive chickpea (*Cicer arietinum*) allergens through an in silico approach. *Comput. Biol. Chem.* **2013**, *47*, 149–155.
10. Rose, J.K.; Bashir, S.; Giovannoni, J.J.; Jahn, M.M.; Saravanan, R.S. Tackling the plant proteome: Practical approaches, hurdles and experimental tools. *Plant J.* **2004**, *39*, 715–733.
11. Agrawal, G.K.; Yonekura, M.; Iwahashi, Y.; Iwahashi, H.; Rakwal, R. System, trends and perspectives of proteomics in dicot plants part I: Technologies in proteome establishment. *J. Chromatogr. B Analyt. Technol. Biomed. Life Sci.* **2005**, *815*, 109–123.

12. Lambert, J.P.; Ethier, M.; Smith, J.C.; Figeys, D. Proteomics: From gel based to gel free. *Anal. Chem.* **2005**, *77*, 3771–3787.

13. Soldi, M.; Sarto, C.; Valsecchi, C.; Magni, F.; Proserpio, V.; Ticozzi, D.; Mocarelli, P. Proteome profile of human urine with two-dimensional liquid phase fractionation. *Proteomics* **2005**, *5*, 2641–2647.

14. Pirondini, A.; Visioli, G.; Malcevschi, A.; Marmiroli, N. A 2-D liquid-phase chromatography for proteomic analysis in plant tissues. *J. Chromatogr. B Analyt. Technol. Biomed. Life Sci.* **2006**, *833*, 91–100.

15. Wu, W.W.; Wang, G.; Yu, M.J.; Knepper, M.A.; Shen, R.F. Identification and quantification of basic and acidic proteins using solution-based two-dimensional protein fractionation and label-free or 18O-labeling mass spectrometry. *J. Proteome Res.* **2007**, *6*, 2447–2459.

16. Batlle, I.; Aletà, N.; Clavé, J.; Ninot, A.; Romero, M.; Rovira, M.; Tous, J.; Vargas, F.J. Bancos de germoplasma de especies de frutos secos y desecados: Avellano, nogal, pistachero y algarrobo. *Fruticultura Profesional* **1999**, *104*, 6–12.

17. Omi, Y.; Kato, T.; Ishida, K.-I.; Kato, H.; Matsuda, T. Pressure-induced release of basic 7S globulin from cotyledon dermal tissue of soybean seeds. *J. Agric. Food Chem.* **1996**, *44*, 3763–3767.

18. Kato, T.; Katayama, E.; Matsubara, S.; Omi, Y.; Matsuda, T. Release of allergenic proteins from rice grains induced by high hydrostatic pressure. *J. Agric. Food Chem.* **2000**, *48*, 3124–3129.

19. Laemmli, U.K. Cleavage of structural proteins during the assembly of the head of Bacteriophage t4. *Nature* **1970**, *227*, 680.

20. Towbin, H.; Staehelin, T.; Gordon, J. Electrophoretic transfer of proteins from polyacrylamide gels to nitrocellulose sheets-procedure and some applications. *Proc. Natl. Acad. Sci. USA* **1979**, *76*, 4350–4354.

21. Sanchez-Madrid, F.; Morago, G.; Corbi, A.L.; Carreira, J. Monoclonal-antibodies to 3 distinct epitopes on human IgE: Their use for determination of allergen-specific IgE. *J. Immunol. Methods* **1984**, *73*, 367–378.

22. AOAC International. *Official Methods of Analysis of AOAC International*, 16th ed.; Association of Official Analytical Chemists: Washington, DC, USA, 1995.

23. Bairoch, A.; Apweiler, R. The SWISS-PROT protein sequence database and its supplement TrEMBL in 2000. *Nucleic Acids Res.* **2000**, *28*, 45–48.

24. Peñas, E.; Gomez, R.; Frias, J.; Baeza, M.L.; Vidal-Valverde, C. High hydrostatic pressure effects on immunoreactivity and nutritional quality of soybean products. *Food Chem.* **2011**, *125*, 423–429.

9

A Cross-Sectional Study: Nutritional Polyamines in Frequently Consumed Foods of the Turkish Population

Nihal Buyukuslu *, Hilal Hizli, Kubra Esin and Muazzez Garipagaoglu

Department of Nutrition and Dietetics, School of Health Sciences, Istanbul Medipol University, Beykoz/Istanbul, 34810, Turkey; E-Mails: hhizli@medipol.edu.tr (H.H.); kesin@medipol.edu.tr (K.E.); mgaripagaoglu@medipol.edu.tr (M.G.)

* Author to whom correspondence should be addressed; E-Mail: nbuyukuslu@medipol.edu.tr

External Editor: Christopher J. Smith

Abstract: Putrescine, spermidine and spermine are the most abundant polycationic natural amines found in nearly all organisms. They are involved in regulation of gene expression, translation, cell proliferation and differentiation. They can be supplied by the endogenous synthesis inside the cell or by the intake from exogenous sources. There is a growing body of literature associated with the effects of bioactive amines on health and diseases, but limited information about polyamine content in foods is available. In the present study, the polyamine content of frequently consumed foods in a typical Turkish diet was estimated for adults, including tea, bread and yoghurt. The estimation of daily intake was defined as 93,057 nmol/day putrescine, 33,122 nmol/day spermidine, 13,685 nmol/day spermine. The contribution of foods to daily intake was: dairy products (47.32%), vegetables and grains (21.09%) and wheat products (12.75%).

Keywords: polyamine; putrescine; spermidine; spermine; daily intake; diet; health

1. Introduction

Putrescine, spermidine, and spermine are the polyamines that have been of main interest among foods due to their biological activities that have beneficial health effects. They are synthesized in all prokaryotic and eukaryotic cells and are known to play some essential roles in cell proliferation, regeneration and differentiation [1]. The metabolic requirement for polyamines is high in rapidly growing tissues, such as normal growth and development, and in tumors [2,3]. Besides many beneficial effects, it was shown that the level of polyamines in cells is also associated with diseases such as cancers and chronic diseases. Since polyamine is involved in the progression of cancer [4], inhibiting polyamine synthesis and reducing its intake through food may have beneficial effects [5].

The body pool of polyamines is supplied by endogenous or *de novo* biosynthesis, intestinal microorganisms, and exogenous sources through the diet [6]. The amount of external dietary polyamines is higher than the endogenous biosynthesis [1]. Nonetheless, dietary polyamines have been thought to support metabolism maintaining optimal health. For instance, they have a potential role in growth and development of digestive system [7]. Intracellular production of polyamines and their concentration in tissues and organs decreases with aging. Therefore, intake of exogenous polyamines was shown to be beneficial in the treatment of some geriatric diseases [8,9]. In the meantime, Binh reported a possible role for the food polyamines that are abundant in the Mediterranean diet in prolonging human life [10]. However, limited information about the polyamine content of food is available to assess dietary polyamine intake and subsequently have dieticians design proper menus. In addition, dietary amines are classified in two categories according to their synthesis; natural polyamines which are formed during *de novo* polyamine biosynthesis, and biogenic amines that are produced by decarboxylation of amino acids. Among natural polyamines, putrescine belongs to both categories. The daily putrescine requirement can be met from either natural or biogenic sources. Inappropriate storage and processing conditions result in biogenic amin formation in foods. Therefore, especially in the case of cancer patients, the daily polyamine intake should be carefully evaluated.

Polyamine contents in foods vary widely between and even within food types [1,11]. This might be due to the origin, processing, storage conditions, seasonal variation and different methodological applications of foods. Thus far, studies on polyamine concentration in food and on the estimation of dietary polyamine were based on the analyses of polyamines in each food by defining the daily intake of foods from a nutrient database and calculating daily intake of polyamines. An inadequate polyamine database seems to be the main limitation for daily estimation of polyamine intake.

Dietary habits in Turkey vary according to region, culture, socioeconomic status, gender and age of the individual. A recent cohort study in Turkey [12] indicated that the most frequently consumed foods were black tea, white bread types, cheese, sunflower oil, sugar, honey, jam and pekmez. The major percentage of energy came from bread. Cheese and yogurt were the most frequently used milk products. Fresh fruits and vegetables were widely consumed throughout the year. Oil and fat consumption showed regional variations. Most frequently used oils were sunflower and olive.

Although there are few studies defining polyamine content in fermented foods such as cheese and meat products, the polyamine content in many widespread foods in Turkey do not exist in a nutrient database. This study aims to identify the main sources of polyamines in frequently consumed foods and to estimate daily putrescine, spermine, and spermidine intake.

2. Experimental Section

2.1. Study Population

In this cross-sectional study, the daily intake of foods of 1218 subjects (45.1% of male; 54.9% of female) was examined by trained interviewers. The subjects were randomly selected from different regions in Turkey. The highest prevalence of distribution among regions was observed in the Mediterranean (42.0%), followed by Central Anatolia (17.5%), Southeastern Anatolia (15.8%), Eastern Anatolia (9.7%), and Marmara (8.7%); it was lowest in the Black Sea region (6.3%).

The mean age and the mean body mass index (BMI) were 40.02 ± 19.15 and 26.08 ± 4.84, respectively. The study took place between July and October 2011.

2.2. Daily Intake of Food

Nutritional information of subjects was collected via a 24 h dietary recall. The quantification of the food and drink was estimated rather than weighed. The interviewers converted these estimates into weights that could be used to calculate food and nutrient intake.

The amounts of daily consumption were registered in grams or milliliters. The dietary records were entered into a computerized nutrient analysis programme BeBis (version 7.2, Pasific Compony, Istanbul, Turkey). Daily intake of foods was calculated as g/person/day. Only in the 81 foods for which the polyamine values were determined have been included in the present study in order to focus on the polyamine content.

Foods were classified as vegetables and grains, fruits, meat products, dairy products, eggs, tea, wheat products, nuts and dried fruits and sweets to define the distribution of polyamines. Foods were ranked from highest to lowest daily consumption of each food item for each group in Table 1.

2.3. Database Development for Polyamine in Foods

Data on content of polyamines in different foods was collected through an extensive literature search using PubMed, Web of Science and the Turkish Academic Network and Information Center (ULAKBIM). "Polyamine", "putrescine", "spermidine", "spermine", "daily intake" and "nutrition" were the key words for the search.

The polyamine contents in foods vary widely between and even within food types. In our research, when an individual food item was obtained from several references or a single reference reported multiple analyses for the same food group, the mean value was calculated. Since the values for polyamine content in some of the studies were reported in nmol/g of food, they were converted into mg/kg, using the appropriate equation (mol = mass/molecular weight). The molecular weights for polyamines were 88.15 g/mol putrescine, 145.25 g/mol spermidine and 202.34 g/mol spermine. The values of polyamines (mg/kg) in food from other publications were placed in Table 1 with references.

Table 1. Daily polyamine intake based on the reference values in the most frequently consumed foods in the Turkish population.

Foods	Daily Intake of Food (g-mL/Person/Day) [a]	A—Daily Polyamine Intake in Turkish Population				B—Reference Values			
		Daily Intake of Each Polyamine (mg-mL/Person/Day) [a]			Daily Intake of Total Polyamine (mg-mL/Person/Day) [a]	Polyamines in Foods (mg-mL/kg) [a]			References
		Putrescine [b]	Spermidine [c]	Spermine [d]		Putrescine [b]	Spermidine [c]	Spermine [d]	
Vegetables and Grains									
Rice [e]	62.30	0.047	0.121	0.287	0.455	0.76	1.95	4.6	[1,13,14]
Tomato [e]	61.94	0.365	0.164	0.000	0.529	5.9	2.65	nd [f]	[1,13–15]
Lentil soup	44.31	0.146	0.000	0.328	0.474	3.3	-	7.4	[16]
Cucumber [e]	38.50	0.307	0.326	0.020	0.653	7.98	8.46	0.53	[1,14,15,17]
Green pepper [e]	9.18	0.572	0.099	0.041	0.712	62.35	10.75	4.5	[14,18]
Eggplant [e]	4.46	0.110	0.020	0.002	0.132	24.57	4.4	0.5	[14,15]
Onion [e]	4.39	0.010	0.036	0.007	0.053	2.34	8.3	1.57	[1,14,15,19]
Chickpea	1.68	0.004	0.000	0.002	0.006	2.6	0.1	1.2	[15]
Potato chips [e]	1.57	0.022	0.027	0.004	0.053	13.84	17.12	2.52	[14,16]
Okra	1.39	0.030	0.026	0.000	0.056	21.7	18.6	Nd [g]	[9]
Tomato puree	1.18	0.031	0.010	0.002	0.043	25.9	8.4	2.1	[20]
Carrot [e]	0.72	0.004	0.005	0.001	0.010	6.03	6.42	2.54	[1,13,14,17,18,21]
Potato [e]	0.58	0.006	0.008	0.001	0.015	10.4	14.02	1.85	[1,13,14,17–19,21]
Cabbage [e]	0.57	0.010	0.006	0.002	0.018	17.58	10.39	2.91	[1,14,19,21]
Lettuce [e]	0.45	0.005	0.008	0.000	0.013	10.08	18.83	0.4	[1,14,17,19]
Green peas [e]	0.34	0.002	0.022	0.018	0.042	5.7	65.2	52.5	[14,22]
Cauliflower [e]	0.23	0.001	0.006	0.001	0.008	4.67	28.1	4.25	[1,17,21]
Ketchup [e]	0.20	0.006	0.001	0.001	0.008	32.42	5.15	2.7	[14,20]
Spring onion	0.19	0.005	0.003	0.000	0.008	24.5	17	0.2	[9]
Broccoli [e]	0.14	0.001	0.005	0.001	0.007	5.75	32.39	9.3	[14,17,21]
Mushroom	0.14	0.001	0.012	0.000	0.013	4	88.6	3.4	[14]

Table 1. *Cont.*

Foods	A—Daily Polyamine Intake in Turkish Population					B—Reference Values			
	Daily Intake of Food (g-mL/Person/Day) [a]	Daily Intake of Each Polyamine (mg-mL/Person/Day) [a]			Daily Intake of Total Polyamine (mg-mL/Person/Day) [a]	Polyamines in Foods (mg-mL/kg) [a]			References
		Putrescine [b]	Spermidine [c]	Spermine [d]		Putrescine [b]	Spermidine [c]	Spermine [d]	
Vegetables and Grains									
Celeriac	0.10	0.001	0.003	0.000	0.004	6.1	26.7	0	[21]
Celery	0.10	0.002	0.001	0.000	0.003	17.1	14.2	3.8	[14]
Maize [e]	0.06	0.003	0.002	0.000	0.005	50.7	32.1	1.4	[9,14]
Garlic	0.05	0.000	0.001	0.000	0.001	2.3	11.1	5.8	[14]
Dill	0.03	0.000	0.001	0.000	0.001	12.7	29.2	8.7	[9]
Spinach [e]	0.02	0.000	0.000	0.000	0.000	8.43	16.03	2.1	[14,19,20]
Mayonnaise [e]	0.18	0.000	0.000	0.000	0.000	0.53	1.6	0.3	[14,15]
Fruits									
Watermelon	69.70	0.000	0.084	0.000	0.084	nd [f]	1.2	nd [f]	[9]
Peach [e]	44.32	0.023	0.194	0.114	0.331	0.52	4.38	2.57	[9,15]
Melon	31.70	0.013	0.371	0.000	0.384	0.4	11.7	nd [f]	[9]
Apple [e]	28.40	0.153	0.045	0.001	0.199	5.39	1.6	0.05	[1,9,14,20]
Grapes	27.10	0.003	0.002	0.000	0.005	0.1	0.06	0.01	[23]
Cherry	5.49	0.009	0.009	0.004	0.022	1.6	1.6	0.8	[14]
Fig	5.69	0.013	0.030	0.000	0.043	2.2	5.2	nd [f]	[9]
Pear [e]	4.98	0.061	0.007	0.000	0.068	12.2	1.5	nd [f]	[1,15]
Banana [e]	3.04	0.042	0.026	0.003	0.071	13.8	8.55	1.05	[14,15]
Tangerine [e]	1.39	0.100	0.003	0.000	0.103	72.28	1.87	0.2	[14,17]
Orange [e]	0.67	0.066	0.003	0.000	0.069	98.28	3.75	0.45	[1,13,14,17]
Strawberry	0.67	0.001	0.001	0.000	0.002	1	2	0.4	[14]
Lemon (lime)	0.58	0.024	0.003	0.001	0.028	41	5	1.8	[9]
Kiwi	0.22	0.000	0.001	0.000	0.001	1.2	5.4	1.5	[15]
Pineapple	0.15	0.000	0.001	0.000	0.001	0.7	4	2.2	[15]
Grapefruit	0.08	0.005	0.001	0.000	0.006	62.1	7.3	nd [f]	[20]

Table 1. *Cont.*

Foods	Daily Intake of Food (g-mL/Person/Day) [a]	A—Daily Polyamine Intake in Turkish Population Daily Intake of Each Polyamine (mg-mL/Person/Day) [a]			Daily Intake of Total Polyamine (mg-mL/Person/Day) [a]	B—Reference Values Polyamines in Foods (mg-mL/kg) [a]			References
		Putrescine [b]	Spermidine [c]	Spermine [d]		Putrescine [b]	Spermidine [c]	Spermine [d]	
Tea									
Black tea [e]	322.2	0.750	2.430	0.050	3.230	0.750	2.430	0.050	[13,24]
Wheat products									
Bread, white [e]	201.4	0.282	1.299	0.624	2.205	1.4	6.45	3.1	[1,15]
Pasta	21.64	0.010	0.072	0.039	0.121	0.44	3.34	1.82	[14]
Pasta, cooked	14.81	0.015	0.107	0.160	0.282	1	7.2	10.8	[1]
Bread, whole grain [e]	5.04	0.013	0.090	0.032	0.135	2.53	17.77	6.3	[1,15,17]
Flour [e]	0.91	0.002	0.007	0.003	0.012	2.16	7.27	3.36	[13,14]
Meat and fish products									
Chicken breast, cooked [e]	4.14	0.000	0.106	0.224	0.330	0	25.5	54.1	[20,25]
Chicken thigh [e]	4.31	0.002	0.037	0.117	0.156	0.4	8.7	27.1	[20,25]
Chicken [e]	1.94	0.007	0.012	0.108	0.127	3.42	6.25	55.74	[1,13,14]
Minced meat, beef [e]	1.59	0.010	0.059	0.054	0.123	6.45	37.05	33.85	[1,17]
Lamb meat	1.46	0.001	0.007	0.069	0.077	1	5	47.1	[13]
Salami	1.44	0.001	0.004	0.013	0.018	0.5	3	9	[15]
Beef meat [e]	1.22	0.005	0.009	0.040	0.054	4.07	7.55	32.95	[1,13,17,26]
Chicken grilled [e]	1.14	0.002	0.020	0.051	0.073	2	17.3	44.4	[20,25]
Chicken breast [e]	1.12	0.000	0.007	0.030	0.037	0.4	6.25	27	[20,25]
Sausages	1.09	0.015	0.007	0.027	0.049	14.2	6.1	25	[1]
Beef liver	0.58	0.001	0.004	0.114	0.119	1	6.8	197	[9]
Tuna, canned	0.17	0.000	0.001	0.003	0.004	1.22	5.75	18.7	[27]
Salmon [e]	0.16	0.000	0.001	0.001	0.002	2.59	4.07	5.72	[9,14,17]
Tuna (in oil)	0.11	0.000	0.000	0.000	0.000	0.01	1.4	0	[27]

Table 1. Cont.

Foods	A—Daily Polyamine Intake in Turkish Population					B—Reference Values			References
	Daily Intake of Food (g-mL/Person/Day) [a]	Daily Intake of Each Polyamine (mg-mL/Person/Day) [a]			Daily Intake of Total Polyamine (mg-mL/Person/Day) [a]	Polyamines in Foods (mg-mL/kg) [a]			
		Putrescine [b]	Spermidine [c]	Spermine [d]		Putrescine [b]	Spermidine [c]	Spermine [d]	
Nuts and dry fruits									
Hazelnut	1.39	0.006	0.029	0.009	0.044	4.2	21	6.5	[15]
Almond	0.24	0.000	0.001	0.003	0.004	1.6	6	13.5	[14]
Raisin	0.52	0.000	0.000	0.000	0.000	0.1	0.4	0.2	[15]
Sweets									
Honey	2.36	0.002	0.000	0.000	0.002	0.7	0.1	nd [f]	[15]
Strawberry marmalade	0.97	0.001	0.002	0.000	0.003	1.4	2	0.4	[15]
Jam	0.48	0.001	0.001	0.000	0.002	1.2	2.2	–	[1]
Apricot marmalade	0.45	0.000	0.001	0.000	0.001	1	1.6	nd [f]	[15]
Prune marmalade	0.44	0.002	0.001	0.000	0.003	4.6	2	0.4	[15]
Chocolate [e]	0.23	0.000	0.001	0.000	0.001	0.38	2.19	1.34	[14,15]
Cacao	0.01	0.000	0.000	0.000	0.000	0.26	0.44	0.2	[14]
Egg									
Egg [e]	20.70	0.007	0.012	0.011	0.030	0.32	0.58	0.53	[13,14,16]
Dairy products									
Yoghurt [e]	92.9	0.019	0.035	0.040	0.094	0.2	0.38	0.43	[17,23]
Cheese, white [g]	37.05	5.039	0.000	0.000	5.039	136	–	–	[28]
Cheddar, fresh [e]	4.22	0.035	0.006	0.003	0.044	8.31	1.39	0.61	[14,15,29–31]
Milk	3.49	0.000	0.000	0.000	0.000	0.013	0.086	–	[32]
Milk, semi-skimmed	3.49	0.001	0.002	0.001	0.004	0.2	0.5	0.3	[1]
Milk, goat [e]	2.65	0.090	0.002	0.001	0.093	33.87	0.6	0.3	[15,32]

[a] Solid food items in mg, liquid food items in mL; [b] molecular weight of putrescine is 88.15 g/mol; [c] molecular weight of spermidine is 145.25 g/mol; [d] molecular weight of spermine is 202.34 g/mol; [e] the mean value of food item (section B) was calculated based on the references given in section A; [f] nd: not determined; [g] there were several polyamine concentrations for various cheeses. We included the result from the most consumed "white cheese" in Turkey.

2.4. Daily Polyamine Intake in Foods

The concentrations of putrescine, spermidine and spermine in foods were reported in [1,9,10,13–15, 17–23,25–31,33,34]. Average daily intakes of putrescine, spermidine and spermine in mg per person per day were calculated; the sum of these three measures yielded the total polyamine intake (Table 1). Foods were categorized as mentioned in Section 2.2. The contribution of foods to the total daily polyamine intake was identified for each group. Based on average values of polyamines that were given for each 100 g of the food item, the daily putrescine, spermidine and spermine intake from the foods were calculated in nmol per person.

2.5. Statistical Analyses

All statistical analyses were performed by using the Statistical Package for Social Sciences (SPSS) version 18.0 (SPSS Inc., Chicago, IL, USA). The values for age and BMI were expressed as means ± standart deviation (SD). Polyamine contents in the individual food items were characterized by arithmetic mean value.

3. Results

3.1. Polyamine Database

Polyamine concentration of foods was taken from the literature. For each food, the reference where the polyamine values were taken from was denoted on the references column in Table 1. The mean value was calculated if there was more than one value for any food item. The concentrations of polyamines were shown in mg per kg food in Table 1. Rice, tomato, tomato puree, ketchup, lentil soup, cucumber, green pepper, eggplant, onion, spring onion, chickpea, potato, okra, carrot, cabbage, lettuce, green peas, cauliflower, broccoli, mushroom, celery, maize, garlic, dill, spinach were placed in the vegetables and grains group. The highest putrescine per kg food was green pepper (62.35 mg), followed by maize (50.7 mg) and ketchup (32.42 mg); the lowest was rice (0.76). Spermidine concentration in foods was as follows: mushroom (88.6 mg/kg), green peas (65.2 mg/kg), and broccoli (32.39 mg/kg). High spermine-containing foods were: green peas (52.5 mg/kg), broccoli (9.3 mg/kg), and dill (8.7 mg/kg), respectively. Among fruits: watermelon, peach, melon, apple, grapes, cherry, fig, pear, banana, tangerine, orange, strawberry, lemon, kiwi, pineapple and grapefruit were placed in the list. The foods including three high-valued polyamines were: orange (98.28 mg), tangerine (72.28 mg), and lemon (41.00 mg) for putrescine; melon (11.7 mg), banana (8.55 mg), and grapefruit (7.3 mg) for spermidine; peach (2.57 mg), pineapple (2.2 mg), and lemon (1.8 mg) for spermine. Tea is the most frequently consumed drink (322.2 mL/person/day) and white bread is a commonly consumed food (201.4 g/person/day) in Turkey. In general, spermidine concentrations of wheat products were higher than putrescine and spermine levels. Poultry, red meat and meat products such as sausages and salami were grouped as meat products. The highest concentration was spermine (4206 nmol/day), the second was spermidine (1886 nmol/day) and the lowest was putrescine (522 nmol/day) in meat products. Tuna (canned and in oil) and salmon were included in the meat products category since the daily frequency intake was higher than other types of fish. Hazelnut, almond and raisin were in the group of nuts and dry fruits. Honey, marmalade, jam, chocolate and cacao were placed in the sweets group. Yoghurt, cheese and milk were grouped as dairy products. Among dairy products, milk and yoghurt were not rich in polyamines, but several types of cheeses had high levels of polyamines due to the fermentation. The polyamine levels in fresh cheeses were lower than ripened cheeses. Our questionnaire concluded that the most frequently consumed cheese among the Turkish population is white cheese; it was therefore the only type of cheese included to the study. The levels of putrescine, spermidine, and spermine in eggs were 0.32 mg/kg, 0.58 mg/kg and 0.53mg/kg, respectively. All the data obtained from published studies were shown in Table 1, section B.

3.2. Frequently Consumed Foods in Turkey

Data from our survey exposed the most frequently consumed foods in Turkey. We picked the ones which have published polyamine values. Foods were listed according to the daily intake of gram food per person per day in Table 1. In g/person/day, rice (62.3), tomato (61.94), and lentil soup (44.31) were the most highly consumed from the vegetables and grains group, whereas watermelon (69.70), peach (44.32) and melon (31.70) were the most consumed fruits. Chicken was consumed more than red meat and fish. Due to lack of polyamine ingredients, most common sweets were not included in the list. Among the sweets, honey intake (2.36 g/person/day) was more than marmalade and jam. Yoghurt (92.9 g/person/day) and cheese consumptions (37.05 g/person/day for ripened; 4.22 g/person/day for fresh) were found higher than daily milk intake. The 15 most frequently consumed foods, from the highest to lowest in g/person/day, were: black tea (322.20), white bread (201.40), yoghurt (92.90), watermelon (69.70), rice (62.30), tomato (61.94), peach (44.32), lentil soup (44.31), cucumber (38.50), ripened cheese (37.05), melon (31.70), apple (28.40), grapes (27.10), pasta (21.64) and egg (20.70).

3.3. Daily Polyamine Intake of Frequently Consumed Foods

The calculated values of polyamine in foods are listed in Table 1. The results are given as mg/person/day. The contribution of polyamines in the top five foods of daily intake (mg/person/day) was as follows: ripened cheese (5.039), black tea (0.750), green pepper (0.572), tomato (0.365) and cucumber (0.282) for putrescine; black tea (2.430), white bread (1.299), melon (0.371), peach (0.194) and tomato (0.164) for spermidine; white bread (0.624), lentil soup (0.328), rice (0.287), chicken breast (0.224) and pasta (0.160) for spermine.

Daily intake of polyamines in food groups among the Turkish population was shown in Table 2. The daily putrescine intake came from dairy products (65.48%), vegetables and grains (20.63%), fruits (6.24%), wheat products (3.91%), black tea (2.95%), meat products (0.56%) and eggs (0.09%). The main spermidine sources were wheat products (32.72), vegetables and grains (19.00%), black tea (16.28%), fruits (16.19%), dairy products (9.12%), meat products (5.70%) and eggs (0.25%). The main spermine sources were wheat products (30.99%), vegetables and grains (25.97%), dairy products (6.32%), and fruits (4.55%). Total daily intake of polyamines per person were 93,057 nmol/day putrescine, 33,122 nmol/day spermidine, and 13,685 nmol/day spermine among the most frequently consumed foods in the Turkish population. The main dietary sources for the total polyamines were dairy products (47.32%), vegetables and grains (21.09%) and wheat products (12.75%). Cheese was the primary food item among dairy products contributing to putrescine intake (5.371 mg/person/day) and also to total intake of polyamines.

Table 2. Daily intake of polyamines in food groups among the Turkish population.

| Foods | Daily Intake of Food | | Daily Intake of Polyamines | | | | | | | | | Total [d] |
| | Group | | Putrescine [a] | | | Spermidine [b] | | | Spermine [c] | | | |
	%	g/person/day	mg/person/day	nmol/person/day	%	mg/person/day	nmol/person/day	%	mg/person/day	nmol/person/day	%	nmol/person/day
Tea	26.55	322.20	0.242	2745	16.28	0.783	5391	0.58	0.016	79	4.95	6725
Wheat products	20.09	243.80	0.321	3642	32.72	1.574	10,839	30.99	0.858	4240	12.75	17,328
Vegetables and grains	19.35	234.82	1.692	19,195	19.00	0.914	6293	25.97	0.719	3553	21.09	28,662
Fruits	18.47	224.18	0.512	5808	16.19	0.779	5363	4.55	0.126	623	7.74	10,525
Dairy products	11.56	140.31	5.371	60,931	9.12	0.439	3022	6.32	0.175	865	47.32	64,306
Eggs	1.71	20.70	0.007	79	0.25	0.012	83	0.40	0.011	54	0.16	214
Meat products	1.69	20.47	0.046	522	5.70	0.274	1886	30.73	0.851	4206	5.69	7735
Sweets	0.41	4.94	0.006	68	0.10	0.005	34	0.04	0.001	5	0.07	100
Nuts and dried fruits	0.18	2.15	0.006	68	0.64	0.031	213	0.43	0.012	59	0.22	304
Total [e]	–	1213.57	8.203	93,057	–	4.811	33,122	–	2.769	13,685	–	135,899

[a] Molecular weight of putrescine is 88.15 g/mol; [b] molecular weight of spermidine is 145.25 g/mol; [c] molecular weight of spermine is 202.34 g/mol; [d] total daily intake of putrescine, spermidine and spermine for each food group; [e] daily intake of each polyamine from all foods and food.

A comparison among three studies which involves total daily polyamine values resulted that the highest contribution to polyamine intake came from putrescine following spermidine and spermine (Table 3). In our study, the contribution to daily intake was from 66.5% putrescine, 23.7% spermidine and 9.8% spermine. When we compare the percentage distribution of polyamines with other studies, putrescine was the highest and spermine was the lowest contributor of daily polyamine intake in the Turkish population.

Table 3. Total daily intake and percentage distribution of polyamines.

Total Daily Intake of Polyamines (nmol/person/day)				Distribution of Polyamines %			Reference
Putrescine	Spermidine	Spermine	Total Polyamine	Putrescine	Spermidine	Spermine	
211,910	86,959	54,704	353,573	59.9	24.6	15.5	[35]
159,133	54,697	35,698	249,528	63.8	21.9	14.3	[36]
93,057	33,122	13,685	135,899	66.5	23.7	9.8	Present study

4. Discussion

In recent years, there has been considerable interest in the influence of ingested polyamines due to their possible functions on cell growth, maintenance and function. Up to date, several research papers were published on the effect of dietary polyamines in health and diseases and on daily consumption of those active bioamines. Multiple abnormalities in the control of polyamine metabolism were shown to be implicated in several pathological processes [37]. In a review by Novotorski et al. it was underlined that the polyamine pathway is a rational target for chemoprevention and chemotherapeutics [38]. Although dietary polyamines have many functional and physiological impacts, their contents in foods are variable. Cooking [21], storage conditions and time [39–41], seasonal variations [42], household recipes [43] and agricultural conditions [44] have to be taken into consideration when polyamine concentrations in foods were analyzed. Moreover, tremendous variation in polyamine values in fermented products such as sausages and cheeses occur due to differences in the fermentation process [17,45]. Analytical methods including ion exchange, thin-layer and high performance liquid chromatography also showed some variation even in the same type of food [13,46]. Seasonal variation may affect the polyamine concentrations in foods. In our study, we made the food consumption questionnaire in between July and October. Therefore, total polyamine intake rose from summertime vegetables and fruits such as tomato, cucumber, green pepper, eggplant, melon, watermelon, apple, peach rather than winter vegetables and fruits such as cauliflower, broccoli, celery, and citrus fruits. Indeed, most of the studies do not mention the season when the samples were collected for polyamine analyses.

The analyses of polyamine concentration in foods and the total amount of polyamine intake for USA [36,47], UK [1,16], Norway [17], Germany [21], Japan [9,13,14], Sweden [34], Czech Republic [20,18], Spain [26,32], France [15], Netherlands [35] have been reported in several papers. In the present study, we reached the polyamine concentrations of 255 food items from a literature survey, and calculated the mean value for the same foods, decreasing the number to 161. Since we selected the most frequently consumed foods, only 81 food items were included to calculate daily polyamine intake. In previous studies, Zoumass [36] included approximately 117 of the 370 foods from the literature that produced polyamine content data. Atiya Ali [34] developed a database using 241 food items. Their polyamine values were obtained from literature to measure food intake derived from dietary surveys. Nishibori [14] dealt with 102 food items among 1000 foods in the study of the polyamine intakes in foods in Japan.

Country-specific food preferences and preferences for their main dietary sources might result in the differences in daily polyamine intake. Daily intake of mostly consumed foods were depicted by Nishibori [14] as beverages (491 mL/day), cereals (480 g/day), vegetables (257 g/day); by Atiya Ali [34] as lentil soup (250 g/portion), grapefruit juice (200 mL/portion), orange juice (200 mL/portion), cooked soybean and red beans (190 g/portion). Zoumas [36] used Fred Hutchinson Cancer Center FFQ programming to calculate the mean nmol/day of polyamine from the samples which was corn (max. putrescine, 902,880 nmol/serving size; max. spermidine, 221,111 nmol/serving size), grapefruit juice, oranges, orange juice, grits, crab, grapefruit, green pea soup (max. spermidine, 36,988 nmol/serving size), pear, peas, lentil soup and chicken breast in serving size. We defined the most frequently consumed three foods in Turkey were black tea (322.2 mL/day), white bread (201.4 g/day) and yoghurt (92.9 g/day). Our results were consistent with a cohort study in Turkey, indicating that tea (92.9%), white bread (88.9%) and yoghurt (55.1%) were among the most frequently consumed foods [12].

Calculation of daily polyamine intake in frequently consumed foods in the Turkish population showed that the top three sources for each polyamine came from ripened cheese, black tea and green pepper for putrescine; from black tea, white bread and melon for spermidine; from white bread, lentil soup and rice for spermine. Our findings were consistent with the previous studies that state a high putrescine level as the main source of polyamine intake. In the present study, the highest contribution to putrescine came from dairy products, to spermidine from wheat products, to spermine from meat and wheat products. However, in Japan and Britain, vegetables were the main source of spermidine; meat and meat products the main source of spermine [14,24]. This could possibly be due to regional variation of food intake and dietary habitats.

The average estimated polyamine intakes for adults in countries including the United Kingdom, Italy, Spain, Finland, Sweden, and the Netherlands [35] were 211,910 nmol/day putrescine, 86,959 nmol/day spermidine, and 54,704 nmol/day spermine. A sudy in the USA by Zoumass [36] reported that the average daily polyamine intake values were 159,133 nmol/day putrescine, 54,697 nmol/day spermidine, and 35,698 nmol/day spermine. Our estimation for daily intake was 93,057 nmol/day putrescine, 33,122 nmol/day spermidine and 13,685 nmol/day spermine. Although the values differ from each other, the percentage distribution of daily polyamine intake resulted in the level of each polyamine having a similar pattern in different populations; from the highest to the lowest, contribution rose from putrescine, spermidine and spermine.

There are several limitations to the present study. It is mainly limited by the scarcity of published polyamine concentration in foods. Second, we had limited information about polyamine content of local foods in Turkey. Therefore, our study is an approximation of the real intake of the Turkish population. Third, the origin of foods and methodology to analyze polyamines vary among studies. Finally, daily intake of foods is substantially associated with seasonal changes.

5. Conclusions

This study gives, for the first time, a complete description of the total polyamine intake and the main food contributors of dietary polyamines in Turkish population. The most frequently consumed three foods were black tea (322.2 mL/day), white bread (201.4 g/day) and yoghurt (92.9 g/day), accounting for approximately 50% of total food intake. The main dietary sources for the total polyamine were dairy products (47.32%), vegetables and grains (21.09%) and wheat products (12.75%). The estimation for daily intake were 93,057 nmol/day putrescine, 33,122 nmol/day spermidine and 13,685 nmol/day spermine.

Acknowledgments

The authors would like to the thank volunteers and students who participated in the study.

Author Contributions

Conception and design of study: N.B.; Data analysis and interpretation: N.B., H.H., K.E., M.G.; Manuscript writing: N.B., H.H.; Critical review and final approval of manuscript: N.B., H.H., K.E., M.G.

Conflicts of Interest

The authors declare no conflict of interest.

References

1. Bardocz, S.; Grant, G.; Brown, D.S.; Ralph, A.; Pusztai, A. Polyamines in food—Implications for growth and health. *J. Nutr. Biochem.* **1993**, *4*, 66–71.
2. Seiler, N.; Atanassov, C.L.; Raul, F. Polyamine metabolism as target for cancer chemoprevention. *Int. J. Oncol.* **1998**, *13*, 993–1006.
3. Gerner, E.W.; Meyskens, F.L., Jr. Polyamines and Cancer: Old molecules, new understanding. *Nat. Rev. Cancer* **2004**, *4*, 781–792.
4. Russell, D.H.; Levy, C.C. Polyamine accumulation and biosynthesis in amouse L1210 leukemia. *Cancer Res.* **1971**, *31*, 248–251.
5. Soda, K. The mechanisms by which polyamines accelerate tumor spread. *J. Exp. Clin. Cancer Res.* **2011**, *30*, 95, doi:10.1186/1756-9966-30-95.
6. Sarhan, S.; Knodgen, B.; Seiler, N. The gastrointestinal tract as polyamine source for tumour growth. *Anticancer Res.* **1989**, *9*, 215–224.
7. Deloyer, P.; Peulen, O.; Dandrifosse, G. Dietary polyamines and non-neoplastic growth and disease. *Eur. J. Gastroenterol. Hepatol.* **2001**, *13*, 1027–1032.
8. Das, R.; Kanungo, M.S. Activity and modulation of ornithine decarboxylase and concentrations of polyamines in various tissues of rats as a function of age. *Exp. Gerontol.* **1982**, *17*, 95–103.
9. Nishimura, K.; Shiina, R.; Kashiwagi, K.; Igarashi, K. Decrease in polyamines with aging and their ingestion from food and drink. *J. Biochem.* **2006**, *139*, 81–90.

10. Binh, P.N.T.; Soda, K.; Maruyama, C.; Kawakami, M. Relationship between food polyamines and gross domestic product in association with longevity in Asian countries. *Health* **2010**, *2*, 1390–1396, doi:10.4236/health.2010.212206.

11. Silla-Santos, M.H. Biogenic amines: Their importance in foods. *Int. J. Food Microbiol.* **1996**, *29*, 213–244.

12. Türkiye Beslenme ve Sağlık Araştırması 2010 (in Turkish). A Report of 2010 National Nutrition Health and Food Consumption Survey. Available online: http://www.sagem.gov.tr/TBSA_Beslenme_Yayini.pdf (accessed on 15 July 2014).

13. Okamoto, A.; Sugi, E.; Koizumi, Y.; Yanagida, F.; Udaka, S. Polyamine content of ordinary food stuffs and various fermented foods. *Biosci. Biotechnol. Biochem.* **1997**, *61*, 1582–1586.

14. Nishibori, N.; Fujihara, S.; Akatuki, T. Amounts of polyamines in foods in Japan and intake by Japanese. *Food Chem.* **2006**, *100*, 491–499.

15. Lavizzari, T.; Teresa Veciana-Nogues, M.; Bover-Cid, S.; Marine-Font, A.; Carmen Vidal-Carou, M. Improved method for the determination of biogenic amines and polyamines in vegetable products by ion-pair high-performance liquid chromatography. *J. Chromatogr. A* **2006**, *1129*, 67–72.

16. Bardocz, S.; Duguid, T.J.; Brown, D.S.; Grant, G.; Pusztai, A.; White, A.; Ralph, A. The importance of dietary polyamines in cell regeneration and growth. *Br. J. Nutr.* **1995**, *73*, 819–828.

17. Eliassen, K.A.; Reistad, R.; Risoen, U.; Ronning, H.F. Dietary polyamines. *Food Chem.* **2002**, *78*, 273–280.

18. Kalac, P.; Krizek, M.; Pelikanova, T.; Langova, M.; Veskrna, O. Contents of polyamines in selected foods. *Food Chem.* **2005**, *90*, 561–564.

19. Moret, S.; Smela, D.; Populin, T.; Conte, L. A survey on free biogenic amine content of fresh and preserved vegetables. *Food Chem.* **2005**, *89*, 355–361.

20. Kalac, P.; Svecova, S.; Pelikanova, T. Levels of biogenic amines in typical vegetable products. *Food Chem.* **2002**, *77*, 349–351.

21. Ziegler, W.; Hahn, M.; Wallnofer, P.R. Changes in biogenic amine contents during processing of several plant foods. *Deutsche Lebensmittel-Rundschau* **1994**, *90*, 108–112. (In German)

22. Valsamaki, K.; Michaelidou, A.; Polychroniadou, A. Biogenic amine production in Feta cheese. *Food Chem.* **2000**, *71*, 259–266.

23. Farriol, M.; Venereo, Y.; Orta, X.; Company, C.; Gomez, P.; Delgado, G.; Rodríguez, R. Ingestion of antioxidants and polyamines in patients with severe burns. *Nutr. Hosp.* **2004**, *19*, 300–304.

24. Bardocz, S. Polyamines in food and their consequences for food quality and human health. *Trends Food Sci. Technol.* **1995**, *6*, 341–346.

25. Silva, C.M.G.; Gloria, M.B.A. Bioactive amines in chicken breast and thigh after slaughter and during storage at 4 ± 1 °C and in chicken-based meat products. *Food Chem.* **2002**, *78*, 241–249.

26. Hernandez-Jover, T.; Izquierdo-Pulido, M.; Veciana-Nogues, M.T.; Marine-Font, A.; Vidal-Carou, M.C. Effect of starter cultures on biogenic amine formation during fermented sausage production. *J. Food Prot.* **1997**, *60*, 825–830.

27. Saaid, M.; Saad, B.; Hashim, N.H.; Ali, A.S.M.; Saleh, M.I. Determination of biogenic amines in selected Malaysian food. *Food Chem.* **2009**, *113*, 1356–1362.

28. Durlu-Ozkaya, F.; Alichanidis, E.; Litopoulou-Tzanetaki, E.; Tunail, N. Determination of biogenic amine content of Beyaz cheese and biogenic amine production ability of some lactic acid bacteria. *Milchwissenschaft* **1999**, *54*, 680–682.

29. Novella-Rodriguez, S.; Veciana-Nogues, M.T.; Vidal-Carou, M.C. Biogenic amines and polyamines in milks and cheeses by ionpair high performance liquid chromatography. *J. Agric. Food Chem.* **2000**, *48*, 5117–5123.

30. Novella-Rodriguez, S.; Veciana-Nogues, M.T.; Izquierdo-Pulido, M.; Vidal-Carou, M.C. Distribution of biogenic amines and polyamines in cheese. *J. Food Sci.* **2003**, *68*, 750–755.

31. Fernandez, M.; Linares, D.M.; del Rio, B.; Ladero, V.; Álvarez, M.A. HPLC quantification of biogenic amines in cheeses: Correlation with PCR-detection of tyramine-producing microorganism. *J. Dairy Res.* **2007**, *74*, 276–282.

32. Novella-Rodriguez, S.N.; Veciana-Nogues, M.T.; Roig-Sagues, A.X.; Trujillo-Mesa, A.J.; Vidal- Carou, M.C. Evaluation of biogenic amines and microbial counts throughout the ripening of goat cheeses from pasteurized and raw milk. *J. Dairy Res.* **2004**, *71*, 245–252.

33. Stratton, J.E.; Hutkins, R.W.; Taylor, S.L. Biogenic amines in cheese and other fermented food: A review. *J. Food Prot.* **1991**, *54*, 460–470.

34. Ali, M.A.; Poortvliet, E.; Stromberg, R.; Yngve, A. Polyamines in foods: Development of a food database. *Food Nutr. Res.* **2011**, *55*, doi:10.3402/fnr.v55i0.5472.

35. Ralph, A.; Englyst, K.; Bardocz, S. Polyamine content of the human diet. In *Polyamines in Health and Nutrition*; Bardócz, S., White, A., Eds.; Kluwer Academic Publishers: London, UK, 1999; pp. 123–137.

36. Zoumas-Morse, C.; Rock, C.L.; Quintana, E.L.; Neuhouser, M.L.; Gerner, E.W.; Meyskens, F.L. Development of a polyamine database for assessing dietary intake. *J. Am. Diet. Assoc.* **2007**, *107*, 1024–1027.

37. Moinard, C.; Cynober, L.; de Bandt, J.L. Polyamines: Metabolism and implications in human diseases. *Clin. Nutr.* **2005**, *24*, 184–197.

38. Nowotarski, S.L.; Woster, P.M.; Casero, R.A. Polyamines and cancer: Implications for chemotherapy and chemoprevention. *Expert Rev. Mol. Med.* **2013**, *15*, e3, doi:10.1017/erm.2013.3.

39. Valero, D.; Martinez-Romero, D.; Serrano, M. The role of polyamines in the improvement of the shelf life of fruit. *Trends Food Sci. Technol.* **2002**, *13*, 228–234.

40. Kozova, M.; Kalac, P.; Pelikanova, T. Contents of biologically active polyamines in chicken meat, liver, heart and skin after slaughter and their changes during meat storage and cooking. *Food Chem.* **2009**, *116*, 419–425.

41. Veciana-Nogues, M.T.; Mariné-Font, A.; Vidal-Carou, M.C. Biogenic amines in fresh and canned tuna. Effects of canning on biogenic amine contents. *J. Agric. Food Chem.* **1997**, *45*, 4324–4332.

42. Wang, S.Y.; Faust, M. Comparison of seasonal growth and polyamine content in shoots of orchard-grown standard and genetic dwarf apple trees. *Physiol. Plant.* **1993**, *89*, 376–380.

43. Working Group on Monitoring Scottish Dietary Targets Workshop. A Short Review of Dietary Assessment Methods Used in National and Scottish Research Studies Briefing Paper Prepared for: September 2003. Available online: http://multimedia.food.gov.uk/multimedia/pdfs/scotdietassessmethods.pdf (accessed on 3 June 2014).

44. Motyl, T.; Ploszaj, T.; Wojtasik, A.; Kukulska, W.; Podgurniak, M. Polyamines in cow's and sow's milk. *Comp. Biochem. Physiol. B Biochem. Mol. Biol.* **1995**, *111*, 427–433.

45. Kalac, P.; Spicka, J.; Krizek, M.; Steidlova, S.; Pelikanova, T. Concentrations of seven biogenic amines in sauerkraut. *Food Chem.* **1999**, *67*, 275–280.

46. Slocum, R.D.; Flores, H.E.; Galston, A.W.; Weinstein, L.H. Improved method for HPLC analysis of polyamines, agmatine and aromatic monuamines in plant tissue. *Plant Physiol.* **1989**, *89*, 512–517.

47. Weiger, T.M.; Aichberger, S.; Wallace, H.M. A comparison of dietary polyamine uptake by humans in Europe, Asia and the USA. In Proceedings of the COST 922 Workshop Health Implications of Dietary Amines, Coimbra, Portugal, 3–6 November 2005.

10

An Investigation of the Complexity of Maillard Reaction Product Profiles from the Thermal Reaction of Amino Acids with Sucrose Using High Resolution Mass Spectrometry

Agnieszka Golon [1], **Christian Kropf** [2], **Inga Vockenroth** [2] and **Nikolai Kuhnert** [1,*]

[1] School of Engineering and Science, Jacobs University Bremen, Campus Ring 1, 28759 Bremen, Germany; E-Mail: a.golon@jacobs-university.de

[2] Henkel AG & Co. KGaA, Henkelstr. 67, 40589 Düsseldorf, Germany; E-Mails: christian.kropf@henkel.com (C.K.); inga.vockenroth@henkel.com (I.V.)

* Author to whom correspondence should be addressed; E-Mail: n.kuhnert@jacobs-university.de

Abstract: Thermal treatment of food changes its chemical composition drastically with the formation of "so-called" Maillard reaction products, being responsible for the sensory properties of food, along with detrimental and beneficial health effects. In this contribution, we will describe the reactivity of several amino acids, including arginine, lysine, aspartic acid, tyrosine, serine and cysteine, with carbohydrates. The analytical strategy employed involves high and ultra-high resolution mass spectrometry followed by chemometric-type data analysis. The different reactivity of amino acids towards carbohydrates has been observed with cysteine and serine, resulting in complex MS spectra with thousands of detectable reaction products. Several compounds have been tentatively identified, including caramelization reaction products, adducts of amino acids with carbohydrates, their dehydration and hydration products, disproportionation products and aromatic compounds based on molecular formula considerations.

Keywords: mass spectrometry; Maillard reaction; carbohydrates; amino acids; complex mixture

1. Introduction

The chemical composition of the raw materials of food is rather well understood; its chemical composition changes, however, completely, upon processing. Most food consumed is processed prior to consumption by thermal treatment, including cooking, frying, roasting, baking and storage, including drying, pickling or fermentation [1]. Upon these processing steps, the original food constituents undergo remarkable chemical changes producing a myriad of novel compounds. For roasted coffee, only 50% of green bean components remain chemically unchanged, whereas the rest are being chemically modified, resulting in a material composed of an unknown structure, usually referred to as coffee melanoidins [2]; while for cocoa powder, around 75% of the material produced through food treatment is unidentified. The situation is similar for most other processed foods, and nowadays, the majority of chemical structures resulting from such a processing is yet to be identified. In 1912, Louis Camille Maillard, for the first time, discovered chemical changes during food processing, describing the reaction between amino acids and sugars at elevated temperatures, typical for food processing conditions [3]. Since then, only a few defined reaction products of food processing have been isolated and structurally characterized [4,5].

The Maillard reaction is one of the most important processes that takes place in food processing and storage. In the first step of the reaction, Amadori compounds are formed and considered as precursors of aroma, color and flavor. In the next steps, series of rearrangements, dehydrations and cyclizations occur to produce advanced glycation end (AGE) products. The Maillard reaction products (MRPs), especially Amadori compounds and melanoidins (high molecular weight compounds), are currently receiving a great deal of interest due to their reported health-promoting properties and their potentials as functional food ingredients [6]. Melanoidins, for example, represent an important part of human diet, with an average intake of several grams per day [7]. Many studies are focused on the high antioxidant capacity of MRPs in model systems and food materials, such as beer, coffee and bakery products [8,9]. Moreover, the antioxidants, antimicrobial and cytotoxic properties of MRPs have been reported [10–12].

The MRPs of cysteine are responsible for meat-like aromas, for example, 2-methyl-3-furanthiol, 2-furfurylthiol or 3-mercapto-2-butanone [8,13]; those of lysine and arginine have been detected in many bakery products, in the crust or crumb of bread, and in roasted coffee. Despite several investigations, the great complexity of Maillard reaction products is still a challenge for food chemists. Much attention has been given to the reactions of sugars, such as glucose and fructose and other monosaccharides, whereas food constituents important to browning include disaccharides, for example sucrose, extensively used in confectionery and pastry products [14,15].

To obtain a comprehensive picture of the Maillard chemistry underlying food processing, further investigations and the development of suitable analytical tools are necessary. Fourier Transform Ion Cyclotron mass spectrometry (FT ICR-MS) has been used only on a few occasions in the analysis of food materials. It has been successfully applied in our group to characterize the composition of black tea thearubigins, roasted coffee beans and the thermal decomposition products of starch [16–19].

The objective of this study is to better understand the composition of unresolved complex mixtures of food materials upon heating by employing the analytical technology and data interpretation approaches. The study focuses on the reaction products formed, when lysine (**1**), arginine (**2**), aspartic acid (**3**), tyrosine (**4**), serine (**5**) and cysteine (**6**) are heated with sucrose (**7**). High resolution mass spectrometry as a direct infusion was applied for all of the samples. Moreover, FT ICR-MS was used for the reactions of cysteine and serine with sucrose. Thus, generated molecular formula lists were subjected to graphical interpretation tools, such as the van Krevelen analysis, in order to provide more information about structural trends.

2. Experimental Section

2.1. Sample Preparation

All chemicals (analytical grade) were purchased from Sigma-Aldrich (Germany). All components of Maillard reaction mixtures were ground in a mortar, mixed together, dissolved in 1 mL of water and heated at 200 °C in the oven with a power of 1.2 kW. The heated samples were then stored at room temperature. Heated products were dissolved in methanol/water (1:1, v/v, 1 mL) and used for mass spectrometry analyses.

2.2. Methods

2.2.1. ESI-TOF-MS

High-resolution mass spectra were recorded using a Bruker Daltonics micrOTOF instrument (Bruker Daltonics, Bremen, Germany) employing both negative and positive electrospray ionization modes. The micrOTOF Focus mass spectrometer (Bruker Daltonics) was fitted with an ESI source, and internal calibration was achieved with 10 mL of 0.1 M sodium formate solution. Calibration was carried out using the enhanced quadratic calibration mode. All MS measurements were performed in both negative and positive ion modes.

2.2.2. FT ICR-MS

Ultra high resolution mass spectra were acquired using a Bruker (Bremen, Germany) solarix Fourier Transform Ion Cyclotron Resonance mass spectrometer (FTICR-MS) with a 12 T refrigerated superconducting cryo-magnet. The instrument was equipped with a dual electrospray ion source with ion funnel technology. The spectra of the samples were recorded in electrospray ionization positive and negative ion modes using direct infusion with a syringe pump with a flow rate of 120 µL/h.

3. Results and Discussion

3.1. Reactions of Amino Acids with Disaccharides

We have chosen different amino acids, namely arginine, lysine, aspartic acid, tyrosine, serine and cysteine, to probe their reactivity with disaccharides under thermal treatment. As disaccharides, the two most common and relevant derivatives in food, sucrose and lactose, were chosen. The mixtures of two selected disaccharides with selected amino acids were heated at various temperatures ranging from 150 to 200 °C. Heating parameters were optimized based on color formation and the recorded mass spectra of products heated at different temperatures. Experiments performed at different heating conditions showed that around a half an hour reaction time at 200 °C are the most suitable parameters to obtain the desired brown material with in excess of 90% of the starting materials being consumed in the reaction. Carbohydrates were mixed with amino acids with different proportions (1:1, 2:1 and 1:2, v/v), which correspond to their composition in typical food products. The reactions were performed with sucrose and lactose, although in this contribution, we focus mainly on the reactions with sucrose. The significant difference between two carbohydrates was observed, and lactose in contrast to saccharose in the reactions with amino acids resulted in more complex mass spectra as judged by the number of resolved peaks observed. This finding might be explained by the stereochemistry of lactose having the axial C4-OH substituent of galactose. Upon heating, this can cause a higher level of dehydration and, consequently, an increased number of intermediates. Mass spectra for heated mixtures of amino acids with sucrose are displayed in Figure 1. All of the amino acids exhibit different reactivities. Mass spectra from the reactions of sugars with lysine and tyrosine are relatively simple, but more complex for arginine and extremely complex for cysteine and serine. The term simple hereby signifies the observation of a small number of intense observed signals ($\leq$100), whereas the term complex refers to a large number of signals ($\geq$400) usually at lower intensities. The complexity can be easily visualized by the number of signals with an S/N ratio higher than 10 in MS spectra, and the number of signals for the reactions are: 221-Lys:Suc (1:1), 495-Cys:Suc (1:1) and 691-Ser:Suc (1:2) (Table 1).

Figure 1. Mass spectra for heated: (**a**) lysine:sucrose (1:2), (**b**) arginine:sucrose (1:2), (**c**) aspartic acid:sucrose (1:2), (**d**) tyrosine:sucrose (1:2), (**e**) serine:sucrose (1:2) and (**f**) cysteine:sucrose (1:1) in negative ion mode using a direct infusion into an ESI-TOF-MS instrument.

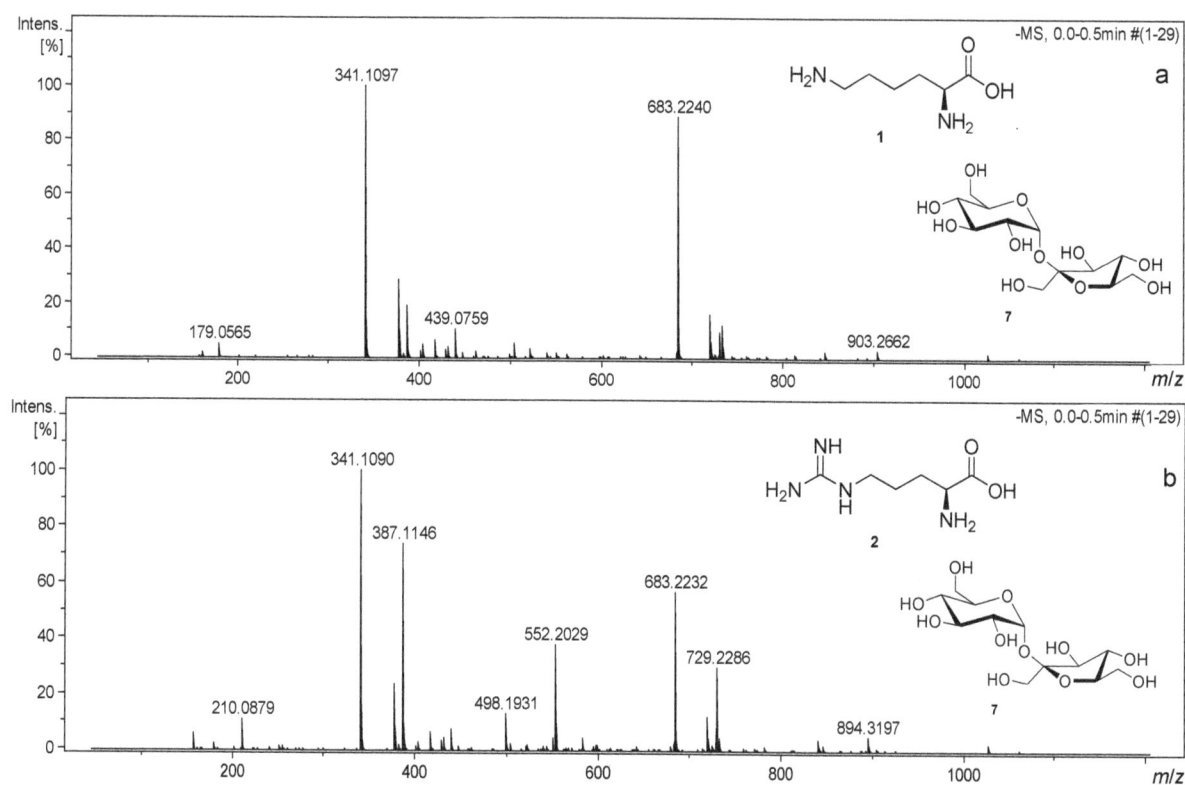

Figure 1. *Cont.*

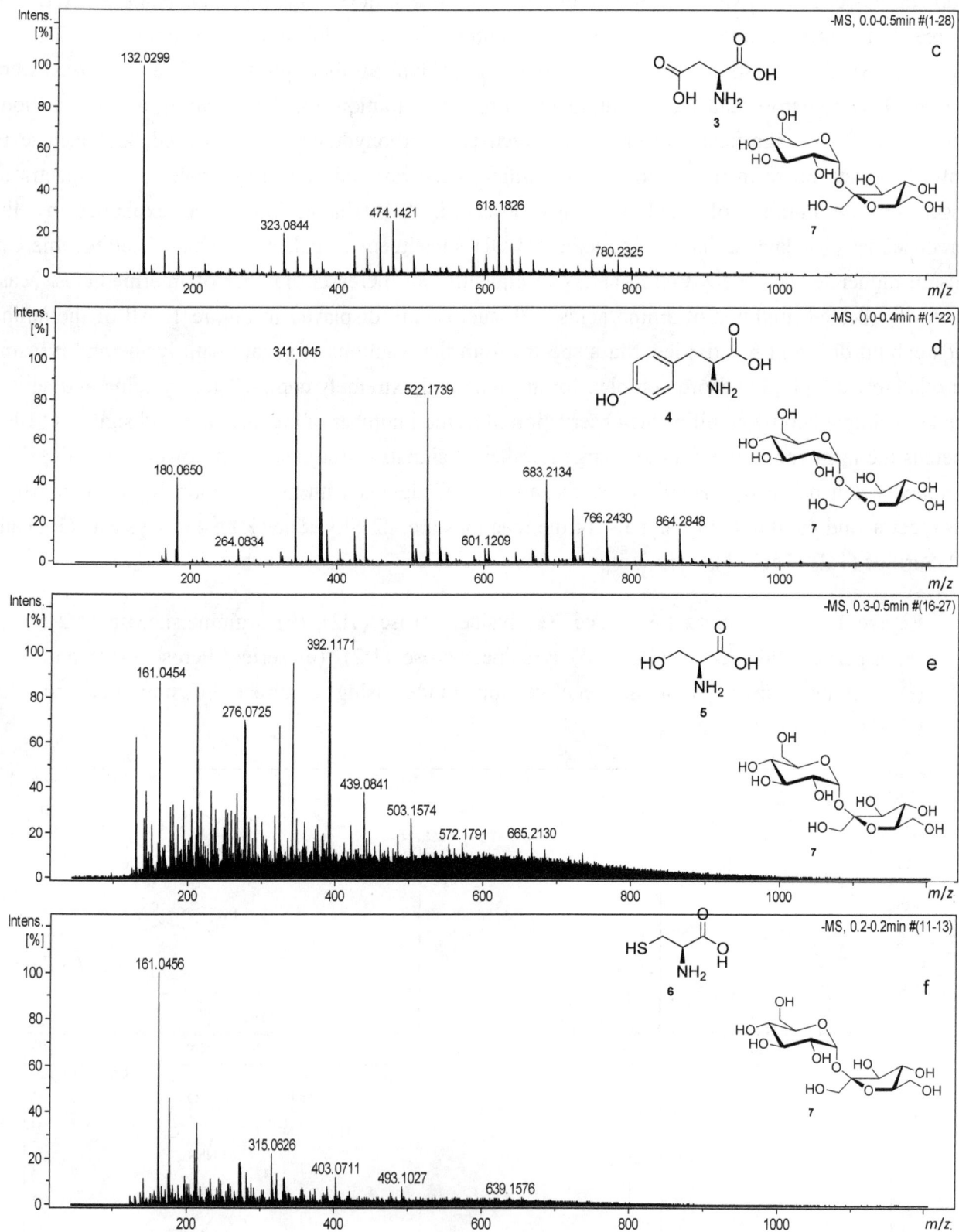

Table 1. The number of peaks observed for samples in ESI-TOF-MS and FT ICR-MS in negative ion mode.

No.	Sample	Number of Signals with S/N > 10	Number of Signals with S/N > 3
1	Arginine-sucrose [a]	126	714
2	Lysine-sucrose [a]	221	670
3	Aspartic acid-sucrose [a]	297	796
4	Tyrosine-sucrose [a]	420	916
5	Serine-sucrose [a]	691	885
6	Cysteine-sucrose [a]	495	791
7	Serine-sucrose [b]	2682	4698
8	Cysteine-sucrose [b]	5193	8527

[a] ESI-TOF-MS; [b] FT ICR-MS.

The examination of the spectra aimed initially at the identification of typical caramelization reactions products [20,21]. From the experimental mass lists, molecular formula lists were generated and compared to those obtained for typical caramelization reactions in the absence of amino acids. The comparison of peaks with identical accurate mass values and, therefore, molecular formulae, if compared to previous studies on caramelization processes, allowed the tentative identification of around 70 caramelization products, including oligomers of hexoses, with up to a maximum of six monomeric units, both dehydration products of monomeric hexoses and oligomeric hexoses showing successive a loss of up to seven water molecules, depending on the number of monomers. Hydration products with up to two water molecules added to an oligomeric carbohydrate were detected [19,20]. Fragmentation products after the redox disproportionation reaction were found and also aromatic compounds after excessive dehydration. The reaction products between amino acids and sucrose gave oligomers of amino acid conjugated to hexoses after the hydrolysis of sucrose in the aqueous reaction medium (either fructose or glucose; MS^n data revealed exclusively neutral losses for hexose fragments and did not allow distinction between fructose and glucose) with a maximum of four carbohydrate units, the dehydration products loosing up to six water molecules and hydration products with up to two water molecules added to the oligomeric products. In Table 2, a mass list of some of the compounds for one of the studied mixtures, the reaction between aspartic acid with sucrose, is shown. In this case, the reaction products of aspartic acid with sucrose with a maximum of four carbohydrate units, their dehydration products loosing up to four water molecules and hydration products with up to one water molecule were found. In most of the cases, signals corresponding to the molecular formulae of dehydration products of amino acid conjugated to carbohydrates with the loss of two water molecules appeared to have the highest intensities.

The studies on processed food components by many research groups are focused on single compounds, such as 5-HMF, acrylamide or heteroaromatic cyclic amines (MeIQ, PhIP) formed during cooking or baking [22]. Our work presented here allows a more global view on the many products so far neglected in thermal processing. It clearly shows that many additional compounds of related chemical structures are present in processed food, with a crude and tentative structure classification based on molecular formulae considerations possible.

Table 2. ESI-TOF-MS data in negative ion mode of a reaction of aspartic acid with sucrose.

No.	Assignment	Molecular Formula	Theoretical m/z $[M - H]^-$	Experimental m/z $[M - H]^-$	Error (ppm)
1	Asp-Glu/Fru	$C_{10}H_{17}NO_9$	294.0831	294.0831	−0.1
2	Asp-Glu/Fru + H_2O	$C_{10}H_{19}NO_{10}$	312.0936	312.0935	0.2
3	Asp-Glu/Fru-2 × H_2O	$C_{16}H_{23}NO_{12}$	420.1147	420.1167	−4.6
4	Asp-Glu/Fru-H_2O	$C_{16}H_{25}NO_{13}$	438.1253	438.1236	3.9
5	Asp-Suc	$C_{16}H_{27}NO_{14}$	456.1359	456.1357	0.3
6	Asp-Suc + H_2O	$C_{16}H_{29}NO_{15}$	474.1464	474.1472	−1.6
7	Asp-3 × Glu/Fru-3 × H_2O	$C_{22}H_{31}NO_{16}$	564.1570	564.1583	−2.3
8	Asp-3 × Glu/Fru-2 × H_2O	$C_{22}H_{33}NO_{17}$	582.1676	582.1680	−0.7
9	Asp-3 × Glu/Fru-H_2O	$C_{22}H_{35}NO_{18}$	600.1781	600.1789	−1.3
10	Asp-3 × Glu/Fru	$C_{22}H_{37}NO_{19}$	618.1887	618.1879	1.4
11	Asp-3 × Glu/Fru + H_2O	$C_{22}H_{39}NO_{20}$	636.1993	636.1990	0.4
12	Asp-4 × Glu/Fru-4 × H_2O	$C_{28}H_{39}NO_{20}$	708.1993	708.2048	−7.7
13	Asp-4 × Glu/Fru-3 × H_2O	$C_{28}H_{41}NO_{21}$	726.2098	726.2139	−5.5
14	Asp-4 × Glu/Fru-2 × H_2O	$C_{28}H_{43}NO_{22}$	744.2204	744.2236	−4.3
15	Asp-4 × Glu/Fru-H_2O	$C_{28}H_{45}NO_{23}$	762.2310	762.2303	0.8
16	Asp-4 × Glu/Fru	$C_{28}H_{47}NO_{24}$	780.2415	780.2385	3.9
17	Asp-4 × Glu/Fru + H_2O	$C_{28}H_{49}NO_{25}$	798.2521	798.2507	1.7

3.2. FT ICR-MS Measurement of Maillard Reactions of Serine and Cysteine with Sucrose

The reactions of serine and cysteine with sucrose resulted in extremely complex MS spectra with signal numbers exceeding several thousand and were therefore analyzed by FT ICR-MS, providing ultimate resolution. The samples were measured as a direct infusion in aqueous methanol solution using a 12T FT ICR-MS instrument with ESI ionization in both positive and negative ion mode. The data obtained provides a comprehensive overview of all products containing the reaction products detectable by ESI-MS. Representative mass spectra are shown in Figure 2. Further expanded mass spectra are available in the Supplementary Materials (Figures S1–S3).

From the data, mass lists were generated with an S/N ratio above four and a relative intensity higher than or equal to 0.1%. A detailed discussion on the cut-off level of the signal intensity and the distinction between noise and signals corresponding to real food processing products has been given in earlier references [1,16]. In the case of serine, 5340 signals were present, while for cysteine 8530. This confirms that the presence of sulfur in the molecule leads to the formation of several new MRPs. It is worth mentioning that the number of detected ions is the minimum number of compounds present in the sample, which must be multiplied by the number of potential isomers, consequently leading to thousands of compounds.

From the mass lists, molecular formula lists were generated, accepting an error below 1 ppm and the presence of C, H, O and N for serine and C, H, O, N and S for cysteine for all signals with a relative intensity of 0.1% of the base peak. The data were then subjected to interpretation tools.

Molecular formulae have been assigned for almost 600 of the most intense peaks for the reactions with serine and cysteine. The majority of assigned signals contain nitrogen atoms. In the case of cysteine, around 90 molecular formulae contain one or more sulfur atoms, for example an adduct at

m/z 264.054607 with the molecular formula $C_9H_{15}NO_6S$, which corresponds formally to the reaction products between cysteine and hexose after dehydration, or at m/z 282.065162 with the molecular formula $C_9H_{17}NO_7S$, derived from the reaction between cysteine and hexose.

Figure 2. Direct infusion ESI-FT-ICR mass spectrum of a reaction of (**a**) sucrose with serine and (**b**) sucrose with cysteine in negative ion mode.

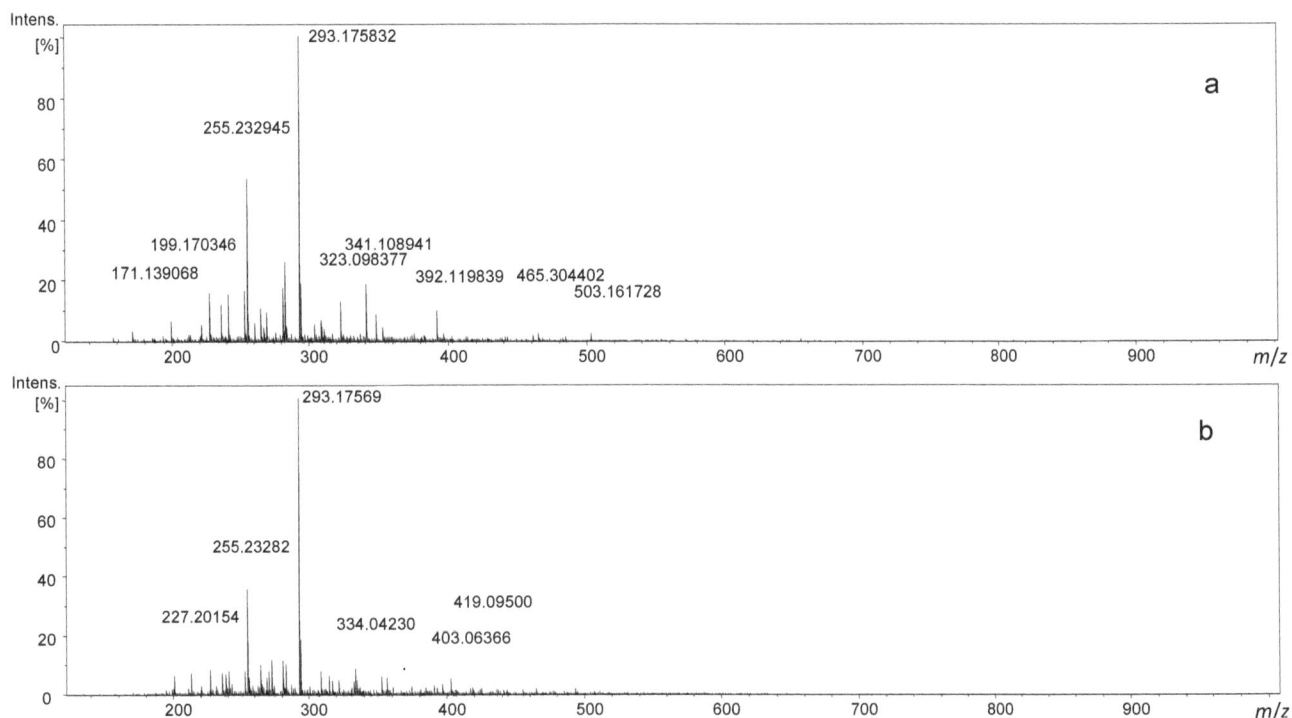

3.3. Graphical Data Interpretation

To visualize differences in product formation, we generated graphs showing a number of compounds and a second graph displaying the sum of ion intensities summed up over all signals with the respective elemental composition, CHO, CHON and $CHON_{1+x}$ for the reaction of serine with sucrose and CHO, CHON, CHOS, CHONS, $CHOS_{1+x}$ and $CHON_{1+x}$ for the reaction of cysteine with sucrose (Figure 3). Each column represents the number of compounds with a defined elemental composition. For serine, the number of nitrogen-containing molecules is higher than nitrogen-free molecules, and more than a half of the latter contains more than one nitrogen atom; although the total intensities of CHO molecules are higher than for nitrogen-containing molecules (Figure 4). The sum of intensities of pseudomolecular ions corresponding to molecular formulae with one nitrogen atom are higher than for the molecules containing two and more nitrogen atoms. For cysteine, the number of compounds with the CHON elemental composition is higher than for CHO. The number of molecules with nitrogen and sulfur atoms is lower than that of CHON molecules. The highest intensities correspond to CHO molecules, followed by CHON, $CHON_{1+x}$, CHONS, $CHOS_{1+x}$ and CHOS. The sum of intensities of molecules with one nitrogen atom are slightly higher than for ones with more than one nitrogen atom (Figure 4).

Figure 3. The number of compounds with the elemental composition (**a**) CHO, CHON and $CHON_{1+x}$ for the reaction of serine with sucrose and (**b**) the elemental composition CHO, CHON, CHOS, CHONS, $CHOS_{1+x}$ and $CHON_{1+x}$ for the reaction of cysteine with sucrose, analyzed by negative ion mode FT ICR-MS.

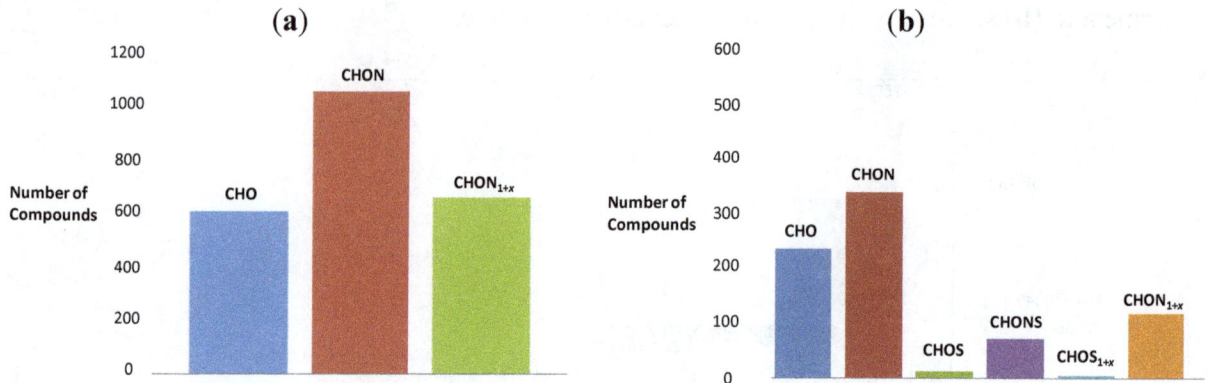

Figure 4. The sum of intensities for the compounds with the elemental composition (**a**) CHO, CHON and $CHON_{1+x}$ for the reaction of serine with sucrose and (**b**) the elemental composition CHO, CHON, CHOS, CHONS, $CHOS_{1+x}$ and $CHON_{1+x}$ for the reaction of cysteine with sucrose analyzed by negative ion mode FT ICR-MS.

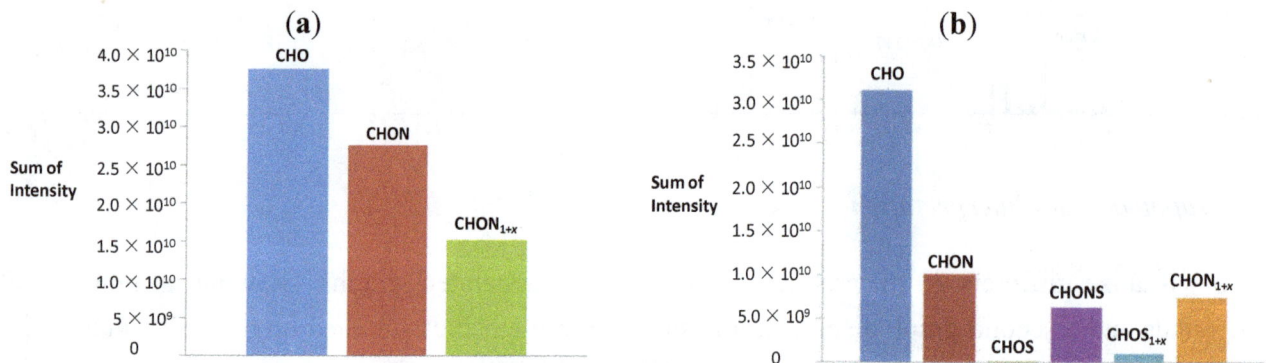

3.4. The van Krevelen Diagrams for the Reaction of Serine with Sucrose

The van Krevelen diagrams were generated from the molecular formulae lists. In this diagram, elemental ratios, such as H/C or O/C, were calculated from the molecular formulas and plotted in two-dimensional graphs, with every point on the graph corresponding to one analyte with a defined elemental composition in the sample. These diagrams allow the tentative classification of classes of compounds based on their characteristic elemental ratio boundaries and the identification of possible reaction trends [1,23].

The van Krevelen diagrams were generated in several variations, with one plot showing all analytes, one bubble plot showing intensity coded data points and two-three plots showing Maillard products with color coding according to different elemental compositions. The van Krevelen diagram of the reaction of sucrose with serine is shown in Figure 5a. In typical elemental ratio boundaries for carbohydrates (right top corner), products can be observed; in the middle of the graph are reaction products between amino acid and carbohydrate and carbohydrate after dehydration. Many compounds after gradual dehydration lay on the diagonal of the plot towards its origin. The second graph shows the same points with their intensities (Figure 5b), and thus, the three compounds with the highest intensities of ions observed correspond to two lipid-type compounds and one carbohydrate. The observation of ions with a lipid-type elemental ratio comes as a surprise, and we cannot exclude an erroneous molecular formulae assignment or other measurement artefact at this point. The reaction products between amino acid and sugar are created in moderate intensities. Moreover, many homologous series can be observed in the graph, such as dehydration products of amino acid-sugar reaction products. In Figure 5c, red points represent nitrogen-free compounds, while blue nitrogen-containing compounds. Within the mixture, 45% of the ions are CHO compounds free of nitrogen and, therefore, originate from caramelization reactions, whereas the remaining 55% contain nitrogen and must be considered as real Maillard products. The data points, which belong to carbohydrates, can be easily seen by following dehydration products on the line with the negative slope toward the origin in the middle of the graph. The next group of nitrogen-free compounds appear in the range of 1.5–2.0 for the H/C ratio and 0.0–0.6 for the O/C ratio, corresponding to disproportionation reaction products. The nitrogen-containing products are located mainly in the middle of the graph and belong to the reaction products between serine and sucrose. Figure 5d displays the distribution of nitrogen-containing compounds with respect to the nitrogen numbers in molecules. The red points correspond to nitrogen-free compounds, while the blue points are the molecules with one nitrogen atom, green with two and yellow with three nitrogen atoms. The distribution shows that one-nitrogen-containing molecules are placed next to the carbohydrate-type molecules on the diagonal line towards point zero, suggesting the dehydration products. From these data, it must be concluded that oligopeptide-like structures are formed with three and more nitrogen atoms, with oligomerization playing an important role in the Maillard reaction, as shown earlier for caramelization [19–21] Compounds with two nitrogen atoms are mainly around 0.5 and 1.5 for the O/C and H/C ratio, respectively, and go toward the more aromatic region. On the other hand, compounds with three nitrogen atoms are placed mainly around the point of 0.3 for O/C and 1.2 for H/C, forming one circle.

3.5. The van Krevelen Diagrams for the Reaction of Cysteine with Sucrose

The profile of the van Krevelen diagram for the reaction of cysteine with sucrose is similar to the one with serine and saccharose, although the number of points in the aromatic region is higher (Figure 6a). The intensities of carbohydrate-type compounds are not as high as in the case of serine, while the signals of some of the reaction products between the amino acid and carbohydrate appear with relatively high intensities (Figure 6b). This can suggest the intensive conversion of sucrose into various products. In Figure 6c, with red nitrogen-free and blue nitrogen-containing compounds, most of the nitrogen-free compounds appearing in the left upper part are characteristic of lipid-type compounds. Some of them are seen in the aromatic region after the successive dehydration reaction. The nitrogen-containing compounds are almost evenly spread over the whole graph. In Figure 6d, we can observe the distribution of nitrogen-containing compounds. The N-containing compounds show a pattern resembling the van Krevelen plot of serine-derived Maillard products. As for serine, the products with one nitrogen atom are distributed in the whole diagram and are placed in the upper left side, in the aromatic region and in the middle of the graph. The last corresponds to the reaction products between cysteine and carbohydrate. The products with three nitrogen atoms are located mainly around the point of 0.2 for O/C and 1.5 for H/C, which is in agreement with the reaction of serine. In the same region, most of the compounds with four nitrogen atoms are observed. In Figure 6e, the allocation of sulfur-containing molecules is shown. Most of these molecules are located in the middle of the graph, being the reaction products between cysteine and sucrose. Some of them are placed in the aromatic region with a low O/C ratio after successive dehydration.

Figure 5. The van Krevelen diagrams of the reaction of sucrose with serine: (**a**) with colored groups of compounds; (**b**) with the intensities; (**c**) with nitrogen-free (red points) and nitrogen-containing compounds (blue points); (**d**) with nitrogen-free (red points) compounds and one nitrogen atom (blue points), two (green), three (yellow) and four (violet) nitrogen atoms in negative ion mode.

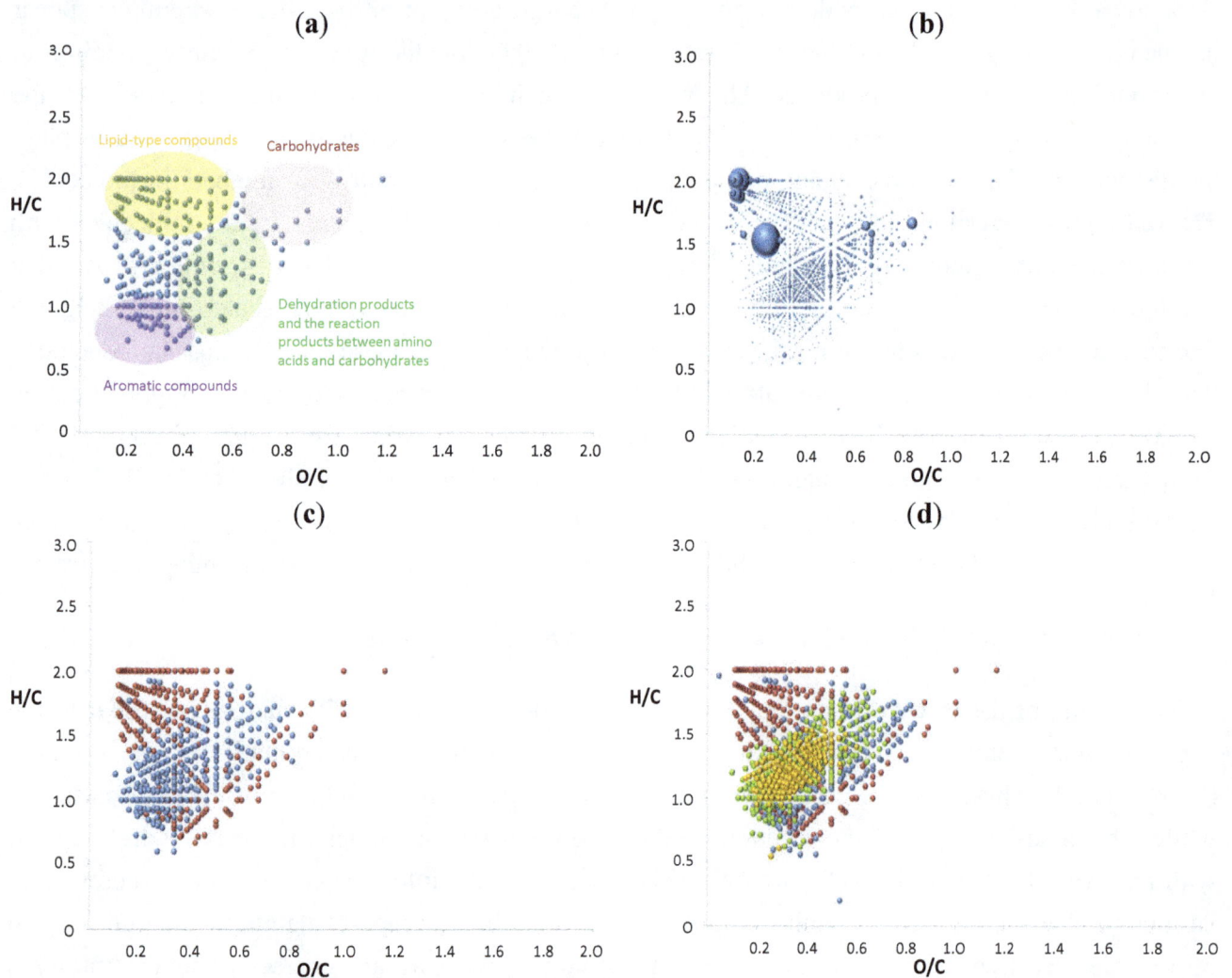

Figure 6. The van Krevelen diagrams of the reaction of sucrose with cysteine: (**a**) with colored groups of compounds; (**b**) with the intensities; (**c**) with nitrogen-free (red points) and nitrogen-containing compounds (blue points); (**d**) with nitrogen-free (red points) compounds and one nitrogen atom (blue points), two (green), three (yellow) and four (violet) nitrogen atoms; (**e**) with sulfur-free (red points) and sulfur-containing (green points) compounds in negative ion mode.

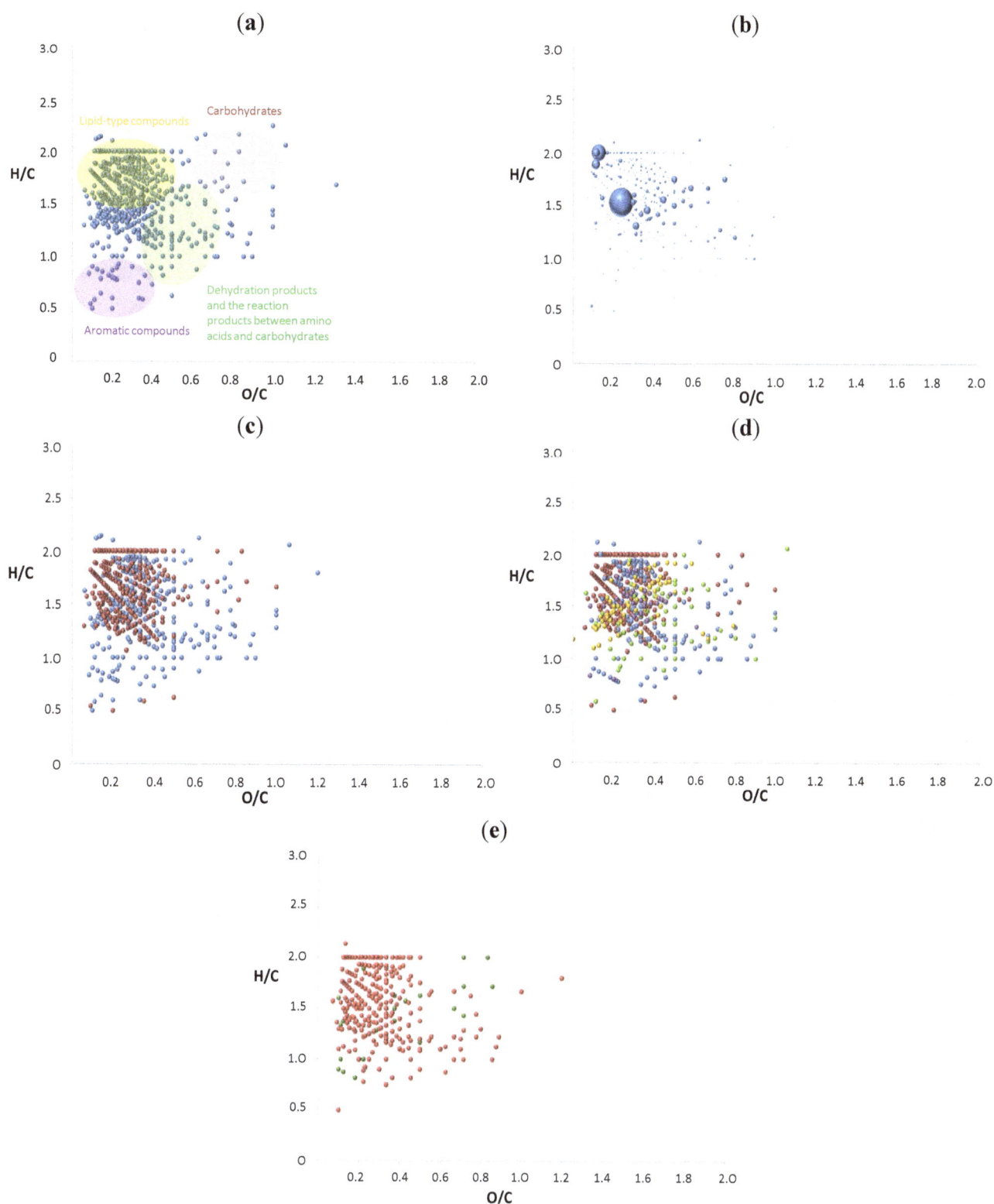

4. Conclusions

To conclude, we have characterized complex mixtures formed upon the heating of different amino acids with sucrose using powerful mass spectrometry techniques. We have shown that thermal treatment of reactive amino acids with carbohydrates results in several thousand reaction products. The different reactivity of free amino acids has been demonstrated. In particular, lysine and tyrosine in the presence of sucrose yielded the simple spectra, followed by arginine and aspartic acid, while cysteine and serine produced the highest number of compounds. We have identified the whole range of caramelization products, including oligomers of carbohydrates with up to six carbohydrate units, dehydration products of oligomers loosing up to a maximum of seven water molecules, hydration products of sugar oligomers, disproportionation products and aromatic compounds. Except caramelization reaction products, many reaction products between amino acids and carbohydrates have been found in the samples, including adducts of amino acids with carbohydrates up to four carbohydrate units. The formation of the reaction products between amino acids and sucrose without two water molecules was favored. The compounds with up to four nitrogen atoms incorporated in the structures have been formed and illustrated in the graphs. Several sulfur-containing MRPs have been identified in the reaction of sucrose with cysteine (e.g., the addition products of cysteine and hexose). Because of the high accuracy of FT ICR-MS, the formula assignment for nitrogen- and sulfur-containing molecules was successful. The van Krevelen diagrams generated form FT ICR-MS data turned out to be useful tools in complex mixture characterization. The work described herein provides a comprehensive overview on Maillard chemistry, without which our life would be tasteless.

Acknowledgments

The authors thank Jacobs University Bremen and Henkel AG & Co. KGaA for the financial support. We are grateful to Nadim Hourani and Matthias Witt for recording the FT ICR-MS spectra. Technical assistance by Anja Müller and graphical support by Borislav Milev are acknowledged.

Author Contributions

Agnieszka Golon carried out all experimental work and data interpretation jointly with Nikolai Kuhnert. All authors contributed to the writing of the manuscript.

Conflicts of Interest

The authors declare no conflict of interest.

References

1.   Kuhnert, N.; Dairpoosh, F.; Yassin, G.; Golon, A.; Jaiswal, R. What is under the hump? Mass spectrometry based analysis of complex mixtures in processed food—Lessons from the characterisation of black tea thearubigins, coffee melanoidines and caramel. *Food Funct.* **2013**, *4*, 1130–1147.

2. Fogliano, V.; Morales, F.J. Estimation of dietary intake of melanoidins from coffee and bread. *Food Funct.* **2011**, *2*, 117–123.

3. Delgado-Andrade, C.; Seiquer, I.; Haro, A.; Castellano, R.; Navarro, M.P. Development of the Maillard reaction in foods cooked by different techniques. Intake of Maillard-derived compounds. *Food Chem.* **2010**, *122*, 145–153.

4. Knerr, T.; Pischetsrieder, M.; Severin, T. 5-Hydroxy-2-methyl-4-(alkylamino)-2H-pyran-3(6H)-one: A New Sugar-Derived Aminoreductone. *J. Agric. Food Chem.* **1994**, *42*, 1657–1660.

5. Du, Q.-Q.; Liu, S.-Y.; Xu, R.-F.; Li, M.; Song, F.-R.; Liu, Z.-Q. Studies on structures and activities of initial Maillard reaction products by electrospray ionisation mass spectrometry combined with liquid chromatography in processing of red ginseng. *Food Chem.* **2012**, *135*, 832–838.

6. Wang, H.-Y.; Qian, H.; Yao, W.-R. Melanoidins produced by the Maillard reaction: Structure and biological activity. *Food Chem.* **2011**, *128*, 573–584.

7. Silván, J.M.; van de Lagemaat, J.; Olano, A.; del Castillo, M.D. Analysis and biological properties of amino acid derivates formed by Maillard reaction in foods. *J. Pharm. Biomed. Anal.* **2006**, *41*, 1543–1551.

8. Cerny, C.; Davidek, T. Formation of aroma compounds from ribose and cysteine during the Maillard reaction. *J. Agric. Food Chem.* **2003**, *51*, 2714–2721.

9. Borrelli, R.C.; Mennella, C.; Barba, F.; Russo, M.; Russo, G.L.; Krome, K.; Erbersdobler, H.F.; Faist, V.; Fogliano, V. Characterization of coloured compounds obtained by enzymatic extraction of bakery products. *Food Chem. Toxicol.* **2003**, *41*, 1367–1374.

10. Lindenmeier, M.; Hofmann, T. Influence of baking conditions and precursor supplementation on the amounts of the antioxidant pronyl-l-lysine in bakery products. *J. Agric. Food Chem.* **2003**, *52*, 350–354.

11. Lindenmeier, M.; Faist, V.; Hofmann, T. Structural and functional characterization of pronyl-lysine, a novel protein modification in bread crust melanoidins showing *in vitro* antioxidative and phase I/II enzyme modulating activity. *J. Agric. Food Chem.* **2002**, *50*, 6997–7006.

12. Hofmann, T. AcetylformoinA chemical switch in the formation of colored Maillard reaction products from hexoses and primary and secondary amino acids. *J. Agric. Food Chem.* **1998**, *46*, 3918–3928.

13. Cerny, C.; Briffod, M. Effect of pH on the Maillard reaction of [13C5]xylose, cysteine, and thiamin. *J. Agric. Food Chem.* **2007**, *55*, 1552–1556.

14. Mundt, S.; Wedzicha, B.L. Role of glucose in the Maillard browning of maltose and glycine: A radiochemical approach. *J. Agric. Food Chem.* **2005**, *53*, 6798–6803.

15. Hwang, I.G.; Kim, H.Y.; Lee, S.H.; Woo, K.S.; Ban, J.O.; Hong, J.T.; Yu, K.W.; Lee, J.; Jeong, H.S. Isolation and identification of an antiproliferative substance from fructose-tyrosine Maillard reaction products. *Food Chem.* **2012**, *130*, 547–551.

16. Kuhnert, N.; Drynan, J.W.; Obuchowicz, J.; Clifford, M.N.; Witt, M. Mass spectrometric characterization of black tea thearubigins leading to an oxidative cascade hypothesis for thearubigin formation. *Rapid Commun. Mass Spectrom.* **2010**, *24*, 3387–3404.

17. Kuhnert, N.; Clifford, M.N.; Muller, A. Oxidative cascade reactions yielding polyhydroxy-theaflavins and theacitrins in the formation of black tea thearubigins: Evidence by tandem LC-MS. *Food Funct.* **2010**, *1*, 180–199.

18. Jaiswal, R.; Matei, M.F.; Golon, A.; Witt, M.; Kuhnert, N. Understanding the fate of chlorogenic acids in coffee roasting using mass spectrometry based targeted and non-targeted analytical strategies. *Food Funct.* **2012**, *3*, 976–984.

19. Golon, A.; González, F.J.; Dávalos, J.Z.; Kuhnert, N. Investigating the thermal decomposition of starch and cellulose in model systems and toasted bread using domino tandem mass spectrometry. *J. Agric. Food Chem.* **2012**, *61*, 674–684.

20. Golon, A.; Kuhnert, N. Unraveling the chemical composition of caramel. *J. Agric. Food Chem.* **2012**, *60*, 3266–3274.

21. Golon, A.; Kuhnert, N. Characterisation of "caramel-type" thermal decomposition products of selected monosaccharides including fructose, mannose, galactose, arabinose and ribose by advanced electrospray ionization mass spectrometry methods. *Food Funct.* **2013**, *4*, 1040–1050.

22. Pais, P.; Knize, M.G. Chromatographic and related techniques for the determination of aromatic heterocyclic amines in foods. *J. Chromatogr. B Biomed. Sci. Appl.* **2000**, *747*, 139–169.

23. Wu, Z.; Rodgers, R.P.; Marshall, A.G. Two- and three-dimensional van Krevelen diagrams: A graphical analysis complementary to the Kendrick mass plot for sorting elemental compositions of complex organic mixtures based on ultrahigh-resolution broadband fourier transform ion cyclotron resonance mass measurements. *Anal. Chem.* **2004**, *76*, 2511–2516.

11

Effect of Pre-Harvest Sprouting on Physicochemical Properties of Starch in Wheat

Senay Simsek [1,*], Jae-Bom Ohm [2], Haiyan Lu [1], Mory Rugg [1], William Berzonsky [3], Mohammed S. Alamri [4] and Mohamed Mergoum [1]

[1] Department of Plant Sciences, North Dakota State University, P.O. Box 6050, Department #7670, Fargo, ND 58108-6050, USA; E-Mails: haiyan.lu22@outlook.com (H.L.); mory.O.P.Rugg@gmail.com (M.R.), mohamed.mergoum@ndsu.edu (M.M.)

[2] USDA-ARS Hard Red Spring and Durum Wheat Quality Laboratory, Harris Hall, North Dakota State University, Fargo, ND 58108, USA; E-Mail: jae.ohm@ars.usda.gov

[3] Department of Plant Sciences, South Dakota State University, Brookings, SD 57007-2141, USA; E-Mail: bill.berzonsky@bayer.com

[4] Nutrition and Food Sciences Department, College of Food and Agricultural Sciences; King Saud University, P.O. Box 2460, Riyadh 11451, Saudi Arabia; E-Mail: msalamri@ksu.edu.sa

* Author to whom correspondence should be addressed; E-Mail: senay.simsek@ndsu.edu

Abstract: Pre-harvest sprouting (PHS) in wheat (*Triticum aestivum* L.) occurs when physiologically mature kernels begin germinating in the spike. The objective of this study was to provide fundamental information on physicochemical changes of starch due to PHS in Hard Red Spring (HRS) and Hard White Spring (HWS) wheat. The mean values of α-amylase activity of non-sprouted and sprouted wheat samples were 0.12 CU/g and 2.00 CU/g, respectively. Sprouted samples exhibited very low peak and final viscosities compared to non-sprouted wheat samples. Scanning electron microscopy (SEM) images showed that starch granules in sprouted samples were partially hydrolyzed. Based on High Performance Size Exclusion Chromatography (HPSEC) profiles, the starch from sprouted samples had relatively lower molecular weight than that of non-sprouted samples. Overall, high α-amylase activity caused changes to the physicochemical properties of the PHS damaged wheat.

Keywords: pre-harvest sprout; wheat; starch; HPSEC

---

1. Introduction

Pre-harvest sprouting (PHS) occurs as a result of the germination of kernels within the wheat spike before harvest, and it can often occur when harvest coincides with relatively high humidity due to untimely rain [1]. PHS generally occurs after maturation of the wheat kernel, however germination may occur as early as 18 days after anthesis [2]. Symptoms of PHS include premature kernel swelling, germ discoloration, seed-coat splitting, and root and shoot emergence [1]. The severity of PHS can also be classified on a continuum from very minor to very severe, which differs from year-to-year, depending on the weather [3,4]. PHS can be measured by calculating the percentage of sprouted wheat kernels and by measuring starch degradation, which is commonly quantified by the falling number test as an indirect measurement of α-amylase activity in grain [4]. PHS may result in monetary losses to growers, millers and bakers [3]. The price of sprouted wheat could be reduced by 20%–50%, and the grain is classified as unacceptable for human food if it contains more than 4% damaged (sprouted) kernels [5]. Severely sprouted grain is often used only as animal feed. Millers experience losses because of an associated reduction in flour yield and quality as well as functional quality. Bakers can experience problems during the bread-making process and suffer losses because the end-product is of poor quality. Products made from flour containing sprouted wheat are porous, sticky, and off-color and generally of poor bake quality [5].

In this study, we investigated physicochemical properties of starch from non-sprouted and sprouted hard red spring (HRS) and hard white spring (HWS) wheat samples. Though, there are many complex factors, such as amylases, amylase genes, abscisic acid (ABA), ABA genes and other genes (VP1) [6,7], dormancy and susceptibility to PHS are also related to seed coat color. Most of the HRS wheat genotypes released from breeding programs display a high level of tolerance to PHS [4,8]. Although many studies have shown that hard white wheat genotypes are more susceptible to PHS, there is considerable variability in the level of tolerance to PHS in wheat [4,8]. Pleiotropic effect of *R* (*Red grain colour*) genes conferring red pericarp color and other dormancy genes (has a major effect on the embryo) affect the wheat grain dormancy [8,9]. Thus, PHS in wheat is controlled by coat-imposed and embryonic pathways regulated by separate genetic systems [6,9,10]. Red wheat tends to have higher levels of PHS resistance than white wheat; although, some studies have shown that sources of resistance to PHS can be acquired in white wheat germ plasm [4,8,11].

A new plant requires energy and nutrients, which is the reason why a sprouting wheat kernel produces amylases, lipases and proteases. These enzymes break down starch, oil and protein to feed the developing embryo [12,13]. The impact of PHS on the final food product depends on the amount of enzymes present and their break down of kernel starches, oils, and proteins [5]. Partial hydrolysis of starch leads to a decrease in the size of the starch molecules and a reduction in the water-holding capacity of the dough. Few studies have documented the physicochemical changes to wheat starch due to PHS. However, the altered pasting, gelation, and retrogradation properties of starch [14], can significantly affect the quality of products such as Asian noodles, sponge cake, and breads.

Starch is the main storage polysaccharide and energy source for plants. Wheat flour is composed of 70%–80% dry matter of starch. Compared to other carbohydrates, starch possesses unique physical and chemical properties [15]. In wheat, there are two main types of starch granules: large, lenticular (A type) and small, spherical (B type). Starch granules are comprised of two constituent polymers: amylose a basically linear polysaccharide (α-1→4 linked glucose), and a highly branched polysaccharide termed amylopectin (α-1→4 linked glucose and α-1→6 linked glucose) [15]. Amylopectin is the major component of starch and normal wheat starch is comprised of about 75% amylopectin. The distribution of starch granule types and structure of amylose and amylopectin and their relative ratios in starch granules play an important part in determining pasting, gelation, and retrogradation properties of starch and end-product quality and stability [16].

2. Materials and Methods

2.1. Materials

Wheat samples were kindly provided by the Spring Wheat Breeding Programs in the Plant Sciences Department at North Dakota State University 24 genotypes, including 12 hard red spring (HRS) and 12 hard white spring (HWS) wheat genotypes grown at three locations (Table 1). "Hanna" and "AC Snowbird" have previously demonstrated a high level of seed dormancy [4]. Therefore, Hanna was considered a HRS and AC Snowbird a HWS control for tolerance to PHS. "Ingot" and "Lolo", previously exhibiting susceptibility to PHS, were considered PHS susceptible HRS wheat and HWS wheat types, respectively. The lines selected as check samples were chosen because they are well characterized and their response to PHS has been determined and previously documented in other research [4].

Experimental plots were grown at Prosper, Carrington, and Casselton, ND in 2008. In each location, the experiment was laid out in a randomized block design with four replicates following the procedure of Rugg [4]. Environmental conditions with respect to temperature and rain in the 2008 growing season were as follows: Each location received lower than average precipitation through the entire growing season. Temperatures were below normal at both environments in the early part of the growing season. Prosper experienced higher precipitation and higher temperatures compared to Carrington, which resulted in higher disease pressure in Prosper than in Carrington [4].

In this study, sprouted and non-sprouted wheat samples were analyzed in two replicates. Two samples from 4 field replicates were combined to produce the 2 replicates used for analysis. Thus, physicochemical characterization of starch from a total of 288 samples (24 entries × 2 replicates × 3 locations × 2—sprouted and non-sprouted) was analyzed in the present research.

2.2. Sample Sprouting and Sprout Score

Wheat samples were evaluated and scored for tolerance to PHS by the Department of Plant Science, North Dakota State University. At plant physiological maturity, 30 wheat spikes were randomly harvested from each experiment unit. There was no PHS damage evident at the time of harvest. To inhibit additional α-amylase activity the spikes were immediately stored at 10 °C. The spikes were placed in a mist chamber and misted for a period of 48 h. Following the misting, a humidifier was placed in the chamber for 3 days. Visual observations of the spikes were made to assess the degree of sprouting induced by maintaining high moisture in the misting chamber. Spikes were scored visually 0–9; whereby 0 represented no visible sprouting and a score of 9 represented very severe sprouting with average coleoptiles lengths greater than 2 cm [4].

Table 1. Sprouting score and α-amylase activity of sprouted wheat, non-sprouted wheat and their difference (ΔD).

Genotype	Sprout Score	α-Amylase Activity (CU/g)		
		Sprouted	Non-Sprouted	ΔD
RSW	H	-	-	-
Hanna	2.8	1.32	0.11	1.20
Ingot	7.0	2.37	0.10	2.27
Alsen	4.8	1.82	0.09	1.57
Briggs	5.7	2.16	0.11	2.06
Freyr	4.4	1.79	0.09	1.70
Glenn	4.0	1.68	0.08	1.61
Granite	5.3	2.09	0.13	1.96
Kelby	3.4	1.56	0.13	1.43
Norpro	6.0	2.18	0.09	2.09
Reeder	4.4	1.76	0.08	1.68
Steele-ND	5.0	1.93	0.08	1.85
Knudson	5.4	2.12	0.10	2.02
Mean	4.8	1.90	0.10	1.79
HWSW				
99S0155-14W	2.5	1.36	0.12	1.24
Otis	7.8	2.47	0.11	2.37
AC Snowbird	2.8	1.39	0.08	1.31
AC Vista	5.8	2.13	0.09	2.04
Argent	4.8	1.98	0.16	1.83
CS3100L	6.8	2.33	0.18	2.16
CS3100Q	6.8	2.44	0.14	2.3
Explorer	6.9	2.37	0.16	2.21
Lolo	5.7	2.17	0.12	2.05
MT9420	6.9	2.33	0.14	2.19
NDSW0602	6.3	2.37	0.12	2.24
Pristine	5.0	1.99	0.18	1.81
Mean	5.7	2.11	0.13	1.98
LSD	1.4	0.39	0.05	0.40

ΔD: Difference between non-sprouted and sprouted wheat; LSD: least significant difference (α = 0.05), used to detect difference between genotypes.

2.3. α-Amylase Activity

Wheat samples were dried and ground in a cyclone sample mill (Udy, Fort Collins, CO, USA) equipped with a 1 mm sieve. Samples of ground wheat (0.5 g) were weighed into test tubes containing stir bars. The test tubes were placed into a stirring heat block at 60 °C and stirred at medium high speed.

Sodium maleate buffer (5 mL, 100 mM, pH 6.0) was heated to 60 °C and added to each tube, stirred for 5 min and then an amylazyme tablet (Megazyme, International, Ireland) was added. The reaction was stopped by adding 6 mL Trizma base (2% w/v, pH 9.5) after 5 min. Subsequently, the sample was left at room temperature for 5 min, then stirred and filtered. The absorbance of the filtrate at 590 nm was measured against the reaction blank and α-amylase activity was calculated by reference to a standard curve.

2.4. Pasting Properties

Pasting properties of samples were evaluated by using a RVA (Newport Scientific, Narrabeen, Australia) according to AACC approved method 76-21.01 [17]. Samples of ground wheat (3.5 g, 14% moisture basis) were added to pre-weighed, de-ionized distilled water samples in a RVA canister. Parameters of Peak Viscosity (PV), breakdown (BD), Hot Paste Viscosity (HPV), setback (SB) and Final Viscosity (FV) were recorded.

2.5. Scanning Electron Microscopy (SEM) Analysis of Starch Granules

Two PHS samples with the highest sprout scores were chosen from both HRS and HWS genotypes. The other two were from non-sprouted samples. Wheat kernels were cracked open longitudinally through the crease and fixed to microscope stubs with Dotite silver paint. The samples were coated with gold using a Hummer II sputter coater (Technics/Anatech Ltd., Alexandria, VA, USA) [18]. Images were obtained using a JEOL JSM-6490LV Scanning Electron Microscope (SEM) (JEOL, Peabody, MA, USA).

2.6. High Performance Size Exclusion Chromatography (HP-SEC) Analysis of Starch

Eight samples of ground wheat were chosen to conduct the HPSEC analysis. Among the 8 samples, 4 samples with each of the HWS and HRS (samples were chosen to give a representation of HWS and HRS with high and low sprout scores) were included. Starch samples were analyzed as described by Simsek et al. [19].

2.7. Statistical Analysis

Statistical analyses were performed using the SAS System for Windows (V. 9.2, SAS Institute, Cary, NC, USA). Bartlett's test was used to analyze the homogeneity of error variance across the three locations. When error variances were homogenous, analysis of variance (ANOVA) was performed using the "Mixed" procedure in SAS, assuming location as a random effect and genotype as a fixed effect. The difference between the HRS and HWS mean value was analyzed using the "Contrast" option. Bartlett's test indicated that error variance of pasting properties for sprouted samples were heterogeneous across the three locations. Thus, each location was analyzed separately for pasting properties. Correlation coefficients were calculated across genotype means using the "Corr" procedure in SAS.

3. Results and Discussion

3.1. α-Amylase Activity in Non-Sprouted and Sprouted HRS and HWS Wheat Samples

In non-sprouted kernels, α-amylase activity is mostly localized to the seed coat, aleurone layer, and scutellum [20]. Based on Bartlett's test, variances across the samples from the three locations were homogeneous for the properties being measured. Thus, the three locations were combined for analyses.

Alpha-Amylase activity of the wheat samples are given in Table 1. The mean value of α-amylase for sprouted samples was 2.00 CU/g; while this value for non-sprouted was 0.12 CU/g, and all of the sprouted samples had higher α-amylase than non-sprouted. These results were in agreement with the findings of Ichinose *et al.* [13] who reported that the α-amylase activity of wheat increases rapidly as germination progresses. A previous report also indicated that the starch in sprouted wheat samples degraded rapidly as the α-amylase activity increased during PHS [13].

Genotypes tested within sprouted samples had sprout scores ranging from 2.5 to 7.8 (Table 1). In this study, the sprout tolerant and susceptible HRS check samples corresponded to the samples with sprout scores and α-amylase activities at the extremes of the evaluation range. Among sprouted samples, Hanna had the lowest sprout score (2.8), while Ingot had the highest sprout score (7.0). The α-amylase activities of sprouted samples of Hanna and Ingot were 1.32 CU/g and 2.37 CU/g, respectively. The check samples for the sprout tolerant (AC Snowbird) and susceptible (Lolo) HWS genotypes were at the low and high end of the sprout score and α-amylase activity range, respectively, but they did not exhibit the highest or lowest values. Breeding line 99S0155-14W had the lowest sprout score at 2.5, while Otis had the highest sprout score (7.8) among sprout damaged HWS samples. Among sprouted samples, the α-amylase activity of 99S0155-14W and Otis were 1.36 CU/g and 2.47 CU/g, respectively. These results agreed with the findings of Huang [21], who reported that genotypes with lower susceptibility to PHS also had lower α-amylase activity.

Wheat genotypes have high α-amylase activity without visible sprouting. The relationship between sprout score and α-amylase activity of each genotype was determined (Table 1). There was a significant difference in α-amylase activity and ΔD among genotypes within sprouted samples ($\alpha = 0.05$). This indicated that varietal differences were highly significant for α-amylase activity. α-amylase activity may be a better indicator—than sprout score—as to which genotypes have low tolerance to PHS. The differences in α-amylase activity between the three locations were not as great as those between genotypes. Interactions between genotypes and location were significant, suggesting that environments affect PHS and enzyme activity, but not as much as the genotype.

3.2. Pasting Properties of Non-sprouted and Sprouted HRS and HWS Wheat

The pasting parameters for the check samples and the averages of all varieties in each wheat class for each location of sprouted and non-sprouted wheat samples and their ΔD are shown in Table 2. All RVA properties changed dramatically due to PHS, which suggests PHS also significantly impacted starch pasting properties. All sprouted samples had much lower peak viscosity, hot paste viscosity, and final viscosity than non-sprouted samples. Peak viscosity of non-sprouted samples ranged from 138.9 to 289 RVU (data not shown). Peak viscosity of sprouted wheat ranged from 5 to 18 RVU. These results suggest that the water binding capacity of starch and starch paste stability decreased due to PHS. The pasting profile of starch determined by RVA has a direct relationship to the microstructure of the starch. Amylose, which is important to high gel consistency upon cooling, may contribute to determining the initial rigidity of swollen starch granules during germination [22]. The branch chain length of amylopectin has an effect on the gelatinization, retrogradation, and pasting properties of starch [16].

Starch granules can lose their resistance to swelling due to higher activity of α-amylase [12], and the reduced resistance to swelling may in turn have lowered the paste viscosity of the sprouted samples. Genotypes with higher tolerance to PHS had higher water binding capacity and higher starch paste stability.

Table 3 shows the correlations per location between pasting characteristics, sprout score, and α-amylase activity among the 24 genotypes. The correlations between sprout score and peak viscosity were negative and highly significant for sprouted samples from Casselton and negative and very highly significant for sprouted samples from Carrington and Prosper. The correlations for sprouted samples between sprout score and peak viscosity were −0.64, −0.56 and −0.76 for Carrington, Casselton and Prosper, respectively. Furthermore, correlations between sprout score and the ΔD of hot paste viscosity was significant for samples grown in Carrington (−0.48) and Casselton (−0.43), which is a consequence of genotypes with high sprout scores also exhibiting low peak viscosity. Significant and negative correlations were obtained between the α-amylase of non-sprouted samples from Carrington. For non-sprouted samples from Casselton and Prosper, correlations between α-amylase activity and peak viscosity and α-amylase and hot paste viscosity were negative and very highly significant. The correlation between α-amylase activity and final viscosity for non-sprouted samples from Casselton and Prosper was negative and highly significant. There were significant negative correlations between α-amylase activity of sprouted samples and peak viscosity of sprouted samples from Carrington (−0.61), Casselton (−0.54) and Prosper (−0.69). Thus, the pasting viscosity of sprouted wheat had also been decreased as a result of PHS.

Table 2. Mean value of pasting profile of check varieties; Hanna, Ingot, Lolo and AC Snowbird; and means of wheat classes from Carrington, Casselton and Prosper for sprouted wheat, non-sprouted wheat and their difference (ΔD).

Wheat Class	Genotype	Peak Viscosity			Hot Paste Viscosity			Final Viscosity		
		Sprouted	Non-Sprouted	ΔD	Sprouted	Non-Sprouted	ΔD	Sprouted	Non-Sprouted	ΔD
						Carrington				
HRSW	Hanna	11.2	222.7	212.8	2.3	134.8	133.2	3.5	258.5	255.4
	Ingot	3.5	233.4	230.4	2.3	139.9	137.6	2.9	256.7	253.8
	Mean	4.4	224.4	221.4	1.4	126.5	125.6	1.7	244.5	243.1
HWSW	AC Snowbird	11.4	252.4	239.3	1.9	125.2	124.2	3.0	237.8	235.5
	Lolo	2.2	210.4	205.4	0.6	93.0	90.8	0.6	202.9	200.3
	Mean	3.1	209.3	207.1	2.0	99.2	97.8	2.4	215.7	213.9
LSD		1.8	19.3	19.5	1.3	18.1	18.6	1.3	28.2	28.6
						Casselton				
HRSW	Hanna	20.5	197.9	177.4	3.1	128.6	125.4	4.8	237.5	232.4
	Ingot	10.6	210.1	198.6	2.5	133.6	130.6	3.1	233.2	229.6
	Mean	12.4	204.5	192.3	2.3	122.0	119.5	3.1	223.1	219.8
HWSW	AC Snowbird	14.8	242.0	228.6	1.9	129.8	128.3	3.1	236.1	233.8
	Lolo	10.6	213.9	200.0	2.8	112.0	108.4	3.5	216.2	212.0
	Mean	10.6	188.9	177.5	2.6	94.0	91.3	3.3	193.7	190.1
LSD		5.3	25.5	25.4	2.1	20.4	19.9	2.8	37.9	37.4
						Prosper				
HRSW	Hanna	14.7	200.2	183.2	1.8	129.1	125.6	3.0	240.8	236.2
	Ingot	6.8	236.2	227.6	2.0	153.7	150.4	2.6	274.4	270.6
	Mean	8.4	201.7	192.9	2.0	119.9	117.8	2.4	228.2	225.9
HWSW	AC Snowbird	17.9	289.0	268.2	1.9	163.7	160.9	3.3	294.6	290.5
	Lolo	8.6	203.5	193.8	3.5	103.5	100.8	3.7	209.6	206.7
	Mean	8.1	189.8	180.8	2.3	95.1	92.4	2.8	203.3	200.3
LSD		2.3	24.1	24.3	1.2	19.5	19.3	1.3	30.9	30.8

Only the two check varieties are shown with the means for each wheat type at each location and the LSD for each location; The unit is expressed by Rapid Viscosity Unit (RVU); ΔD: Difference between non-sprouted and sprouted wheat; LSD: least significant difference (α = 0.05).

Table 3. Correlation coefficients [a] between α-amylase activity and pasting characteristics among 24 genotypes from Carrington, Casselton and Prosper.

Pasting Characteristics		Sprout Score	Sound α-Amylase	PHS α-Amylase	ΔD α-Amylase
Carrington					
Non-Sprouted	Peak Viscosity	NS	−0.45 *	NS	NS
	Hot Paste Viscosity	−0.47 *	−0.50 *	−0.51 *	−0.45 *
	Final Viscosity	NS	NS	−0.43 *	NS
Sprouted	Peak Viscosity	−0.64 ***	NS	−0.61 **	−0.60 **
ΔD	Peak Viscosity	NS	−0.45 *	NS	NS
	Hot Paste Viscosity	−0.48 *	−0.50 *	−0.51 *	−0.45 *
	Final Viscosity	NS	NS	−0.43 *	NS
Casselton					
Non-Sprouted	Peak Viscosity	−0.41 *	−0.67 ***	NS	NS
	Hot Paste Viscosity	−0.44 *	−0.68 ***	NS	NS
	Final Viscosity	NS	−0.54 **	NS	NS
Sprouted	Peak Viscosity	−0.56 **	NS	−0.54 **	−0.54 **
ΔD	Peak Viscosity	NS	−0.68 ***	NS	NS
	Hot Paste Viscosity	−0.43 *	−0.69 ***	NS	NS
	Final Viscosity	NS	−0.55 **	NS	NS
Prosper					
Non-Sprouted	Peak Viscosity	NS	−0.64 ***	NS	NS
	Hot Paste Viscosity	NS	−0.68 ***	NS	NS
	Final Viscosity	NS	−0.57 **	NS	NS
Sprouted	Peak Viscosity	−0.76 ***	NS	−0.69 ***	−0.68 ***
ΔD	Peak Viscosity	NS	−0.66 ***	NS	NS
	Hot Paste Viscosity	NS	−0.69 ***	NS	NS
	Final Viscosity	NS	−0.57 **	NS	NS

[a] Correlation coefficient is significant at * $p < 0.05$, ** $p < 0.01$ and *** $p < 0.001$, respectively; NS, not significant; ΔD: difference between non-sprouted and sprouted wheat; variables with no significant correlation to any other variable were omitted from the table.

3.3. Starch Granule Morphology of Non-sprouted and Sprouted HRS and HWS Wheat

Starch is deposited in the endosperm of the wheat kernel and is comprised of discrete granules [23]. The endosperm of wheat kernels contains two types of granules—a larger type, mostly about 20–35 mm in diameter (A-starch) and lenticular in shape; and a smaller spherical type, ranging from 2 to 8 mm in diameter (B-starch) [23]. Scanning electron microscopy (SEM) can be used to determine the distribution of the A and B type granules as well as their degradation due to high α-amylase activity. In the endosperm of non-sprouted wheat kernels, starch granules are usually embedded in a dense protein matrix. Artificial sprouting of barley and corn for seven days demonstrated that extensive damage to starch granules can take place due to an increase in α-amylase activity [24]. Dronzec *et al.* [25] and Bean *et al.* [14] have shown extensive damage to wheat starch granules due to PHS. Two genotypes, Steele-ND and Pristine, were chosen to conduct SEM analyses because of their high sprout scores (Table 1) under conditions of PHS. The SEM images of the sprouted and non-sprouted samples from the two genotypes are shown in Figure 1. The non-sprouted samples (Figure 1A,C) exhibited intact starch granules embedded in a very dense protein matrix. Conversely, starch granules had been degraded, and the protein matrix was absent in the sprouted samples (Figure 1B,D). There was also pitting observed on the starch granules of the sprouted samples, most likely due to an increase in α-amylase activity. Similarly, Huang *et al.* [21] reported that the protein matrix was missing or compromised in sprouted wheat samples. The same study demonstrated that proteolytic enzymes broke down the protein matrix, thereby producing a loose structure around the starch granules, which made them more accessible to α-amylase.

Figure 1. Scanning Electron Microscopy (SEM) images of starch from genotypes Steel-ND and Pristine. (**A**) Image from hard red spring (HRS) wheat genotype Steel-ND non-sprouted sample; (**B**) image from HRS wheat genotype Steel-ND sprouted sample; (**C**) image from hard white spring (HWS) wheat genotype Pristine non-sprouted sample; and (**D**) image from HWS wheat genotype Pristine sprouted sample. Starch granules are circled and protein matrix is identified by arrows.

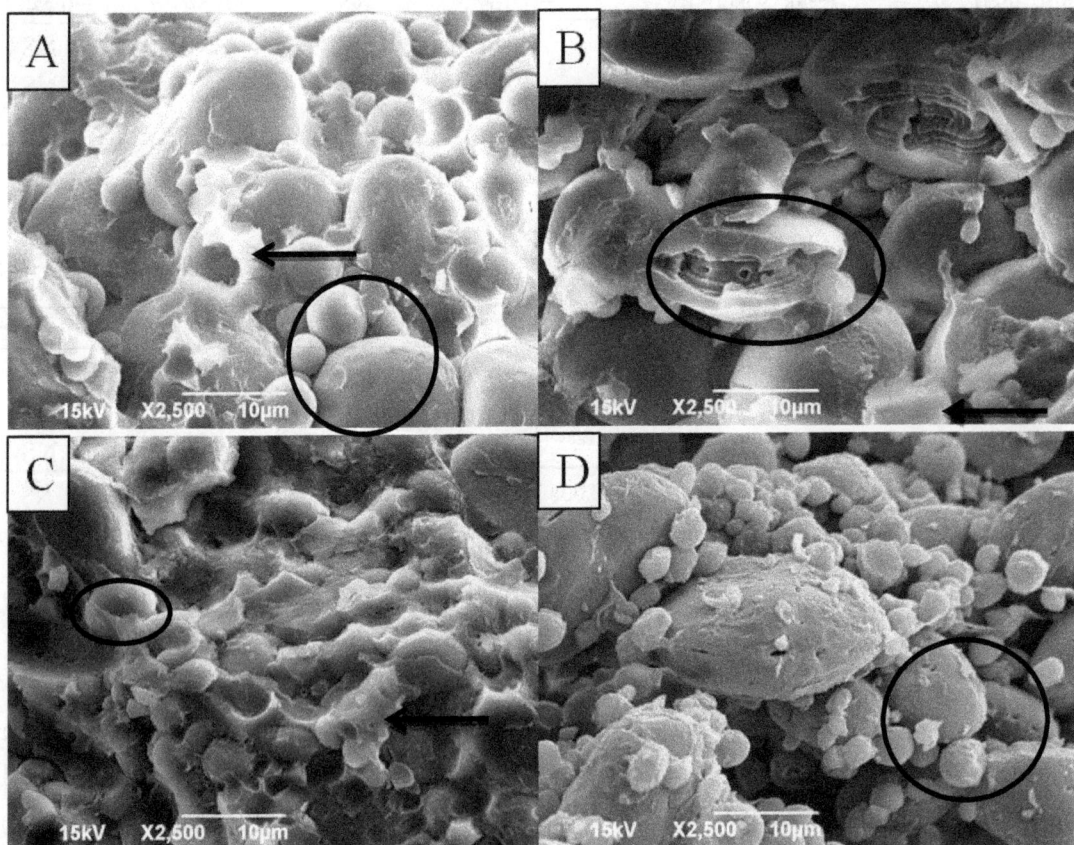

3.4. HPSEC of Starch in Non-sprouted and Sprouted HRS and HWS Wheat

There were three peaks detected in the HPSEC chromatogram of starch in sprouted and non-sprouted samples (Figure 2), a high molecular weight amylopectin (HMW-AP) peak, a low molecular weight amylopectin (LMW-AP) peak, and an amylose (AM) peak [19]. A comparison of the HPSEC chromatography profiles of sprouted and non-sprouted samples of Hanna, Ingot, 99S0155-14W and Otis, showed that sprouted samples had lower HMW-AP than non-sprouted wheat samples. There was a shift from HMW-AP to LMW-AP and AM, and an overall shift from amylopectin to AM starch in the endosperm of sprouted samples. Thus, the apparent AM content seemed to increase in sprouted samples (Figure 2).

Figure 2. High Performance Size Exclusion Chromatograph (HPSEC) profiles of sound and PHS damaged wheat samples of genotypes Hanna and Ingot (hard red spring wheat genotypes), Otis and 99S0155-14W (hard white spring wheat genotypes). Dashed line represents the HPSEC chromatography of pre-harvest sprouting (PHS) damaged sample; solid line represents the HPSEC chromatography of sound sample.

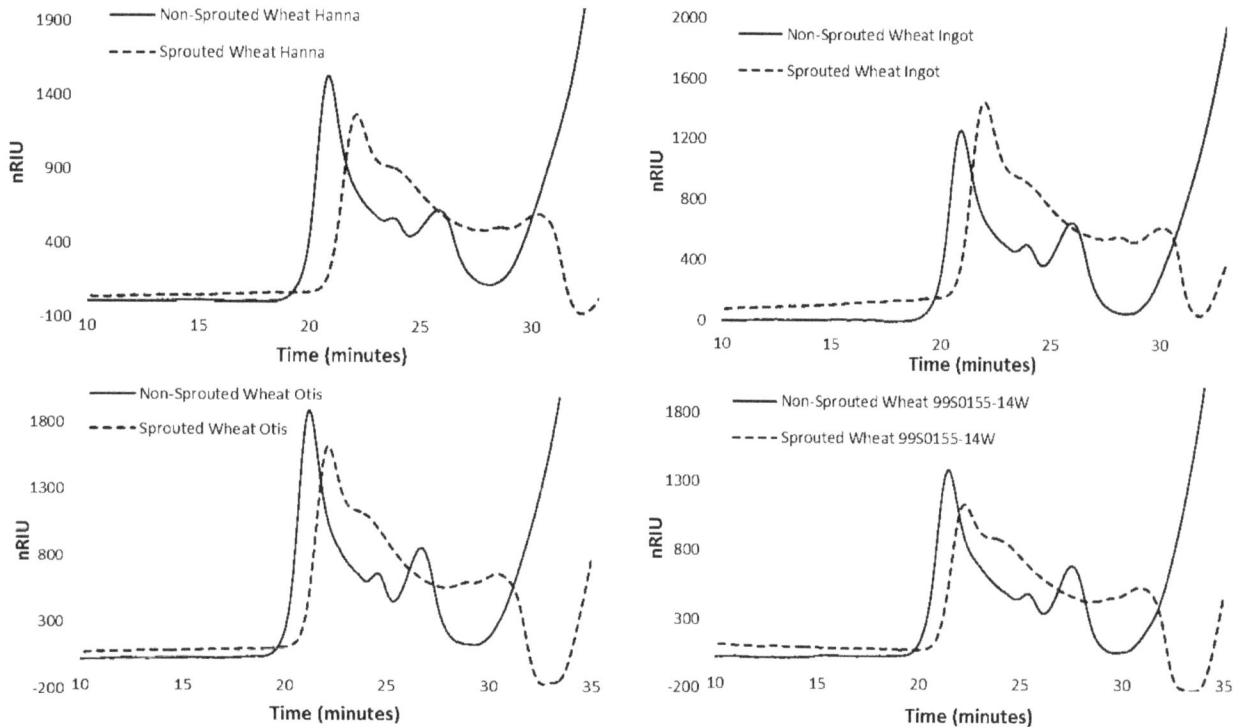

The percentages of each starch fraction and apparent average molecular weights determined by HPSEC are given in Table 4. The large starch molecules were degraded to smaller molecules due to the higher α-amylase activity in sprouted samples. Furthermore, more starch degradation occurred in genotypes that received higher PHS scores. Among sprouted samples, there was a significant decrease in HMW-AP for Hanna (61.5%–31.6%), Ingot (60.6%–34%), Otis (62.8%–32.9%) and 99S0155-14W (64.5%–28.9%), and this hydrolysis of amylopectin resulted in an apparent increase in the percentages of LMW-AP and amylose in the sprouted samples (Table 4). Based on these changes, we presume that starch had been hydrolyzed during PHS due to an increase in α-amylase activity, causing some portion of HMW-AP starch to be converted into LMW-AP and amylose. The structure of amylose and amylopectin and their relative ratios in starch granules play an important part in determining pasting, gelation, and retrogradation properties of starch and end-product quality and stability [16]. Degradation of the amylopectin changes the functional properties of the starch by altering the molecular weight and chain length distribution. The overall decrease in apparent average molecular mass of the starch and the increase of LMW-AP and amylose contents likely caused the reduction in paste viscosity [14] seen in Table 2.

Table 4. Percent of starch fractions and molecular weight distribution (MWD) of HMW-AP, LMW-AP and AM from genotypes Hanna and Ingot (HRSW), Otis and 99S0155-14W (HWSW).

Treatment	Genotype	HMW-AP		LMW-AP		AM	
		(%)	MWT	(%)	MWT	(%)	MWT
Non-Sprouted	Hanna	61.5	1.68×10^7	11.8	4.61×10^6	26.8	1.82×10^6
	Ingot	60.6	1.57×10^7	12.8	4.08×10^6	26.6	1.60×10^6
	Otis	62.8	1.41×10^7	11.1	2.99×10^6	26.1	1.15×10^6
	99S0155-14W	64.5	1.26×10^7	11.0	2.10×10^6	24.6	0.80×10^6
	LSD	1.2	1.03×10^6	1.1	6.25×10^5	0.1	1.71×10^5
Sprouted	Hanna	31.6	0.95×10^7	40.8	5.21×10^6	27.5	0.24×10^6
	Ingot	34.0	1.01×10^7	38.2	5.46×10^6	27.8	0.27×10^6
	Otis	32.9	0.95×10^7	37.5	4.90×10^6	29.7	0.24×10^6
	99S0155-14W	28.9	0.92×10^7	42.2	4.88×10^6	28.9	0.19×10^6
	LSD	0.8	7.60×10^4	1.7	1.89×10^5	1.1	7.71×10^3
ΔD	Hanna	29.9	0.73×10^7	29.0	0.60×10^6	0.7	1.58×10^6
	Ingot	26.6	0.56×10^7	25.4	1.38×10^6	1.2	1.33×10^6
	Otis	29.9	0.46×10^7	26.4	2.09×10^6	3.6	0.91×10^6
	99S0155-14W	35.6	0.34×10^7	31.2	2.78×10^6	4.3	0.61×10^6
	LSD	1.5	1.01×10^6	2.0	4.80×10^5	1.1	1.66×10^5

HMW-AP: High Molecular Weight Amylopectin; LMW-AP: Low Molecular Weight Amylopectin; AM: Amylose; MWT: Molecular weight, measured as apparent average molecular weight by HPSEC; LSD: least significant difference (α = 0.05); ΔD: difference between non-sprouted and sprouted wheat.

4. Conclusions

Overall, PHS damage resulted in significant changes in physicochemical properties of the starch. Genotype and wheat seed coat color had significant effects on PHS score, α-amylase activity starch pasting and degradation of starch molecules. Mean α-amylase activities for HRS genotypes were lower than that for HWS genotypes, suggesting that in general, the HRS genotypes were less susceptible to PHS and starch degradation than the HWS genotypes. However, some HWS genotypes, including the AC Snowbird and 99S0155-14W exhibited a level of tolerance to PHS similar to HRS genotypes.

Acknowledgments

We would like to thank Kristin Whitney for technical support and valuable input. This research (in part) was supported by ND-SBARE: Wheat Fund #FARGO90148, North Dakota Wheat Commission and North Dakota State University Agricultural Experiment Station.

Author Contributions

The authors Mory Rugg, William Berzonsky, Mohammed S. Alamri and Mohamed Mergoum produced the wheat used in this study, as well as completing the artificial sprouting of the sprouted wheat samples and evaluation of sprout score for the sprouted wheat samples. Jae-Bom Ohm assisted with experimental procedures, statistical analysis and interpretation of data. Senay Simsek and Haiyan

Lu had a major part in the experimental procedures and conducted the analysis (excluding sprout score) of the samples. Senay Simsek and Haiyan Lu also contributed to interpretation of data. All authors contributed to revision and writing of the manuscript.

Conflicts of Interest

The authors declare no conflict of interest.

References

1. Groos, C.; Gay, G.; Perretant, M.R.; Gervais, L.; Bernard, M.; Dedryver, F.; Charmet, G. Study of the relationship between pre-harvest sprouting and grain color by quantitative trait loci analysis in a white × red grain bread-wheat cross. *Theor. Appl. Genet.* **2002**, *104*, 39–47.
2. Mares, D.J. Temperature dependence of germinability of wheat (*Triticum aestivum* L.) grain in relation to pre-harvest sprouting. *Crop Pasture Sci.* **1984**, *35*, 115–128.
3. Wahl, T.I.; O'Rourke, A.D. The Economics of Sprout Damage in Wheat. In *Pre-harvest Sprouting in Cereals*; Walker-Simmons, M.K., Ried, J.L., Eds.; AACC International Press: St. Paul, MN, USA, 1992; pp. 10–17.
4. Rugg, M.O.P. Evaluating Hard Red and White Spring Wheat (*Triticum Aestivum* L.) Genotypes for Tolerance to Pre-harvest Sprouting. Master's Thesis, North Dakota State University, Fargo, ND, USA, 2011.
5. Sorenson, B.; Wiersma, J. Sprout damaged wheat, crop insurance and quality concerns. *Minn. Crop News Arch.* **2004**; Available online: http://web.archive.org/web/20130205033819/ http://www. extension.umn.edu/cropEnews/2004/04MNCN31.htm (accessed on 2 April 2014).
6. Flintham, J.E. Different genetic components control coat-imposed and embryo-imposeddormancy in wheat. *Seed Sci. Res.* **2000**, *10*, 43–50.
7. Flintham, J.; Adlam, R.; Bassoi, M.; Holdsworth, M.; Gale, M. Mapping genes for resistance to sprouting damage in wheat. *Euphytica* **2002**, *126*, 39–45.
8. Mares, D.; Mrva, K.; Cheong, J.; Williams, K.; Watson, B.; Storlie, E.; Sutherland, M.; Zou, Y. A QTL located on chromosome 4A associated with dormancy in white- and red-grained wheats of diverse origin. *Theor. Appl. Genet.* **2005**, *111*, 1357–1364.
9. Mares, D.; Mrva, K.; Tan, M.K.; Sharp, P. Dormancy in white-grained wheat: Progress towards identification of genes and molecular markers. *Euphytica* **2002**, *126*, 47–53.
10. Himi, E.; Mares, D.J.; Yanagisawa, A.; Noda, K. Effect of grain colour gene (*R*) on grain dormancy and sensitivity of the embryo to abscisic acid (ABA) in wheat. *J. Exp. Bot.* **2002**, *53*, 1569–1574.
11. Anderson, J.A.; Sorrells, M.E.; Tanksley, S.D. RFLP analysis of genomic regions associated with resistance to preharvest sprouting in wheat. *Crop Sci.* **1993**, *33*, 453–459.
12. Morad, M.M.; Rubenthaler, G.L. Germination of soft white wheat and its effect on flour fractions, breadmaking, and crumb firmness. *Cereal Chem.* **1983**, *60*, 413–417.
13. Ichinose, Y.; Takata, K.; Kuwabara, T.; Iriki, N.; Abiko, T.; Yamauchi, H. Effects of increase in alpha-amylase and endo-protease activities during germination on the breadmaking quality of wheat. *Food Sci. Technol. Res.* **2001**, *7*, 214–219.

14. Bean, M.M.; Keagy, P.M.; Fullington, J.G.; Jones, F.T.; Mecham, D.K. Dried Japanese noodles I. Properties of laboratory prepared noodle doughs from sound and damaged wheat flours. *Cereal Chem.* **1974**, *51*, 416–426.

15. Whistler, R.L.; BeMiller, J.N. Starch. In *Carbohydrate Chemistry for Food Scientists*; Whistler, R.L., BeMiller, J.N., Eds.; American Association of Cereal Chemists: St. Paul, MN, USA, 1997; pp. 117–152.

16. Jane, J.; Chen, J. Effects of amylose molecular size and amylopectin branch chain length on paste properties of starch. *Cereal Chem.* **1992**, *69*, 60–65.

17. AACC. General Pasting Method for Wheat or Rye Flour or Starch Using the Rapid Visco Analyzer. In *International Approved Methods of Analysis*, 11th ed., Method 76-21.01; AACC International: St. Paul, MN, USA, 1999.

18. MacGregor, A.W.; Matsuo, R.R. Starch degradation in endosperm of barley and wheat kernels during initial stages of germination. *Cereal Chem.* **1972**, *59*, 210–216.

19. Simsek, S.; Whitney, K.; Ohm, J.B. Analysis of cereal starches by high-performance size exclusion chromatography. *Food Anal. Methods* **2013**, *6*, 181–190.

20. Rani, K.U.; Prasada Rao, U.J.S.; Leelavathi, K.; Haridas Rao, P. Distribution of enzymes in wheat flour mill streams. *J. Cereal Sci.* **2001**, *34*, 233–242.

21. Huang, G.R. A Study of Alpha Amylase Activity in Kansas Hard White Wheats. Master's Thesis, Kansas State University, Manhattan, NY, USA, 1979.

22. Tsai, M.L.; Li, C.F.; Lii, C.Y. Effects of granular structures on the pasting behaviors of starches. *Cereal Chem.* **1997**, *74*, 750–757.

23. Cornell, H. The Functionality of Wheat Starch. In *Starch in Food*; Eliasson, A.C., Ed.; Woodhead Publishing Limited: Cambridge, UK, 2004; pp. 221–238.

24. Lorenz, K.; Kulp, K. Sprouting of cereal grains—Effects on starch characteristics. *Starch* **1981**, *33*, 183–187.

25. Dronzec, B.L.; Hwang, P.; Bushuk, W. Scanning electron microscopy of starch from sprouted wheat. *Cereal Chem.* **1972**, *49*, 232–239.

12

A Review on the Role of Vibrational Spectroscopy as An Analytical Method to Measure Starch Biochemical and Biophysical Properties in Cereals and Starchy Foods

D. Cozzolino *, S. Degner and J. Eglinton

School of Agriculture, Food and Wine, The University of Adelaide, Waite Campus, PMB 1 Glen Osmond, SA, 5064, Australia; E-Mails: sophia.degner@adelaide.edu.au (S.D.); jason.eglinton@adelaide.edu.au (J.E.)

* Author to whom correspondence should be addressed; E-Mail: d.cozzolino@adelaide.edu.au

External Editor: Glen Fox

Abstract: Starch is the major component of cereal grains and starchy foods, and changes in its biophysical and biochemical properties (e.g., amylose, amylopectin, pasting, gelatinization, viscosity) will have a direct effect on its end use properties (e.g., bread, malt, polymers). The use of rapid and non-destructive methods to study and monitor starch properties, such as gelatinization, retrogradation, water absorption in cereals and starchy foods, is of great interest in order to improve and assess their quality. In recent years, near infrared reflectance (NIR) and mid infrared (MIR) spectroscopy have been explored to predict several quality parameters, such as those generated by instrumental methods commonly used in routine analysis like the rapid visco analyser (RVA) or viscometers. In this review, applications of both NIR and MIR spectroscopy to measure and monitor starch biochemical (amylose, amylopectin, starch) and biophysical properties (e.g., pasting properties) will be presented and discussed.

Keywords: gelatinization; pasting properties; starch; near infrared spectroscopy; mid infrared spectroscopy

1. Introduction

The starch stored in the seeds and tubers of various agricultural crops including maize, wheat, rice, barley, potato and cassava provides the main source of energy in the human diet [1–4]. Starch is the major component of cereal grains, and changes in its biophysical and biochemical properties are related with the amount and ratio of amylose and amylopectin, that influence and affect properties such as viscosity, gelatinization, that will determine its end use properties (e.g., bread, malt, beer, polymers) [1,4–7].

Current chemical and physical methods used in research and by the industry to measure starch properties are slow, destructive, with many based on empirical relationships [8–12]. Overall knowledge of the pasting properties of the sample can help to improve starch content in cereals and starchy foods as well as will allow understanding the biophysical and structural properties of the starch to be used in foods (e.g., bread, beer, whisky) and in industrial applications (e.g., polymer production) [13]. This can also provide with useful information about the specific application of the starch in the selection or screening of new genotypes or lines in breeding programs [6,14].

Analytical methods currently used to determine biophysical properties in cereals and starchy foods include instrumental methods such as Differential Scanning Calorimetry (DSC) and the Rapid Visco Analyzer (RVA) [1,9–11,13,15–17]. Instruments such as the RVA are also used in the routine analysis of cereals (e.g., wheat, barley) to determine the effects of rain damage on grain quality at the delivery point [17–19].

During the last 20 years, methods based on vibrational spectroscopy in combination with chemometric techniques have resulted in the development of rapid methods to predict and monitor starch biochemical and biophysical properties [8–11]. Desirable characteristics of these tools include speed, ease-of-use, minimal or no sample preparation, and in some case the avoidance of sample destruction. Most of these features are characteristic of mid-infrared (MIR) and near-infrared (NIR) spectroscopy [8].

In a recent report, Kaddour and Cuq [20,21] highlighted three main ways in which NIR spectroscopy is currently applied in the analysis of cereals and starchy products: (1) as the straightforward analysis and very rapid determination of composition; (2) as screening tool in plant-breeding (for the selection of cross-breeds with the desired qualities); and (3) as an in-line tool to monitor physical and chemical changes during processing.

In this review, applications of both NIR and MIR spectroscopy to measure and monitor starch biochemical composition (e.g., amylose, amylopectin and starch) and biophysical properties (e.g., pasting properties, viscosity) will be presented and discussed.

2. Applications of NIR and MIR Spectroscopy

Several authors reported the ability of both NIR and MIR spectroscopy to measure and monitor starch biochemical and biophysical characteristics in a wide range of starchy foods and cereals. Examples of these applications are discussed and summarised below.

2.1. Determination of Amylose, Amylopectin, Starch and Granule Structure

The feasibility of using NIR reflectance spectroscopy to analyse starch and amylose, in buckwheat flours was reported [22]. Samples from different cultivars of buckwheat harvested in 12 different countries were analysed using NIR and multiple linear regression (MLR). The authors reported correlation coefficients (R) higher than 0.93 for starch, while workable standard error of predictions (SEP) in relation to the reference standard deviation data were reported. In contrast, the MLR models for amylose were judged as unstable due to the poor SEP values obtained. These authors attributed this issue to the standard deviation of the reference data [22]. The main wavelengths reported by the authors were found around 1925 nm associated with water (combination of the stretching and bending vibrations of hydroxyl group), around 2057 nm associated with protein (combination of NH and amide II or III), and around 2100 nm associated with starch (combination of OH and CO) [22]. The authors concluded that NIR spectroscopy can only be used for the measurement of the amounts starch in buckwheat flours [22].

Near infrared analysis was used to predict starch, moisture, and sugar content in sliced and fresh samples of sweetpotato (*Ipomoea batatas* (L.) Lam.). Samples were collected during three growing seasons, where the best calibration equation for starch was developed from the combination of samples from the three harvests [23]. An R of 0.95, standard error of calibration (SEC) of 2.01, and SEP of 1.91 was reported by these authors [23] for the determination of starch content. Calibrations based on samples from a given year adequately predicted the variables but could not account for variances introduced by samples from other years or harvests [23].

The use of NIR transmittance (NIT) spectroscopy was assessed to measure amylose content in corn (*Zea mays*) [24]. Calibrations were developed using a set of genotypes having endosperm mutations in single and double-mutant combinations, ranging in starch-amylose content (SAC) from 8.5% to 76% [24]. Prediction models for SAC yield an R of 0.96 and SEP of 5.1% [24]. According to the authors, narrowing the amylose range of the calibration set generally did not improve performance statistics except when partial least squares regression (PLS) was used, in which a decrease in the SEP values were observed [24]. Although the amylose calibrations were of limited precision, the authors stated that the calibration models developed may be useful when a rough screening method is needed for SAC [24].

The determination of starch in potato tubers sourced from different varieties and breeding lines was achieved using NIR spectroscopy [25]. The authors reported an R^2 of 0.90 and SECV values ranging between 0.74% and 0.79% [25,26]. Amylose content was measured in sorghum samples using differential scanning calorimetry and predicted using NIR spectroscopy [27]. The R^2 and SECV for the determination of amylose in ground and whole sorghum grain were 0.75 (SECV = 0.77) and 0.70 (SECV = 0.75), respectively [27].

The potential of NIR reflectance spectroscopy was investigated as an alternative method for predicting the major constituents in yam tuber samples (*Dioscorea pp*) [28,29]. Two hundred and sixty-five samples, belonging to seven different yam accessions were analysed for starch, amylose, sugars, proteins, minerals and cellulose [28,29]. The reported coefficients of determination (R^2) were 0.84 for starch, 0.86 for sugars and 0.88 for proteins [28,29]. Calibrations developed by combining both the calibration and validation sets determined an improvement of the calibration statistics [29].

Discriminant analysis (DA) was also used by the authors to classify samples according to the amylaceous fraction of the chemotype [29].

A methodology for the rapid estimation of taro quality (*Colocasia esculenta*) (starch, total sugars, cellulose, proteins, and minerals) was reported using NIR spectroscopy [30]. NIR calibration models reported a R^2 of 0.89 for starch, 0.90 for sugars, 0.44 for amylose and 0.61 for cellulose [30]. The predictions were tested on an independent set of 58 randomly selected accessions and the R^2 in prediction for starch and sugars were 0.76 and 0.74 respectively [30].

Wheat (*Triticum aestivum* L.) samples sourced from breeding programs that developed genotypes free of amylose (waxy wheat), as well as genetically intermediate (partial waxy) types were analysed using NIR spectroscopy [31–34]. Linear discriminant analysis (LDA) identifies the fully waxy genotype samples (greater than 90% accuracy) [34]. However, accuracy was reduced for partial and wild-type genotypes. It was suggested by the authors that the spectral sensitivity to waxiness is due to the lipid-amylose complex which diminishes with waxiness. These physical differences in endosperm affect light scatter as well as determine changes in starch crystallinity [34].

Interactions between carbohydrate monomers and polymers and their effects on NIR spectra were explored as well as the implications of such effects with regards to the development of NIR calibrations were evaluated [35]. The effects of the presence of amylopectin, amylose, cellulose and starch during the drying of glucose and sucrose on the resulting spectra were investigated [35]. Sugars in various molar ratios with polymers were dried in a rotating mixer and then reground, and spectra from 10,000 to 4000 cm^{-1} were recorded [35]. Although simple mixing of sugars with cellulose, amylose, amylopectin or starch caused some changes in the spectra of the sugars, in general the spectra obtained by spectral subtraction were considerably more like those of the pure sugars than were those obtained for the materials dried together [35].

The use of Fourier transform infrared (FTIR) spectroscopy was used as tool to differentiate between patterns of amylose in different granule types [36]. According to the authors, the IR spectrum of starch samples was described by peaks near 3500, 3000, 1600, 1400, 1000, 800 and 500 cm^{-1} [36]. The peaks at 3405 and 2930 cm^{-1} could be attributed to O–H and C–H bond stretching, respectively, while the peaks at 1420 and 1366 cm^{-1} were attributed to the bending modes of H–C–H, C–H and O–H. The peaks at 1300~1000 cm^{-1} were attributed to C–O–H stretching. The peaks at 1155, 1097 and 1019 cm^{-1} were assigned as the C–O bond stretching. The bands at 1047 and 1022 cm^{-1} were associated with the ordered and amorphous structures of starch, respectively. The ratio between absorbances 1047/1022 cm^{-1} was also used to quantify the degree of order in starch samples [36].

The structure of the A-and B-type granules of wheat starch was measured using polarized light microscopy, X-ray diffraction (XRD), and FTMIR spectroscopy in order to study the granular, crystalline, and short-range structures [37,38]. The A-and B-type granules displayed a typical A-type crystalline structure with the degrees of crystallinity of 31.95% and 29.38% respectively. A second order reflection was found in both A-and B-type granules, which was proposed due to the crystalline lamellae of the semicrystalline lamellae. The A-and B-type granules had mass and surface fractal structures respectively [37,38]. Table 1 shows the validation statistics for the analysis of amylose and starch content in different cereals and starchy products using NIR spectroscopy as well as different reference methods.

Table 1. Validation statistics for the measurement of amylose and starch, using near infrared spectroscopy and different reference analytical methods as reported by various authors.

Chemical parameter	Reference method	Sample	SECV/SEP	Reference
Amylose (%)	Iodine—colorimetric method	Rice	0.30–0.31	[39]
Amylose (g kg^{-1})	Iodine—colorimetric method	Beans [1]	11.4–12.8	[40]
Amylose (%)	Enzymatic	Barley	0.93–1.09	[41]
Starch (%)	Enzymatic	Barley	0.78–0.98	[41]
Amylose (%)	Iodine—colorimetric method	Yam	3.71	[29]
Starch (%)	Iodine—colorimetric method	Yam	1.78	[29]
Crude starch (%)	Acetic acid-calcium chloride and polarization	Maize	0.72–0.96 *	[42]

Notes: SECV: standard error of cross validation; SEP: standard error of prediction; [1] handheld and bench instruments were compared; * RMSEP: root mean square error of prediction.

2.2. Gelatinization, Pasting Properties and Retrogradation of Starch

The first attempt to get information correlated to rice RVA pasting data by NIR spectroscopy was only partially successful and reported by Delwiche and collaborators [33]. Later few studies were found in the literature that attempted to characterise the biophysical properties of starchy samples (e.g., pasting properties, viscosity) using either NIR or MIR spectroscopy [43–48]. Calibration and validation statistics reported by other authors on the use of NIR spectroscopy to predict pasting properties in cereals are summarised in Table 2.

Table 2. Calibration and validation statistics for the determination of pasting properties using near infrared spectroscopy in different cereal and starchy samples as reported by several authors.

Sample	Method and wavelength range	Parameter (RVU)	R^2	SECV/SEP	Reference
Rice	NIR (400–2500 nm)	PV	0.35	17.5	[49]
		BD	0.88	10.2	
		SB	0.92	13.6	
		HPV	0.55	16.7	
Rice non-waxy	NIR (1100–2500 nm)	PV	0.37	32.44	[50]
		BD	0.58	13.36	
		SB	0.60	25.07	
Rice	NIR (1100–2500 nm)	PV	0.63	23.7	[33]
		BD	0.72	14.2	
		SB	0.73	20.2	
Rice	NIR (1100–2500 nm)	PV	0.38–0.42		[43]
		BD	0.057–0.060		
		SB	0.57–0.59		
Rice	NIR (1100–2500 nm)	PV	0.74	20.99	[51]
		BD	0.80	21.47	
		SB	0.97	22.23	
		TH	0.80	7.37	
		FV	0.95	13.2	
Sweet potato	NIR (1100–2500 nm)	PV	0.91	13.1	[46,47]
		BD	0.81	10.67	
		SB	0.92	1.82	
Maize	NIR (1100–2500 nm)	PV	0.92	183	[44]
		BD	0.92	232	
		SB	0.92	412	

Notes: RVU: rapid visco units; R^2: coefficient of determination; SECV: standard error of cross validation; SEP: standard error of prediction; PV: peak viscosity; BD: breakdown; SB: setback; HPV: hot pasting viscosity; TH: trough; FV: final viscosity.

The ability of NIR spectroscopy to analyse changes in structure of starch due to gelatinization, and determination of degree of gelatinization was reported. The second derivative NIR spectra of rice starch samples having different degrees of gelatinization, showed large deviations at wavelengths around 1204, 1368, 1436, 1700, 1748, 1784, 1924, 2088, 2280, 2320 and 2348 nm [52]. Especially, NIR spectra in the wavelength region around 2100 and 2280 nm changed complexly according to progress of gelatinization. According to the authors particle size effects could explain some of the high correlations obtained (high correlation existed between degree of gelatinization and particle size) [52].

The gelatinization of rice starch was reported as a function of temperature and pressure from the changes in the IR spectrum [53]. These authors proposed that the re-entrant shape for starch is not only due to hydrogen bonding but also to the imperfect packing of amylose and amylopectin chains in the starch granule [53].

The RVA instruments are widely used in assessing cooking and processing characteristics in rice [43]. The ability to predict RVA parameters by NIR spectroscopy would be useful in rapidly determining rice pasting qualities, but NIR spectroscopy does not correlate with the traditional parameters such as peak viscosity (PV), final viscosity (FV), breakdown (BD), consistency, and setback (SB) [43]. Alternative RVA parameters were sought by collecting RVA and NIR data for a total of 86 short, medium, and long grain rice cultivars. The amylose contents were 0.41%–24.90% (w/w) and protein concentrations were 8.47%–11.35% (w/w). PLS regression models generated for the entire NIR spectrum against the RVA curve showed viscosity varied linearly with the NIR spectra between 1100 to 2500 nm [43].

The use of NIR reflectance spectroscopy for the rapid and accurate measurement of starch gelatinization degree was explored in fresh pasta made with eggs [54,55]. The samples ($n = 48$) were analysed using a FT-NIR spectrophotometer [54,55]. Modified PLS (MPLS) model was developed to predict the gelatinization degree. The model was able to accurately predict gelatinization degree with SEP values of 0.24 and R of 0.97 [54,55].

Lu and co-workers [46,47] reported good calibration statistics using sweet potato. The results obtained in this study proved that predictions of parameters derived from the RVA by NIR spectroscopy are sufficiently accurate to be applied as indicators of quality in breeding, especially for early selection of materials in breeding [46,47,56]. However, because NIR spectroscopy is a secondary technique relying on calibration against a reference analysis (RVA profile), and the quality of the calibration process is critical, for accurate testing of advanced breeding lines, instrumental testing with RVA will still require for confirmation of differences in quality parameters [46,47,56]. Precision and reliability of NIR calibration and prediction may be affected by many factors, such as sample representativeness, genetic variability for traits, accuracy of reference data, as well as other factors [46,47]. The genetic variability available for some traits was the most limiting factor in achieving high R values. A rapid predictive method based on NIR spectroscopy was developed to measure sweet potato starch physiochemical quality and pasting properties [46,47]. The results reported by the authors showed that NIR analysis was sufficiently accurate and effective for rapid evaluation of starch physicochemical properties in sweet potato. According to the authors the NIR based protocol developed in this study can be used for screening large number of starch samples in food enterprises and sweet potato breeding programs [46,47].

Bao and collaborators [49] reported that the PV and HPV derived from the RVA profiles were poorly predicted, while BD and SB achieved better prediction with low SEP (<20.8 RVU; RVU, rapid visco units). The same authors also reported differences between NIR calibrations for RVA parameters when waxy and non-waxy rice samples were analysed [49,50].

Three RVA profile parameters of rice were predicted with NIR spectroscopy [57,58]. The coefficients of determination of calibration (R^2) for BD, SB and CSV were 0.97, 0.99 and 0.99, respectively [58]. The root mean square errors of calibration (RMSEE) were 4.12, 2.41 and 1.72 [58]. The R^2 of NIR models of brown rice RVA profile parameters was 0.94, 0.99 and 0.99, respectively. The RMSECV was 5.4, 2.87 and 1.99, respectively [58].

Calibration models based on NIR spectra were developed using modified PLS regression with different mathematical treatments based on the grain and flour spectra of non-waxy rice alone or in combination with waxy rice [50]. The results showed that calibration models built with flour spectra are more robust than those with grain spectra, and with total rice including waxy rice are superior to those with only non-waxy rice [50]. However, for accurate assay of the pasting viscosity and gel textural parameters, direct instrumental measurement should be employed in later generations [50]. Wu and collaborators [58,59] also reported the prediction of amylose content using NIR spectroscopy. They reported differences in the calibrations obtained using brown, whole and rice flour [58,59].

The performance of NIR spectroscopy as method was examined for the non-destructive determination of major components of potato (*Solanum tuberosum*) and its pasting viscosity properties [60]. Good calibration models with reasonable accuracy for moisture, carbohydrate, protein and amylase contents and RVA parameters were determined using PLS regression [60]. The SEP values obtained were 0.87% for moisture, 0.95% for carbohydrates, 0.6% for amylose, and 0.15% for protein, as well as 30 RVU for PV, 34 RVU for FV, 24 RVU for BD, and 22 RVU for ST. The calibration model for ash content was not significant; further research is needed to improve calibration for ash content [60]. Thus, NIR spectroscopy is a useful method to non-destructively measure the major components and pasting viscosity of intact potato [60].

Retrogradation of gelatinised starch is the main phenomenon that influences the texture of MiGao (rice cake) [61]. The hardness of the MiGao increased during stored at 25 °C for 5 days. The RVA, FTIR spectroscopy and X-ray were used in order to quantify and analyze the retrogradation behaviour of MiGao [61].

Three rice starches with different amylose contents (Glutinous: 1.4%, Jasmine: 15.0% and Chiang: 20.2%) were pregelatinized in a double drum dryer at 110, 117 and 123 °C. Starch crystallinity was determined by XRD and FTIR spectroscopy [62]. Rheological properties were assessed using RVA and a rheometer. Pre-gelatinized starches obtained from Glutinous (PGS) and Jasmine rice (PJS) gave an RVA pasting profile with cold PV [62].

Glutinous flour derived from rice varieties such as RD6, Sakonnakorn and Niew Ubon as well as Horn-Mali rice flour samples were analysed using both RVA and NIR spectroscopy [51]. The models for determining the pasting properties using the PT, ST, FV, BD and trough (TH) yield all had a similar level of accuracy with R^2 values of 0.99, 0.97, 0.95, 0.80 and 0.80, respectively [51]. Furthermore, the thermal properties indicated that the models for onset temperature, peak temperature, and enthalpy produced moderate correlation values with R^2 values of 0.82, 0.75 and 0.73, respectively, whereas pasting temperature could not be predicted by NIR spectroscopy ($R^2 = 0.06$) [51].

Twenty-two diverse sorghum landraces, classified as normal and opaque types obtained from Ethiopia, were characterised for grain quality parameters using NIR spectroscopy, chemical and the RVA method [19]. Protein content ranged from 77 to 182 g kg^{-1}, and starch content from 514 to 745 g kg^{-1}. The normal sorghums had higher digestible energy than the opaque sorghums, which exhibited lower RVA viscosities, and higher pasting temperatures and ST ratios [19]. The RVA parameters were positively correlated with the starch content and negatively correlated with the protein content. Landraces were different for the various grain quality parameters with some landraces displaying unique RVA and NIR profiles [19]. The authors concluded that this study might guide utilisation of the sorghum landraces in plant improvement programs, and provides a basis for further studies into how starch and other constituents behave in and affect the properties of these landraces [19].

In recent years, the use of attenuated total reflectance and mid infrared (ATR-MIR) spectroscopy was evaluated and used to understand the gelatinization and retro-gradation of flour barley samples and the relationship with malting quality. Samples were sourced from commercial barley varieties, analysed using RVA and ATR-MIR instrument. These results showed that ATR-MIR spectroscopy is capable of characterising gel samples derived from barley flour samples having different malting characteristics. Infrared spectra can effectively represent a "fingerprint" of the sample being analysed and can be used to simplify and reduce analytical times in the routine methods currently used [11].

2.3. Monitoring of Starch Gelatinization and Processing

Currently a large proportion of analytical and routine process steps or controls in the cereal industry are performed in the laboratory. Vibrational spectroscopy have proven to be very versatile in dealing with either inorganic or organic constituents within a broad range from environmental to industrial applications [63,64]. The main advantages of these techniques are their multiplexing capability: more than ten different and locally separated measurement positions can be monitored with only one instrument, require minimal sampling; hence, in conjunction with chemometric modelling, it often replaces analytical methods like chromatography [63,64]. In recent years, the term process analytical technologies (PAT) describes the field of process analysis and measurement technologies that have been expanded to include several physical, chemical, mathematical and other analytical tools used to characterize chemical and biological processes [63–65]. The so called PAT technologies demonstrated to be one of the most efficient and advanced tools for continuous monitoring, as well as controlling the processes and the quality of raw ingredients and products in several applications among food processing, petrochemical and pharmaceutical industries [63–65].

According to Kaddour and collaborators [20,21] NIR spectroscopy is probably one of the physical methods best adapted to analyse starch products in cereals and starchy foods. The use of NIR spectroscopy as an in-line monitoring tool in cereal processing has opened a new field of applications, whereby NIR spectroscopy can provide information on changes in chemical and physico-chemical characteristics of foods during manufacturing processes and has been reported by various authors [65–77]. Recently, the use of dynamic NIR spectroscopy provides information that is relevant to calculate the reaction rate constants for the process [20,21,78–80]. Dynamic NIR spectroscopy gives complementary information with those that can be obtained using classical analytical methods, such as rheology or chemical analysis [20,21,78–80].

The potential applications of dynamic NIR spectroscopy for cereal products concern in-line monitoring of unit operations involved in transformation of wheat flour to food products (flour agglomeration, dough mixing, pasta dough extrusion and sheeting, dough proofing, thermal treatment and storage) [20,21,78–80]. The changes in NIR spectrum *versus* mixing time were considered to construct the NIR mixing curve and to determine the optimum mixing times. Several research groups [65,78–85] have also confirmed the ability of NIR spectroscopy to monitor dough mixing and extended applications over different mixer systems, in particular in wheat and dough formulations [20,21,78–80] as well as during potato processing [77]. The use of NIR technology in breadmaking process due to its non-invasive aspect would lead to both the industrial automation of the mixing process and the better description of physico-chemical phenomenon implied during dough mixing. A more detailed literature analysis on the NIR monitoring of bread dough mixing was presented recently in a review by Kaddour and Cuq [20,21].

The physicochemical events occurring during batter mixing at different water contents (51.8, 54.4, and 56.7 g of water/100 g of dough) were described and analysed using NIR spectroscopy [84]. An FT-NIR spectrometer over the 1000–2500 nm range with a fibre optic probe was used to record NIR spectra in-line. The analysis of both one-dimensional statistical method such as PCA and generalised two-dimensional correlation spectroscopy (2D COS) was conducted to evaluate the possibilities of NIR spectroscopy to monitor physical and physicochemical modifications observed during mixing of batter [84]. The NIR results were in agreement with the physical and physicochemical analysis traditionally used to study bread dough mixing including consistency and glutenin depolymerisation [84]. The 2D COS method allowed a sequence of chemical events occurring during mixing for the batters at 51.8% and 54.4% water contents to be tentatively proposed. However, the 2D COS did not give clear physicochemical differences between the three batters during mixing according to the authors [84]. The NIR results for the highly hydrated batter (56.7%) were difficult to analyse due to its high water content [20,21,84].

Rheological properties of dough are important for wheat quality characterization [80,84–86]. This research sought to obtain prediction models for rheological characteristics and to characterise the breadmaking quality of whole wheat using NIR spectroscopy, in order to offer a rapid tool to the farmers to know the quality of their product at the harvest moment. Tenacity (P), extensibility (L), deformation energy (W) and ratio P/L of dough were measured using traditional methods [86]. NIR spectra were acquired from these samples and models to predict the values from these parameters were developed. The SEC values achieved for the extensibility, deformation energy and tenacity of dough and the ratio between the two latter parameters were 5.27 mm, 9.97×10^{-4} J, 3.98 mm and 0.025 J mm^{-1}, respectively. The four models were validated by cross-validation, and by independent validation. The precision obtained in these models was enough for being applied in harvesters or at delivering moment [86].

The different quality of rice in simulated cooking process was investigated by using scanning electron microscope (SEM), FTIR, XRD, and DSC methods [87]. The result shows that the starch granules of rice of low amylose content (AC) have more pores and cavities, which directly led to the swelling of rice starch granule [87]. With the change in temperature, the changes in intensity and width of the IR bands showed that the variation in proportion area of crystalline to non-crystalline was related to AC of rice. According to the interpretative sensitivity of the data obtained, XRD method was suggested adequate for characterizing the pasting property of different quality of rice. Besides, DSC method showed the AC was not relevant to pasting temperature and enthalpy [87].

The potential application of NIR, FT-Raman spectroscopy to monitor starch retrogradation in stored bread crumb was investigated [88]. Semolina-based bread was made and cut into slices, which were stored under controlled conditions in sealed plastic bags. The aging of the bread crumb was monitored by both NIR FT-Raman spectroscopy and a texture analysis over a period of 20 days. The use of 2D COS analysis in the spectral range of 390–975 cm^{-1} revealed characteristic differences among the spectra collected over time for bands that peaked at 480, 765, and 850 cm^{-1} [88]. The band at 480 cm^{-1} is studied here in detail. During the storage, the peak frequency of this band shifted towards lower wavenumbers, and its full width at half height decreased. Both of these parameters were highly correlated R^2 of 0.92 and 0.95, respectively to crumb hardness measured by the texture analyser [88].

The gelation process of starch was monitored using IR spectra of starch in water while heating were obtained using ATR-MIR spectroscopy [89]. The authors evaluated the relationships between gelation and spectral changes using factor analysis, evolving factor analysis (EFA) and three-way PCA [89]. The authors reported that absorption values at 3300 and 1610 cm^{-1} decreased with temperature while absorptions at 1000 cm^{-1} increased [89]. The factor score plot patterns of amylose, amylopectin and rice starches were similar however those derived from potato and corn starches were unique [89]. van Velzen and co-workers [90] studied the factors associated with dough stickiness with the use of an ATR sampling module attached to a FTMIR instrument [90]. More recently Bock and collaborators [91] evaluated the use of ATR-MIR spectroscopy to monitor the addition of water on the gluten properties in dough [91,92]. The authors compared the OH stretch band of water in flour dough with that in H_2O-D_2O mixtures having the same water content revealed the formation of two distinct water populations in flour dough corresponding to IR absorption frequencies at 3600 and 3200 cm^{-1} [91,92]. The band intensity at 3200 cm^{-1}, which is related to water bound to the dough matrix, decreased and shifted to lower frequencies with increasing moisture content of the dough [91,92]. The authors concluded that when bran is added to flour dough, water redistribution among dough components promotes partial dehydration of gluten and collapse of β-spirals into β-sheet structures [91,92]. According to the authors this transconformational changes could be related to the physical basis for the poor quality of bread dough containing added bran [91,92].

3. Conclusions

The measurement of starch chemical components (e.g., amylose and amylopectin), and physical characteristics of the endosperm such as granule type (A and B), and biophysical properties such as gelatinization, retrogradation, crystalinity were reported by several authors using either NIR or MIR spectroscopic. However, contradictory results in terms of precision and accuracy of the calibrations obtained were reported. Such differences were attributed to the accuracy of the reference method used (e.g., enzymatic or colorimetric methods), with interferences with other properties (e.g., lipids, proteins), range in composition, with the number of samples used to develop the calibration models, and the methods used (NIR, MIR, transmission, ATR, reflectance).

Infrared (IR) spectroscopy also has demonstrated its capability of in/on and at-line monitoring of different stages during wheat processing. This method is well adapted to identify and describe physical, chemical and physicochemical modifications occurring during wheat processing. Overall, the different data found in the literature demonstrated that the IR and in particular NIR spectroscopy can be used to generate relevant information to monitor wheat processing because it is the least perturbing method of exploring molecular changes, during the input of mechanical energy. It can be considered as a valuable method for the industry to analyse, control and understand the origin and construction of starch based products.

Overall, the main advantages of these techniques are the speed and ease of use in routine operations, determining potential savings, through reduction of analysis time and cost. However, the major disadvantage has been the difficulty of quantification and interpretation of data generated by spectroscopic methods. It is clear that the breadth of these applications, either in routine use or under developed is showing no sign of diminishing. The combination of these techniques with multivariate methods could be used as a tool on a large scale to be used in R & D or industrial applications.

However, the chemical or biophysical basis that determines the corner stone is still not well understood. The biggest challenges in the wider use of these technologies will be the interpretation of the complex data obtained. The interpretation of the models through multivariate methods in particular calibration development, the knowledge of the fundamentals of molecular spectroscopy in rheological analysis by IR spectroscopy is still not well known. The future development of these applications will provide the cereal and starch base food industry with very fast and non-destructive methods to quantify different samples providing the industry with a rapid means of qualitative rather than quantitative analysis.

Acknowledgments

This project is supported by Australia's grain growers through their investment body the Grain Research and Development Corporation, with matching funds from the Australian government.

Conflicts of Interest

The authors declare no conflict of interest.

References

1. Evers, A.D.; Blakeney, A.B.; O'Brien, L. Cereal structure and composition. *Aust. J. Agric. Res.* **1999**, *50*, 629–650.
2. Pérez, S.; Baldwin, P.M.; Gallant, D.J. Structural features of starch granules. In *Starch: Chemistry and Technology*, 3rd ed.; BeMiller, J., Whistler, R., Eds.; Academic Press: New York, NY, USA, 2009.
3. Perez, S.; Bertoft, E. The molecular structures of starch components and their contribution to the architecture of starch granules: A comprehensive review. *Starch* **2010**, *62*, 389–420.
4. Schwartz, D.; Whistler, R.L. History and future of starch. In *Starch: Chemistry and Technology*, 3rd ed.; BeMiller, J., Whistler, R., Eds.; Academic Press: New York, NY, USA, 2009.
5. Xie, S.; Liu, Q.; Cui, S.W. Starch modification and applications. In *Food Carbohydrates*; Cui, S., Ed.; Taylor and Francis: Boca Raton, FL, USA, 2005.
6. Keeling, P.L.; Myers, A.M. Biochemistry and genetics of starch synthesis. *Ann. Rev. Food. Sci. Technol.* **2010**, *1*, 271–303.
7. Willett, J.L. Starch in polymer compositions. In *Starch: Chemistry and Technology*, 3rd ed.; BeMiller, J., Whistler, R., Eds.; Academic Press: New York, NY, USA, 2009.
8. Karoui, R.; Downey, G.; Blecker, C. Mid-infrared spectroscopy coupled with chemometrics: A tool for the analysis of intact food systems and the exploration of their molecular structure-quality relationships—A review. *Chem. Rev.* **2010**, *110*, 6144–6168.
9. Cozzolino, D.; Allder, K.; Roumeliotis, S.; Eglinton, J. Feasibility study on the use of multivariate data methods and derivatives to enhance information from the rapid visco analyser. *J. Cereal Sci.* **2012**, *56*, 610–614.

10. Cozzolino, D.; Roumeliotis, S.; Eglinton, J. Exploring the use of near infrared (NIR) reflectance spectroscopy to predict starch pasting properties in whole grain barley. *Food Biophys.* **2013**, *8*, 256–261.

11. Cozzolino, D.; Roumeliotis, S.; Eglinton, J. Prediction of starch pasting properties in barley flour using ATR-MIR spectroscopy. *Carbohydr. Polym.* **2014**, *95*, 509–514.

12. Reeves, J.B., III. Potential of near- and mid-infrared spectroscopy in biofuel production. *Commun. Soil Sci. Plant Anal.* **2012**, *43*, 478–495.

13. Singh, J.; Dartois, A.; Kaur, L. Starch digestibility in food matrix: A review. *Trends Food Sci. Technol.* **2010**, *21*, 68–180.

14. Hodsagi, M.; Gergely, S.; Galencser, T.; Salgo, A. Investigations of native and resistant starches and their mixtures using near infrared spectroscopy. *Food Bioprocess Technol.* **2012**, *5*, 401–407.

15. Kim, J.-O.; Kim, W.-S.; Shin, M.-S. A comparative study on retrogradation of rice starch gels by DSC, X-ray and alpha-amylase methods. *Starch* **1997**, *49*, 71–75.

16. Booth, R.; Bason, M.L. Principles of operation and experimental techniques. In *The RVA Handbook*; Crosbie, G.B., Ross, A.S., Eds.; AACC International: St Paul, MN, USA, 2007; pp. 1–19.

17. Batey, I.L. Interpretation of RVA curves. In *The RVA Handbook*; Crosbie, G.B., Ross, A.S., Eds.; AACC International: St Paul, MN, USA, 2007; pp. 19–31.

18. Zhou, M.; Mendham, N.J. Predicting barley malt extract with a Rapid Viscoanalyser. *J. Cereal Sci.* **2005**, *41*, 31–36.

19. Shewayrga, H.; Sopade, P.A.; Jordan, D.R. Characterisation of grain quality in diverse sorghum germplasm using a rapid visco-analyzer and near infrared reflectance spectroscopy. *J. Sci. Food Agric.* **2012**, *92*, 1402–1410.

20. Kaddour, A.A.; Cuq, B. Dynamic NIR spectroscopy to monitor wheat product processing: A short review. *Am. J. Food Technol.* **2011**, *6*, 186–196.

21. Kaddour, A.A.; Cuq, B. Dynamic NIR spectroscopy to monitor bread dough mixing: A short review. *Am. J. Food Technol.* **2011**, *6*, 173–185.

22. Hong, J.H.; Ikeda, K.; Kreft, I.; Yasumoto, K. Near-infrared diffuse reflectance spectroscopic analysis of the amounts of moisture, protein, starch, amylose, and tannin in buckwheat flours. *J. Nutr. Sci. Vitaminol.* **1996**, *42*, 359–366.

23. Katayama, K.; Komaki, K.; Tamiya, S. Prediction of starch, moisture, and sugar in sweetpotato by near infrared transmittance. *HortScience* **1996**, *31*, 1003–1006.

24. Campbell, M.R.; Mannis, S.R.; Port, H.A.; Zimmerman, A.M.; Glover, D.V. Prediction of starch amylose content versus total grain amylose content in corn by near-infrared transmittance spectroscopy. *Cereal Chem.* **1999**, *76*, 552–557.

25. Haase, N.U. Rapid estimation of potato tuber quality by near-infrared spectroscopy. *Starch* **2006**, *58*, 268–273.

26. Haase, N.U.; Mintus, T.; Weipert, D. Viscosity measurements of potato starch paste with the rapid visco analyser. *Starch* **1995**, *47*, 123–126.

27. Alencar-Figueredo, L.F.; Davrieux, F.; Fliedel, G.; Rami, J.F.; Chantereau, J.; Deu, M.; Courtois, B.; Mestres, C. Development of NIRS equations for food grain quality traits through exploitation of core collection of cultivated sorghum. *J. Agric. Food Chem.* **2006**, *54*, 8501–8509.

28. Lebot, V.; Champagne, A.; Malapa, R.; Shiley, D. NIR determination of major constituents in tropical root and tuber crop flours. *J. Agric. Food Chem.* **2009**, *57*, 10539–10547.

29. Lebot, V.; Malapa, R. Application of near infrared reflectance spectroscopy for the evaluation of yam (*Dioscorea alata*) germplasm and breeding lines. *J. Sci. Food Agric.* **2013**, *93*, 1788–1797.

30. Lebot, V.; Malapa, R.; Bourrieau, M. Rapid estimation of taro (*Colocasia esculenta*) quality by near-infrared reflectance spectroscopy. *J. Agric. Food Chem.* **2011**, *59*, 9327–9334.

31. Delwiche, S.R.; Weaver, G. Bread quality of wheat flour by near-infrared spectroscopy: Feasibility of modeling. *J. Food Sci.* **1994**, *59*, 410–415.

32. Delwiche, S.R. Protein content of single kernels of wheat by near-infrared reflectance spectroscopy. *J. Cereal Sci.* **1998**, *27*, 241–254.

33. Delwiche, S.R.; McKenzie, K.S.; Webb, B.D. Quality characteristics in rice by near-infrared reflectance analysis of whole-grain milled samples. *Cereal Chem.* **1996**, *73*, 257–263.

34. Delwiche, S.R.; Graybosch, R.A.; St Amand, P.; Bai, G. Starch waxiness in hexaploid wheat (*Triticum aestivum* L.) by NIR reflectance spectroscopy. *J. Agric. Food Chem.* **2011**, *59*, 4002–4008.

35. Reeves, J.B. Solid-state matrix effects on near-infrared spectra: Interactions of glucose and sucrose with amylose, amylopectin, cellulose, and starch—Implications for near-infrared calibrations. *Appl. Spectrosc.* **1996**, *50*, 154–160.

36. Zeng, J.; Li, G.; Gao, H.; Ru, Z. Comparison of A and B starch granules from three wheat varieties. *Molecules* **2011**, *16*, 10570–10591.

37. Zhang, X.; Yu, P. Using ATR-FT/IR molecular spectroscopy to detect effects of blend DDGS inclusion level on the molecular structure spectral and metabolic characteristics of the proteins of hulless barley. *Spectroc. Acta A Mol. Biomol. Spectrosc.* **2012**, *95*, 53–63.

38. Zhang, B.; Li, X.; Liu, J.; Xie, F.; Chen, L. Supramolecular structure of A- and B-type granules of wheat starch. *Food Hydrocoll.* **2013**, *31*, 68–73.

39. Xie, L.H.; Tang, S.Q.; Chen, N.; Luo, J.; Jiao, G.A.; Shao, G.N.; Wei, X.J.; Hu, P.S. Optimisation of near-infrared reflectance model in measuring protein and amylose content of rice flour. *Food Chem.* **2014**, *142*, 92–100.

40. Plans, M.; Siro, J.; Casanas, F.; Sabate, J.; Rodriguez-Saona, L. Characterization of common beans (*Phaseolus vulgaris* L.) by infrared spectroscopy: Comparison of MIR, FT-NIR and dispersive NIR using portable and benchtop instruments. *Food Res. Int.* **2013**, *54*, 1643–1651.

41. Ping, H.; Wang, J.; Ren, G. Prediction of the total starch and amylose content in barley using near infrared reflectance spectroscopy. *Intell. Autom. Soft Comput.* **2013**, *19*, 231–237.

42. Jiang, H.Y.; Zhu, Y.J.; Wei, L.M.; Dai, J.R.; Song, T.M.; Yan, Y.L.; Chen, S.J. Analysis of protein, starch and oil content of single intact kernels by near infrared reflectance spectroscopy (NIRS) in maize (*Zea mays* L.). *Plant Breed.* **2007**, *126*, 492–497.

43. Meadows, F.; Barton, F.E. Determination of rapid visco analyser parameters in rice by near infrared spectroscopy. *Cereal Chem.* **2002**, *79*, 563–566.

44. Juhasz, R.; Gergely, S.; Gelencser, T.; Salgo, A. Relationships between NIR spectra and RVA parameters during wheat germination. *Cereal Chem.* **2005**, *82*, 488–493.

45. Gergely, S.; Salgo, A. Changes in carbohydrate content during wheat-maturation-what is measured by near infrared spectroscopy? *J. Near Infrared Spec.* **2005**, *13*, 9–17.

46. Lu, G.Q.; Huang, H.H.; Zhang, D.P. Application of near-infrared spectroscopy to predict sweet potato starch thermal properties and noodle quality. *J. Zhejiang Univ. Sci.* **2006**, *7*, 475–481.

47. Lu, G.Q.; Huang, H.H.; Zhang, D.P. Prediction of sweet potato starch physiochemical quality and pasting properties using near-infrared reflectance spectroscopy. *Food Chem.* **2006**, *94*, 632–639.

48. Garcia-Rosas, M.; Bello-Perez, A.; Hernani, Y.M.; Gonzalo, R.; Flores-Morales, A.; Mora-Escobedo, R. Resistant starch content and structural changes in maize (*Zea mays*) tortillas during storage. *Starch* **2009**, *61*, 414–421.

49. Bao, J.S.; Cai, Y.Z.; Corke, H. Prediction of rice starch quality parameters by near-infrared reflectance spectroscopy. *J. Food Sci.* **2001**, *66*, 936–939.

50. Bao, J.; Wang, Y.; Shen, Y. Determination of apparent amylose content, pasting properties and gel texture of rice starch by near infrared spectroscopy. *J. Sci. Food Agric.* **2007**, *87*, 2040–2048.

51. Chueamchaitrakun, P.; Chompreeda, P.; Haruthaithanasan, V.; Suwonsichon, T.; Kasemsamran, S. Prediction of pasting and thermal properties of mixed Hom-Mali and glutinous rice flours using near infrared spectroscopy. *Kasetsart J. (Nat. Sci.)* **2011**, *45*, 481–489.

52. Onda, T.; Abe, H.; Matsunaga, A.; Komiyama, Y.; Kawano, S. Analysis of gelatinization of starch by near-infrared spectroscopy. *J. Jpn. Soc. Food Sci. Technol.* **1994**, *41*, 886–890.

53. Rubens, P.; Heremans, K. Pressure-temperature gelatinization phase diagram of starch: An *in situ* fourier transform infrared study. *Biopolymers* **2000**, *54*, 524–530.

54. Zardetto, S. Potential applications of near infrared spectroscopy for evaluating thermal treatments of fresh egg pasta. *Food Control* **2005**, *16*, 249–256.

55. Zardetto, S.; Rosa, M.D. Study of the effect of lamination process on pasta by physical chemical determination and near infrared spectroscopy analysis. *J. Food Eng.* **2006**, *74*, 402–409.

56. Lu, G.Q.; Sheng, J.L. Application of near infrared reflectance spectroscopy (NIRS) in sweet potato quality breeding. *Sci. Agric. Sin.* **1990**, *23*, 76–81.

57. Liu, Y.F.; Lu, Y.X.; Kui, L.M.; Guo, Y.M.; Tan, C.Y.; Liu, X.L. Determination of RVA profile parameters of rice by near infrared spectrometer. *Southwest China J. Agric. Sci.* **2007**, *20*, 974–978.

58. Wu, J.G.; Shi, C.H. Prediction of grain weight, brown rice weight and amylose content in single grains using near infrared reflectance spectroscopy. *Field Crops Res.* **2004**, *87*, 13–21.

59. Wu, J.G.; Shi, C.H. Calibration model optimization for rice cooking characteristics by near infrared reflectance spectroscopy (NIRS). *Food Chem.* **2007**, *103*, 1054–1061.

60. Chen, J.Y.; Zhang, H.; Yagi, Y. Nondestructive determination of major components and pasting viscosity of potato tuber by near infrared spectroscopy. *J. Jpn. Soc. Food Sci. Technol.* **2009**, *56*, 299–306.

61. Ji, Y.; Zhu, K.; Zhou, H.; Qian, H. Study of the retrogradation behaviour of rice cake using rapid visco analyser, fourier transform infrared spectroscopy and X-ray analysis. *Int. J. Food Sci. Technol.* **2010**, *45*, 871–876.

62. Nakorn, K.N.; Tongdang, T.; Sirivongpaisal, P. Crystallinity and rheological properties of pregelatinized rice starches differing in amylose content. *Starch* **2009**, *61*, 101–108.

63. Codgill, R.P.; Anderson, C.A.; Drennen, J.K., III. Using NIR spectroscopy as an integrated PAT tool. *Spectroscopy* **2004**, *19*, 104–109.

64. Jorgensen, P.; Pedersen, J.G.; Jensen, E.P.; Esbensen, K.H. On line batch fermentation process monitoring (NIR)—Introducing biological process time. *J. Chemom.* **2004**, *18*, 81–91.

65. Robert, P.; Devaux, M.F.; Mouhous, N.; Dufour, E. Monitoring the secondary structure of proteins by near-infrared spectroscopy. *Appl. Spectrosc.* **1999**, *53*, 226–232.

66. Millar, S.; Robert, P.; Devaux, M.F.; Guy, R.C.E.; Maris, P. Near-infrared spectroscopy measurements of structural changes in starch-containing extruded products. *Appl. Spectrosc.* **1996**, *50*, 1134–1139.

67. Osborne, B.G. Near infrared spectroscopic studies of starch and water in some processed cereal foods. *J. Near Infrared Spectrosc.* **1996**, *4*, 195–200.

68. Osborne, B.G. NIR measurements of the development of crystallinity in stored bread crumb. *Analusis* **1998**, *26*, M55–M57.

69. Osborne, B.G. Applications of near infrared spectroscopy in quality screening of early-generation material in cereal breeding programmes. *J. Near Infrared Spectrosc.* **2006**, *14*, 93–101.

70. Wesley, I.J.; Larsen, N.; Osborne, B.G.; Skerritt, J.H. Non-invasive monitoring of dough mixing by near infrared spectroscopy. *J. Cereal Sci.* **1998**, *27*, 61–69.

71. Wesley, I.J.; Blakeney, A.B. Investigation of starch-protein water mixtures using dynamic near infrared spectroscopy. *J. Near Infrared Spectrosc.* **2001**, *9*, 211–220.

72. Wellner, N.; Mills, E.N.; Brownsey, G.; Wilson, R.H.; Brown, N.; Freeman, J.; Halford, N.C.; Shewry, P.R.; Belton, P.S. Changes in protein secondary structure during gluten deformation studied by dynamic fourier transform infrared spectroscopy. *Biomacromolecules* **2005**, *6*, 255–261.

73. Wilson, R.H.; Goodfellow, B.J.; Belton, P.S.; Osborne, B.G.; Oliver, G.; Russell, P.L. Comparison of Fourier transform mid infrared spectroscopy and near infrared reflectance spectroscopy with differential scanning calorimetry for the study of the staling of bread. *J. Sci. Food Agric.* **1991**, *54*, 471–483.

74. Chung, H.; Arnold, M.A. Near-infrared spectroscopy for monitoring starch hydrolysis. *Appl. Spectrosc.* **2000**, *54*, 277–284.

75. Xie, F.; Dowell, F.E.; Sun, X.S. Comparaison of near-infrared spectroscopy and texture analyser for measuring wheat bread changes in storage. *Cereal Chem.* **2003**, *80*, 25–29.

76. Xie, F.; Dowell, F.E.; Sun, X.S. Using visible and near-infrared reflectance spectroscopy and differential scanning calorimetry to study starch, protein and temperature effects on bread staling. *Cereal Chem.* **2004**, *81*, 249–254.

77. Haase, N.U. Prediction of potato processing quality by near infrared reflectance spectroscopy of ground raw tubers. *J. Near Infrared Spectrosc.* **2011**, *19*, 37–45.

78. Kaddour, A.A.; Morel, M.-H.; Barron, C.; Cuq, B. Dynamic monitoring of dough mixing using near-infrared spectroscopy: Physical and chemical outcomes. *Cereal Chem.* **2007**, *84*, 70–79.

79. Kaddour, A.A.; Morel, M.-H.; Barron, C.; Cuq, B. Physico-chemical description of bread dough mixing using two-dimensional near-infrared correlation spectroscopy and moving-window two-dimensional correlation spectroscopy. *J. Cereal Sci.* **2008**, *48*, 10–19.

80. Kaddour, A.A.; Cuq, B.; Damiri, H.; Cassan, D.; Menut, P.; Morel, M. Physico-chemical mechanisms involved in the near infrared monitoring of dough mixing. In Proceedings of AACC Annual Meeting, Orlando, FL, USA, 24–28 July 2005; p. 216.

81. Alava, J.M.; Millar, S.J.; Salmon, S.E. The determination of wheat breadmaking performance and bread dough mixing time by NIR spectroscopy for high speed mixers. *J. Cereal Sci.* **2001**, *33*, 71–81.

82. Apruzzese, F.; Balke, S.T.; Diosady, L.L. In-line colour and composition monitoring in the extrusion cooking process. *Food Res. Inter.* **2000**, *33*:621–628.

83. Evans, A.J.; Huang, S.; Osborne, B.G.; Kotwal, Z.; Wesley, I.J. Near infrared on-line measurement of degree of cook in extrusion processing of wheat flour. *J. Near Infrared Spectrosc.* **1999**, *7*, 77–84.

84. Kaddour, A.A.; Morel, M.H.; Cuq, B. Description of batter dough mixing using near-infrared spectroscopy. *J. Cereal Sci.* **2008**, *48*, 698–708.

85. Psotka, J. Utilizing predictive technologies in milling and baking. *Cereal Foods World* **1999**, *44*, 30–31.

86. Arazuri, S.; Arana, I.; Arias, N.; Arregui, L.M.; Torralba, J.G.; Jaren, C. Rheological parameters determination using near infrared technology in whole wheat grain. *J. Food Eng.* **2012**, *111*, 115–121.

87. Huang, T.; Zhu, B.; Du, X.; Li, B.; Wu, X.F.; Wang, S.H. Study on gelatinization property and edible quality mechanism of rice. *Starch* **2012**, *64*, 846–854.

88. Piccinini, M.; Simonetta, F.; Secchi, N.; Sanna, M.; Roggio, T.; Catzeddu, P. The application of NIR FT-Raman spectroscopy to monitor starch retrogradation and crumb firmness in semolina bread. *Food Anal. Methods* **2012**, *5*, 1145–1149.

89. Iizuka, K.; Aishima, T. Starch gelation process observed by FT-IR/ATR spectrometry with multivariate data analysis. *J. Food Sci.* **1999**, *64*, 653–658.

90. Van Velzen, E.J.J.; van Duynhoven, J.P.M.; Pudney, P.; Weegels, P.L.; van der Maas, J.H. Factors associated with dough stickiness as sensed by attenuated total reflectance infrared spectroscopy. *Cereal Chem.* **2003**, *80*, 378–382.

91. Bock, J.E.; Connelly, R.K.; Damodaran, S. Impact of bran addition on water properties and gluten secondary structure in wheat flour doughs studied by attenuated total reflectance fourier transform infrared spectroscopy. *Cereal Chem.* **2013**, *90*, 377–386.

92. Bock, J.E.; Damodaran, S. Impact of bran addition on water properties and gluten secondary structure in model gluten dough studied by attenuated total reflectance Fourier transformed infrared spectroscopy. *Food Hydrocoll.* **2013**, *31*, 146–155.

13

Comprehensive and Comparative Metabolomic Profiling of Wheat, Barley, Oat and Rye Using Gas Chromatography-Mass Spectrometry and Advanced Chemometrics

Bekzod Khakimov *, Birthe Møller Jespersen and Søren Balling Engelsen

Department of Food Science, Faculty of Science, University of Copenhagen, Rolighedsvej 30, Frederiksberg C, 1958 Copenhagen, Denmark; E-Mails: bm@food.ku.dk (B.M.J.); se@food.ku.dk (S.B.E.)

* Author to whom correspondence should be addressed; E-Mail: bzo@food.ku.dk

External Editor: Francesco Capozzi

Abstract: Beyond the main bulk components of cereals such as the polysaccharides and proteins, lower concentration secondary metabolites largely contribute to the nutritional value. This paper outlines a comprehensive protocol for GC-MS metabolomic profiling of phenolics and organic acids in grains, the performance of which is demonstrated through a comparison of the metabolite profiles of the main northern European cereal crops: wheat, barley, oat and rye. Phenolics and organic acids were extracted using acidic hydrolysis, trimethylsilylated using a new method based on trimethylsilyl cyanide and analyzed by GC-MS. In order to extract pure metabolite peaks, the raw chromatographic data were processed by a multi-way decomposition method, Parallel Factor Analysis 2. This approach lead to the semi-quantitative detection of a total of 247 analytes, out of which 89 were identified based on RI and EI-MS library match. The cereal metabolome included 32 phenolics, 30 organic acids, 10 fatty acids, 11 carbohydrates and 6 sterols. The metabolome of the four cereals were compared in detail, including low concentration phenolics and organic acids. Rye and oat displayed higher total concentration of phenolic acids, but ferulic, caffeic and sinapinic acids and their esters were found to be the main phenolics in all four cereals. Compared to the previously reported methods, the outlined protocol provided an efficient and high throughput analysis of the cereal metabolome and the acidic hydrolysis improved the detection of conjugated phenolics.

Keywords: GC-MS; metabolomics; barley; wheat; oat; rye; TMSCN; PARAFAC2

1. Introduction

Cereals such as wheat, barley, rye and oat are amongst the mostly grown agricultural food products worldwide and the most important cereal crops for human consumption in northern Europe. The detailed chemical and functional composition of these crops is defining their use for food and feed as well as their prices. Cereals are the most important study objects in foodomics studies seeking to optimize their health beneficial factors and/or reducing deleterious metabolites. While the gross chemical composition, such as carbohydrates, proteins, dietary fibers and micronutrient contents, are important characteristics of cereal products, recent studies showed that relatively low concentration secondary metabolites such as antioxidant phenolics, organic acids and phytosterols have a significant influence on the health and nutritional values of cereals [1,2]. The beneficial health effects associated with the consumption of cereals have been attributed to dietary fiber content [3] as well as phenolics that possess antioxidant, radical scavenging and cholesterol lowering properties [4–7]. Whole grain barley intake has proven to decrease the low-density lipoprotein (LDL) cholesterol in an intervention study involving hypercholesterolemic patients [8]. Moreover, phenolic acids were found to be important texturizing agents in cooking-extrusion of cereals [9] and recognized as the main antioxidant constituents of cereals [10].

Quantitative and qualitative analysis of both, secondary and primary metabolites (with molecular weight of up to 1500 Da) of grains are studied within cereal metabolomics. Cereal metabolomics offers an insight into the metabolic fluctuations of cereal cultivars that may reveal effects of genetic modifications as well as of biotic and abiotic stresses [11]. Recent studies have illustrated the power of cereal metabolomics to reveal effects of growth temperature [12], salt stress [13], drought stress [14], and biotic stress [15]. Cereal metabolomics is also a promising approach to reveal biochemical and genetic backgrounds of quality traits and may open new possibilities towards targeted breeding [16,17].

Comprehensive metabolomic profiling of cereals requires a reliable protocol that enables extraction of maximum metabolic information in a high-throughput and reproducible manner. Metabolomics studies performed for uncovering single and/or multiple internal and/or external effects on cereals aim to cover as broad range of metabolites as possible. However, due to the great physico-chemical diversity of cereal metabolites, it is in practice impossible to cover the whole cereal metabolome using a single protocol. The phytochemical composition, including phenolics of wheat [18–20], barley [21], oat [22] and rye [23] have been investigated in a number of studies within the HEALTHGRAIN diversity-screening program [24].

This study demonstrates the development of comprehensive GC-MS metabolomics protocol for profiling a broad range of phenolics and organic acids from whole grain flour samples, and applied on wheat, barley, rye and oat. Phenolics of cereals are primarily present in conjugated and bonded forms with carbohydrates, lipids and other cell membrane components that alter their solubility and thus bioavailability [21]. Analysis of phenolic content of cereals is mainly performed by basic hydrolysis of cereal extracts [18], which can only cleave ester bonds and stabilize de-esterification reactions. However, a substantial part of phenolics and other organic acids of cereals are conjugated through glycosidic and/or ether bonds to carbohydrates and other molecules. In contrast to basic hydrolysis, acidic hydrolysis allows the cleavage of not only ester bonds, but also glycosidic and ether bonds at an elevated temperature. The advantages of this approach have been demonstrated in polyphenol analysis of the wheat and rice grains [25,26].

In this study, a standardized, high-throughput and unbiased protocol was developed for GC-MS metabolomic profiling of free and conjugated phenolics and organic acids of whole-grain cereals using hydrochloric acid based hydrolysis followed by trimethylsilyl derivatization. The study demonstrates the first application of a novel trimethylsilylation method based on trimethylsilyl cyanide (TMSCN) for derivatization of cereal metabolites. When compared to other frequently used derivatization methods, the new protocol provides a more unbiased and broad-spectrum derivatization of metabolites and is able to provide reproducible metabolomics profiles of complex biological samples [27]. The obtained raw GC-MS data of cereals were processed by a semi-automated multi-way decomposition method, PARAFAC2 [28]. The PARAFAC2 processing of the raw GC-MS data lead to unambiguous deconvolution of elusive peaks such as, overlapped, retention time shifted and low s/n peaks and enable an automatic estimation of relative concentrations of detected peaks [29,30]. Metabolite extraction and GC-MS analysis of the cereal samples were performed within a bigger study, which involved a larger set of barley samples (manuscript in preparation). The main aim of this study was to demonstrate the performance of the protocol, using new technologies within metabolomics, and to show first results of a comparative application to the four major north European cereals: wheat, barley, rye and oat. To the best of our knowledge, this is the first study illustrating a comprehensive GC-MS profiling of phenolics and organic acids of cereals using exactly the same protocol across different cereals.

2. Experimental Section

Whole grain samples of wheat (*Tr. aestivum*, variety Bussard), barley (*H. vulgare*, variety Bomi), rye (*S. cereal*, variety Petkus) and oat (*A. sativa*, variety Sang) were purchased in Sepetember 2012 from the Danish bread cereal producing company Aurion (Hjørring, Denmark). All four cereals were grown under biodynamical conditions in Jutland during the season 2011/12.

2.1. Metabolite Extraction and Sample Derivatization

Cereal metabolites were extracted from 50 mg of milled grains that were soaked into 600 μL 85% methanol and vortexed for 20 s at 3000 rpm followed by 20 min incubation at 30 °C using a Thermomixer (Model 5436, Eppendorf, Hamburg, Germany) at 1400 rpm. After 3 min of centrifugation at $16,000 \times g$, the supernatant was transferred to a fresh 2 mL Eppendorf tube (Hamburg, Germany) and the remaining flour sample was extracted a second time using the same extraction procedure. Then, the combined extracts were completely dried under nitrogen gas flow at 40 °C and hydrolyzed by using 240 μL of 6 M hydrochloric acid at 96 °C for 1 h by stirring at 1400 rpm. The hydrolyzed extracts were transferred into a fresh 2 mL glass vials and phenolics and organic acids were extracted into diethyl ether. Ether-based extraction of phenolics and organic acids was performed twice, by addition of 800 μL diethyl ether and vortexing for 25 s. The obtained ether fractions were completely dried using nitrogen gas flow and re-solubilized in 200 μL 100% methanol. Aliquots, 90 microliter, of the final extracts were transferred into 200 μL glass inserts and completely dried under nitrogen gas flow, sealed and stored at −20 °C until

GC-MS analysis. Each sample was spiked with an internal standard (IS) (5 μL of 0.2 mg mL^{-1} solution of ribitol). In order to avoid any moisture, the samples stored in the freezer were dried under reduced pressure before derivatization. Sample derivatization and injection were fully automated by using a Multi-Purpose Sampler (MPS, GERSTEL, Mülheim, Germany) with DualRait WorkStation integrated to a GC-MS system from Agilent (CA, USA). Each sample was individually derivatized by addition of 40 μL trimethylsilyl cyanide (TMSCN) and incubated for 40 min at 40 °C. Two replicate samples per cereal were analyzed in randomized order and the MPS autosampler allowed a sequential derivatization of all samples in the same manner by keeping the derivatization time constant, throughout the analysis.

2.2. GC-MS Data Acquisition

The GC-MS consisted of an Agilent 7890A GC and an Agilent 5975C series MSD. GC separation was performed on a Phenomenex ZB 5MSi column (30 m × 250 μm × 0.25 μm). A derivatized sample volume of 1 μL was injected into a cooled injection system (CIS port) using Solvent Vent mode at the vent pressure of 7 kPa until 0.3 min after injection at the vent flow of 100 mL min^{-1}. Detailed information on CIS and MPS parameters are described in Khakimov *et al.* 2013 [27]. Hydrogen was used as carrier gas, at a constant flow rate of 1.2 mL min^{-1}, and the initial temperature of CIS was set to 120 °C for 0.3 min followed by heating at 5 °C s^{-1} until reaching 320 °C and then held for 10 min. The GC oven program was as follows: initial temperature 40 °C, equilibration time 3.0 min, heating rate 12.0 °C min^{-1}, end temperature 300 °C, hold time 8.0 min and post run time 5 min at 40 °C. Mass spectra were recorded in the range of 50–500 *m/z* with a scanning frequency of 3.2 scans s^{-1}, and the MS detector was switched off during the 8.5 min of solvent delay time and after 25.5 min of the run time. The transfer line, ion source and quadrupole temperatures were set to 290, 230 and 150 °C, respectively. The mass spectrometer was tuned according to manufacturer's recommendation by using perfluorotributylamine (PFTBA).

2.3. Data Analysis

Initial analysis and visualization of the GC-MS data was performed using ChemStation software (Agilent, Germany). Retention indices of detected metabolites were calculated using the Van den Dool and Kratz equation and retention times of C10-C40 alkanes that were analyzed using the same GC-MS protocol [31]. The raw GC-MS data was imported from netCDF format to .mat files into Matlab® ver. R2012b (8.0.0.783) and data was manually divided into 121 smaller baseline separated intervals in retention time dimension. Each interval was modeled separately by PARAFAC2 as described previously [30]. PARAFAC2 modeled the three-way raw GC-MS data (elution time × mass spectra × samples) without any prior data pre-processing. The PARAFAC2 model outcomes: the elution profiles, which represent the TIC in the raw data, and spectral profiles, which represent the experimental EI-MS of deconvoluted peaks, were used for metabolite identification. The PARAFAC2 resolved mass spectrum of each peak was extracted and compared against NIST05 library (NIST, USA), Golm Metabolite Database [32]. Finally, PARAFAC2 concentration profiles, which represented relative concentrations of detected peaks were extracted and normalized according to the peak area of the internal standard (ribitol). The obtained metabolite table was used for exploring variations of phenolics in cereals and for principal component analysis (PCA) [33] after autoscaling of the data.

3. Results and Discussion

3.1. GC-MS Metabolomic Profiling and PARAFAC2 Based Data Processing

The total ion current (TIC) chromatograms of the GC-MS data obtained from hydrolyzed extracts of the four cereals are illustrated in Figure 1. Just over 300 peaks with a s/n ratio >10 were detected from GC-MS profiles. Validated PARAFAC2 models of 121 intervals of the raw GC-MS data revealed 389 components including resolved peaks, shoulders of neighbor peaks and baseline. Then, each PARAFAC2 model was individually evaluated and components that represent baseline, artifact peaks such as column bleed and reagent derived peaks and shoulders of neighbor peaks were eliminated, resulting in 247 chromatographic peaks with unique retention indices and mass spectra. The PARAFAC2 modeling of GC-MS intervals representing vanillin, protocatechuic acid and β-resorcylic acid are demonstrated in Figure 2.

Figure 1. The total ion current (TIC) chromatograms of GC-MS data obtained on wheat, barley, rye and oat metabolite extracts.

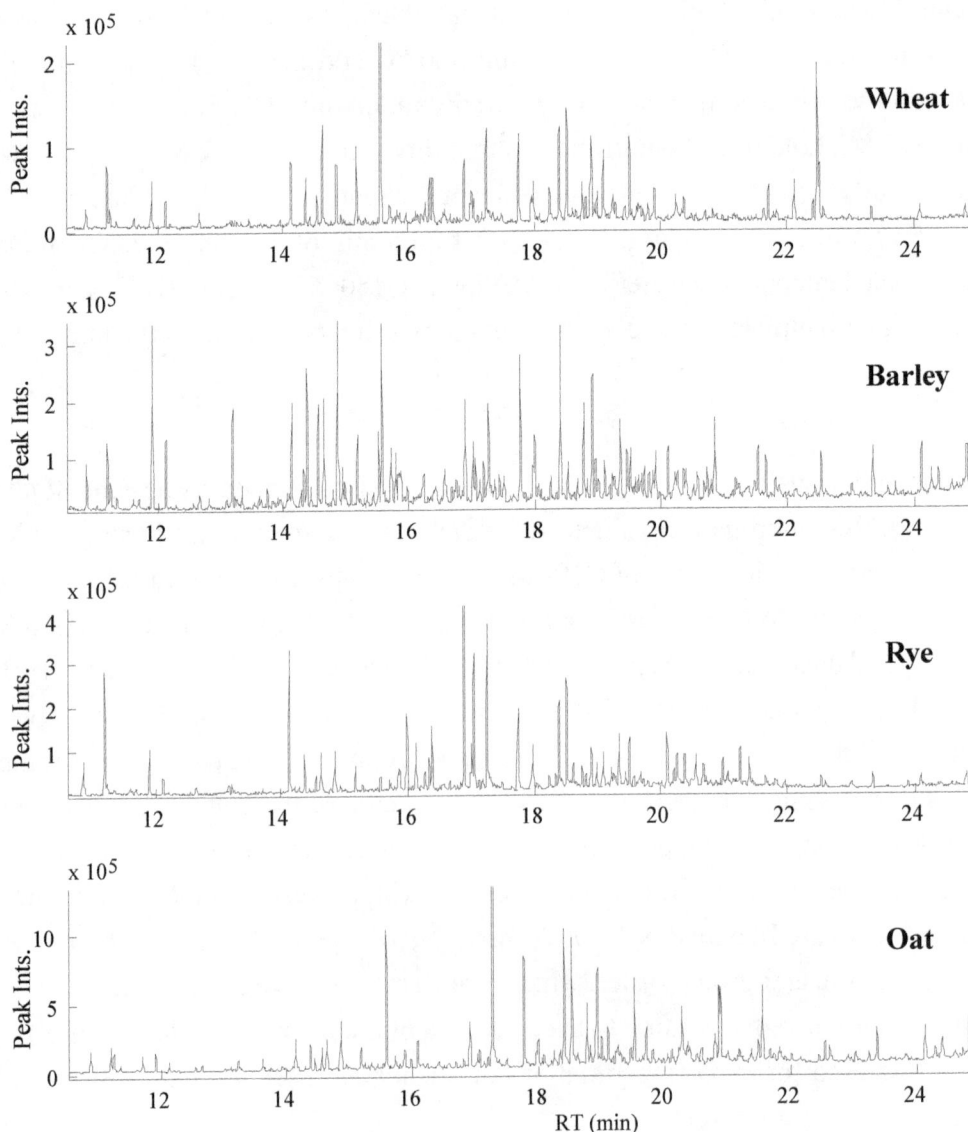

Figure 2. PARAFAC2 based processing of raw GC-MS data intervals. (**A**) and (**E**) are the TIC of raw GC-MS data intervals. (**B**) and (**F**) are the superimposed PARAFAC2 elution profiles of the raw GC-MS data intervals with seven and four components, respectively. (**C**) and (**G**) are subplots of (**B**) and (**F**), respectively. * Numbers of elution profiles correspond to the metabolites represented in Table 1. (**D**) and (**H**) are subplots of PARAFAC2 mass spectral profiles.

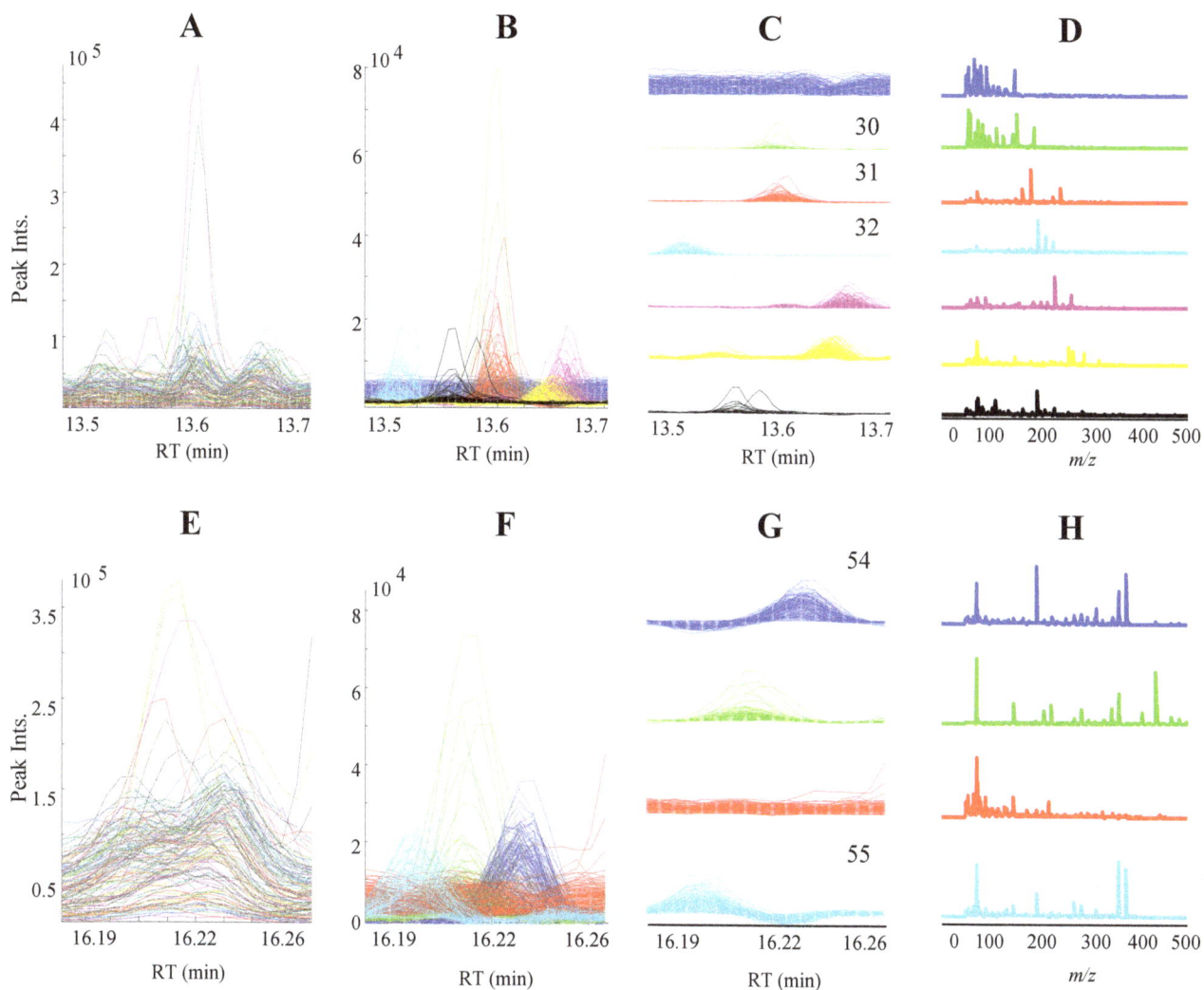

Comparison of RIs and PARAFAC2 resolved mass spectra of 247 resolved peaks against the NIST05 and Golm Metabolite Database resulted in the identification of 89 metabolites (Table 1) at level 2 as described in Metabolomics Standards Initiative report [34]. A total of 32 out of 89 identified metabolites were trimethylsilyl (TMS) derivatives of phenolic acids, their esters and aldehydes. In addition to the previously found phenolic acids from different barley genotypes [21], several other phenolics such as *p*-salicylic, gallic, gentisic, homovanillic and α-resorcylic acids and methyl esters of ferulic, caffeic, protocatechuic and sinapinic acids were identified. Small molecular organic acids, alcohols and their esters constituted 30 out of 89 identified metabolites. These included succinic, glyceric, maleic, fumaric, malic, pyroglutamic, azelaic acids and methyl esters of aconitic and citric acids that are part of the same or different metabolic pathways, and in addition, TMS-derivatives of 10 fatty acids and their esters, 6 sterols and a flavonoid, catechin-*n*TMS.

Table 1. A list of identified metabolites from wheat, barley, rye and oat flour samples by GC-MS. Metabolite identification was performed at level 2 as described in Metabolomics Standards Initiative report [34] and was based on RI and EI-MS library match (>80). [a] Metabolites with more than one isomers and/or TMS-derivatives; [b] tentatively identified.

No	Metabolites	RT min	RI (r)	RI (c)
1.	Laevulic acid-1TMS	9.04	1030	1070
2.	Sorbic acid-1TMS	9.06	1009	1071
3.	Hepta-2,4-dienoic acid, methyl ester	9.28	1000	1080
4.	Octanol-1-1TMS	9.51	1101	1090
5.	Malonic acid-2TMS	9.99	1205	1207
6.	(3,3-Dimethyl-1-cyclohexen-1-yl)oxy]-1TMS	9.97	1110	1206
7.	Benzoic acid-1TMS	10.42	1228	1226
8.	3-Methyl-2-furoic acid-1TMS	10.38	1107	1224
9.	Glycerol-3TMS	10.88	1282	1246
10.	1,3-Dihydroxypropanone-2-2TMS	11.03		1249
11.	Succinic acid-2TMS	11.24	1292	1262
12.	Glyceric acid-3TMS	11.51	1199	1274
13.	Maleic acid-2TMS	11.55	1286	1275
14.	Fumaric acid-2TMS	11.60	1178	1278
15.	p-Hydroxybenzaldehyde-1TMS	11.85	1280	1289
16.	2-Hydroxyheptanoic acid-2TMS	11.83	1312	1288
17.	3-Hydroxybutanoic acid-2TMS	12.12	1403	1401
18.	Resorcinol-2TMS	12.2	1378	1404
19.	Trimethyl aconitate	12.50	1428	1419
20.	Citric acid, trimethyl ester	12.82	1442	1435
21.	3-Hydroxyanthranilic acid, methyl ester-1TMS	12.8		1434
22.	2,4-Dihydroxy-5-methylpyrimidine-2TMS	12.89	1403	1439
23.	5-Hydroxy-2-(hydroxymethyl)-4H-pyran-4-one-2TMS	13.08	1492	1448
24.	Maseptol-1TMS	13.12	1358	1450
25.	Malic acid-2TMS	13.19	1494	1453
26.	2-Hydroxycyclohexanecarboxylic acid-2TMS	13.23	1402	1456
27.	3-Hydroxyoctanoic acid-2TMS	13.35	1452	1462
28.	Pyroglutamic acid-2TMS	13.46	1466	1467
29.	Erythritol-4TMS	13.47		1467
30.	Dimethyl azelate	13.61	1485	1474
31.	4-Hydroxybenzeneacetic acid, methyl ester-1TMS	13.62	1458	1475
32.	Vanillin-1TMS	13.55	1469	1471
33.	Citric acid, trimethyl ester-1TMS	13.76		1482
34.	2-Furancarboxylic acid, 5-[(oxy)methyl]-1TMS	13.72	1540	1480
35.	4-Hydroxyphenylethanol-2TMS	13.92	1475	1490
36.	Anozol	14.15	1603	1601

Table 1. *Cont.*

No	Metabolites	RT min	RI (r)	RI (c)
37.	2-Ketoglutaric acid-3TMS	14.34	1622	1612
38.	3-Methyl-3-hydroxypentanedioic acid-3TMS	14.3	1610	1609
39.	Dodecane-6-hydroxy-1TMS	14.40	1631	1615
40.	4-Hydroxybenzoic acid-2TMS	14.45	1618	1618
41.	Methyl Isovanillate-1TMS	14.66	1547	1629
42.	Suberic acid-2TMS	15.11	1682	1654
43.	Syringaldehyde -1TMS	15.15	1658	1656
44.	β-D-Arabinopyranose-4TMS [a]	15.23	1692	1660
45.	β-D-Xylopyranose-4TMS	15.30	1694	1664
46.	3,5-Dihydroxybenzoic ac. met.est.-2TMS	15.35	1656	1667
47.	2,5-Dimethoxymandelic acid-2TMS	15.38	1867	1669
48.	Vanillic acid-2TMS	15.72	1656	1687
49.	4-Hydroxycinnamic acid, methyl ester -1TMS	15.88	1565	1696
50.	Azelaic acid-2TMS	15.98	1800	1802
51.	2,3-Dihydroxyphosphoric acid, propyl ester-4TMS	15.86	1708	1695
52.	Methyl 2-(oxy)-2-(4-(oxy)phenyl)propanoate-2TMS	16.14	1757	1811
53.	α-D-Galactofuranoside, methyl-2,3,5,6-tetrakis-4TMS [a]	16.11	1845	1810
54.	3,5-Dihydroxy benzoic ac.-3TMS	16.24	1826	1818
55.	3,4-Dihydroxy benzoic ac.-3TMS	16.20	1826	1815
56.	D-Fructose-5TMS	16.41	1867	1828
57.	Isocitric acid-4TMS	16.34	1835	1823
58.	Catechin-*n*TMS [a]	16.44		1830
59.	Homovanilic acid-2TMS	16.4	1867	1827
60.	β-D-Galactopyranoside, methyl 2,3,4,6-tetrakis-4TMS [a]	16.68	1900	1844
61.	Catechin-*n*TMS [a]	16.77		1849
62.	2,5-Dihydroxy benzoic ac.-3TMS	16.78	1796	1850
63.	α-D-Glucopyranoside, methyl 2,3,4,6-tetrakis-4TMS [a]	16.90	1928	1857
64.	Syringic acid-2TMS	16.88	1845	1856
65.	β-D-Glucopyranoside, methyl 2,3,4,6-tetrakis-4TMS [a]	17.05	1928	1866
66.	α-D-Glucopyranose, 1,2,3,4,6-pentakis-5TMS [a]	17.02	1924	1864
67.	Palmitic acid, methyl ester	17.01	1870	1864
68.	D-Galactose, 2,3,4,5,6-pentakis-5TMS [a]	17.12	1970	1871
69.	*p*-Coumaric acid-2TMS	17.18	1924	1874
70.	Ferulic acid, methyl ester-1TMS	17.25	1765	1878
71.	3,4,5-Trihydrozy benzoic ac.-4TMS	17.45	1976	1890
72.	2-Hydroxymandelic acid, ethyl ester-2TMS	17.34	1777	1884
73.	4'-Cyclohexylacetophenone	17.58	1703	1898
74.	Caffeic acid methyl ester-2TMS	17.76	1863	2010
75.	β-D-Glucopyranose-5TMS [a]	17.75	1970	2009

Table 1. *Cont.*

No	Metabolites	RT min	RI (r)	RI (c)
76.	2-Hydroxysebacic acid-3TMS	18.13	2059	2034
77.	Ferulic acid-2TMS	18.40	2076	2052
78.	8,11-Octadecadienoic acid, methyl ester	18.35	2093	2049
79.	Sinapinic acid methyl ester-1TMS	18.51	1943	2059
80.	Methyl vanillactate-2TMS	18.55	2030	2062
81.	Caffeic acid-3TMS	18.76	2114	2076
82.	9-Methoxy-4α-methyl-2,3,7-trihydroxy-4,4a-dihydro-2H-benzo[c]chromen-6(3H)-one [b]	18.85		2082
83.	Linoleic acid-1TMS	19.23	2202	2207
84.	4,8-Dihydroxy-2-quinolinecarboxylic acid-3TMS	19.46	2265	2224
85.	Sinapinic acid-2TMS	19.52	2221	2228
86.	Androsterone type plant sterol [b]	19.89		2254
87.	3-Hydroxyandrostan-17-one-1TMS	19.98	2186	2261
88.	19-Norandrosterone-3-TMS [b]	20.36	2198	2288
89.	9,10-Dihydroxystearic acid-3TMS	20.87	2517	2426
90.	3,7-di-Hydroxy-androstan-17-one-2TMS	21.09	2432	2443
91.	9,10-Dihydroxystearic acid, dimethyl ester-2TMS	21.49	2784	2474
92.	2,3-Dihydroxypalmitic acid, propyl ester-2TMS	21.84	2581	2601
93.	2-Deoxy-6-phosphogluconolactone-5TMS	23.26		2820
94.	2-Hydroxytetracosanoic acid, methyl ester-1TMS	23.69	2894	2858
95.	3,7-Dihydroxycholest-5-ene-2TMS	23.95	2900	2881

3.2. Principal Component Analysis (PCA)

In order to explore the metabolomics data, PCA was performed on the metabolite table, including eight cereal samples in duplicates and 89 identified metabolites. PC1 *versus* PC2 scores plot of the PCA model (Figure 3A) show a clear separation of four different cereals explaining more than 60% variation of the data. The loadings plot of the corresponding model (Figure 3B) demonstrates a large spread of the 89 metabolites and revealed no clear groupings of metabolites classes. However, major part of the benzoic acid derived phenolics such as 3,5-dihydroxybenzoic, 3,4-dihydroxybenzoic and 3,4,5-trihydroxybenzoic acids are grouped on the upper left part of the loadings plot showing greater abundance in barley compared to the other cereals. In contrast to this, cinnamic acid derived phenolics such as ferulic, sinapinic and syringic acids are located on the bottom right corner showing greater concentrations in rye and wheat. Phenolics such as caffeic and 4-hydroxybenzoic acids have the highest concentrations in oat and significantly contribute to its separation from other cereals. However, detailed variations of phenolics and organic acids within and between cereal cultivars require a closer investigation of the data. In the following section, univariate comparisons of some metabolites are represented and the findings are compared to previous results reported in the literature.

Figure 3. (**A**) scores and (**B**) loading plots of the three component PCA model developed using identified metabolite table. * Numbers in loadings plot correspond to the metabolites represented in Table 1.

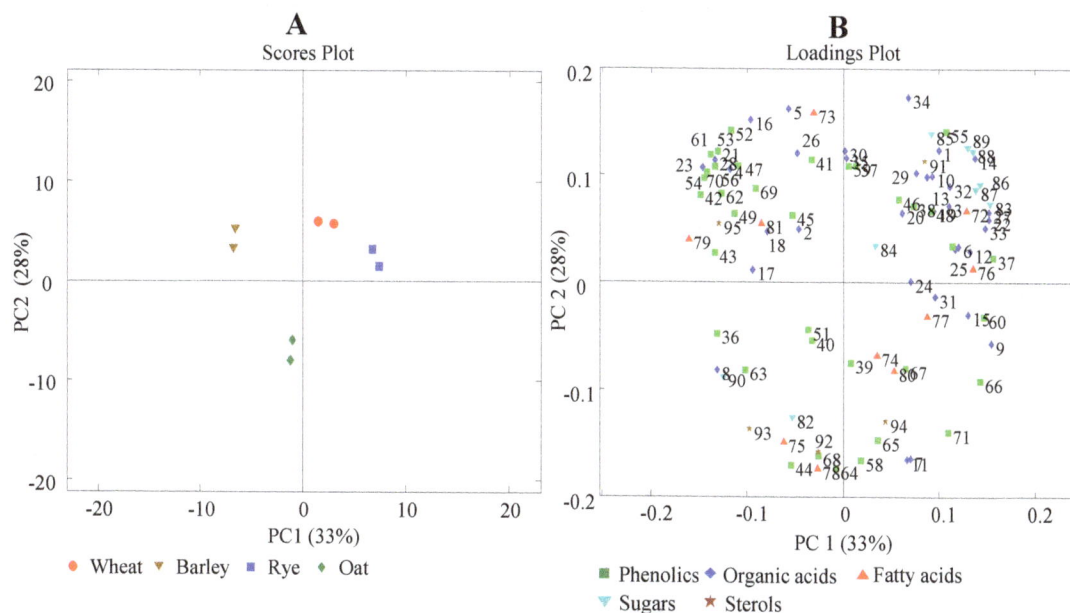

3.3. Variation of Phenolics and Organic Acids in Cereals

Phenolic acid composition of wheat, barley, rye and oat were compared to previously reported data [18,21–23]. Figure 4 shows relative percentages of the nine most abundant, free and conjugated phenolic acids of cereals reported in previous studies and makes comparisons with the data obtained in the current study. In previous studies, the phenolic acids of cereals were extracted using 80% ethanol followed by hydrolysis of conjugated phenolics in 2 M sodium hydroxide solution and analyzed by LC-DAD. In the current study, free and conjugated phenolics were extracted using 85% methanol, hydrolyzed in 2 M solution of hydrochloric acid followed by GC-MS analysis and PARAFAC2 based data processing. These two methodologies in phenolic profiling of cereals result in several apparent compositional differences. However, it should be underlined that the compared cereal genotypes are different in the two studies and the goal of this study is not a comprehensive comparison of phenolics of cereal varieties, but to demonstrate the power of the standardized cereal metabolomics protocol developed.

Nine major phenolics of the cereals investigated in this study were compared with winter wheat (*Triticum aestivum* var. *aestivum*) [18], Dicktoo barley (USA) [21], Grandrieu rye (France) [23] and Bajka oat (Poland) [22] varieties (Figure 4). Figure 4 shows that the relative concentrations of caffeic acid consistently increased (14%–23%) in all cereal cultivars compared to the previous studies where its abundance was below 1%. Similarly, for wheat, barley and oat, concentrations of ferulic acid increased from approximately 20% to 33%, while the comparison is more consistent for the two rye varieties. These results suggest that in grains, a significant amount of caffeic and ferulic acids are present in conjugated forms that cannot be cleaved by alkaline hydrolysis. Thus, the most abundant phenolic acids in previous cereal metabolomics studies were ferulic, sinapinic and 3,5-dihydroxybenzoic acids, while in this study, ferulic, sinapinic and caffeic acids were the most abundant ones.

Figure 4. Comparison of relative percentages of the nine most abundant phenolic acids of cereals reported in the literature (L) with the results of the current study (R). In literature the following genotypes were studied: winter wheat (*Triticum aestivum* var. *aestivum*) [18], Dicktoo barley (USA) [21], Grandrieu rye (France) [23] and Bajka oat (Poland) [22].

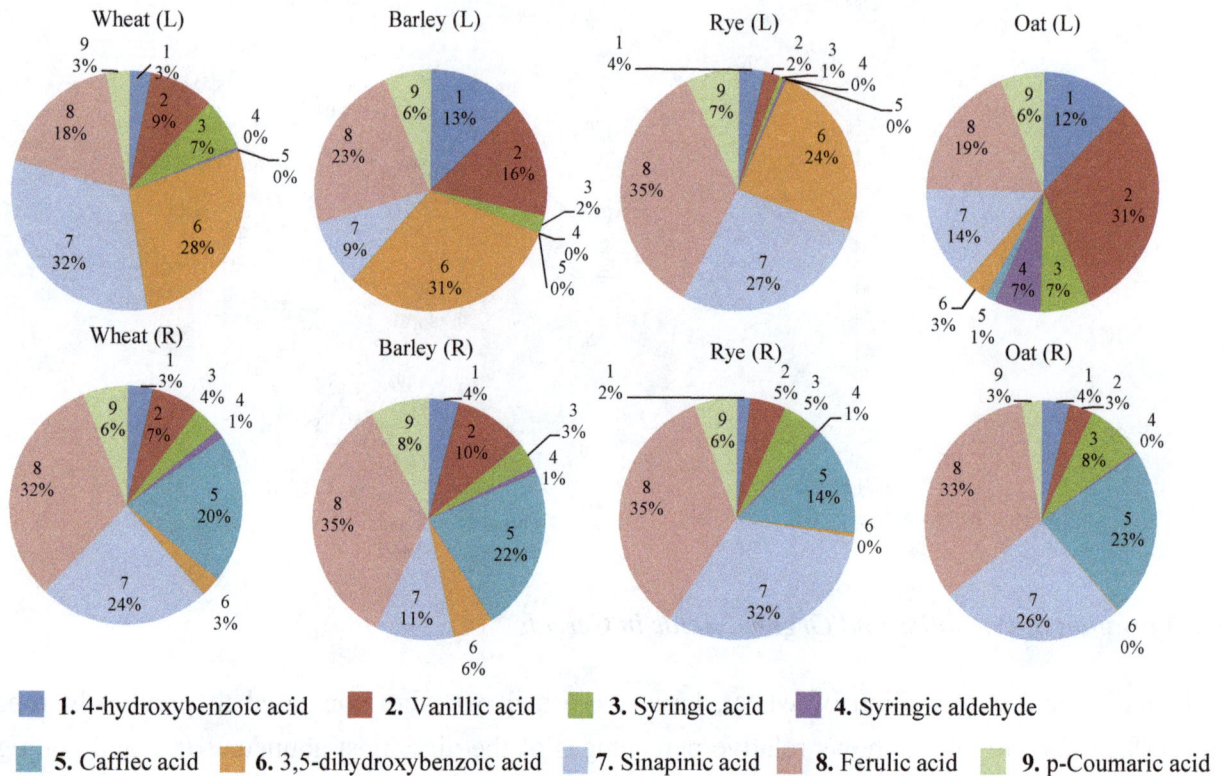

1. 4-hydroxybenzoic acid   2. Vanillic acid   3. Syringic acid   4. Syringic aldehyde
5. Caffiec acid   6. 3,5-dihydroxybenzoic acid   7. Sinapinic acid   8. Ferulic acid   9. p-Coumaric acid

Figures 5 and 6 demonstrate relative concentrations of phenolics and organic acids/alcohols of wheat, barley, rye and oat genotypes investigated in this study. Figure 5 show that ferulic, caffeic and sinapic acids and their methyl esters are the most abundant metabolites among all other phenolics in the cereal samples. Moreover, the relative concentrations of the most abundant phenolics are found to be up to three times greater in rye and oat than in wheat and barley. Succinic and 3-hydroxybutanoic acids were the most abundant metabolites among all organic acids detected in the four different cereals (Figure 5). Relative concentrations of fumaric and 2-hydroxycyclohexanecarboxylic acids were significantly higher in rye, while concentrations of malic and ketoglutaric acids were highest in barley.

Figure 5. Relative concentrations of 32 phenolics detected from wheat, barley, rye and oat. Metabolites are numbered according to the Table 1.

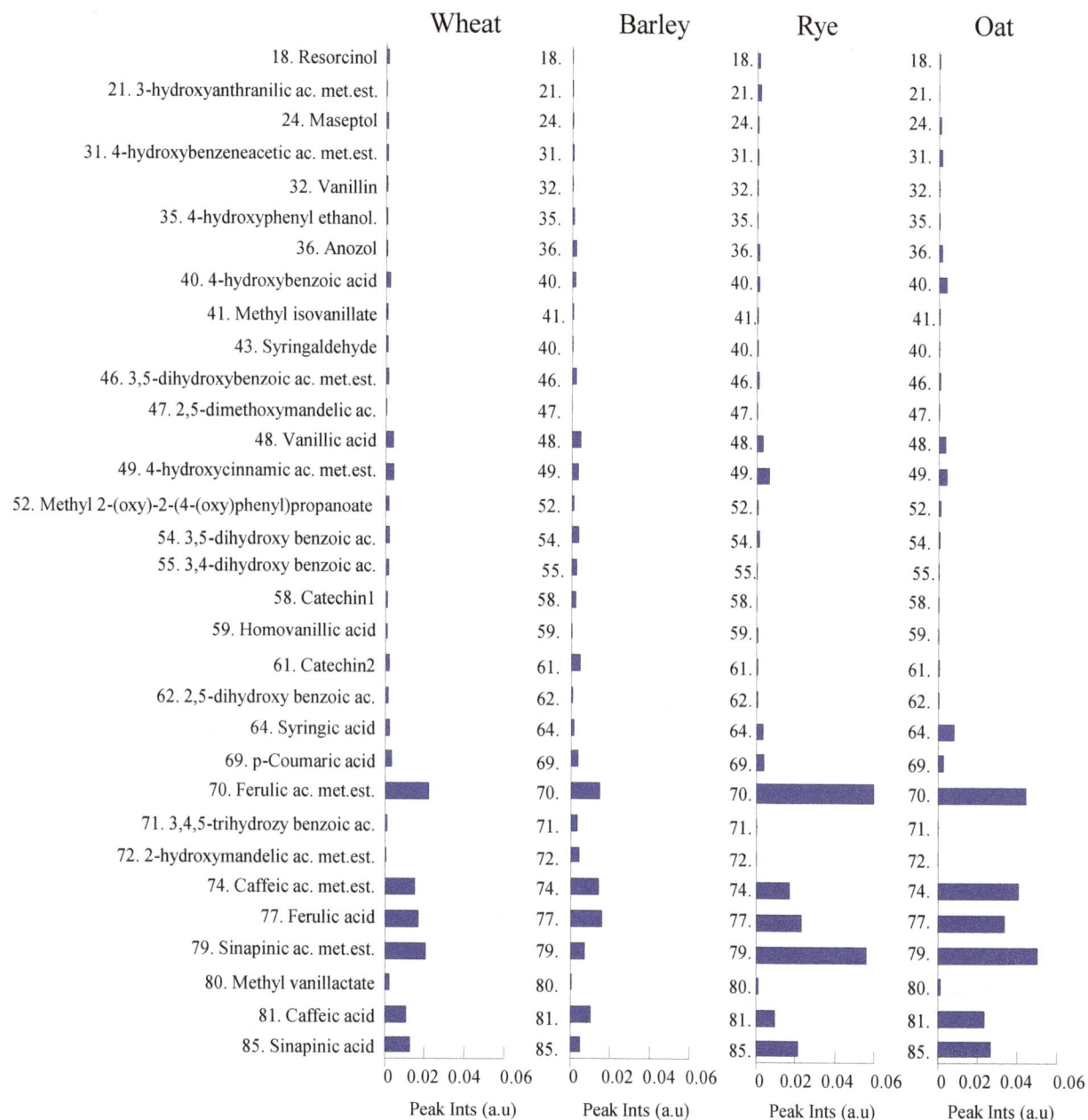

Figure 6. Relative concentrations of 29 organic acids/alcohols detected from wheat, barley, rye and oat. Metabolites are numbered according to the Table 1.

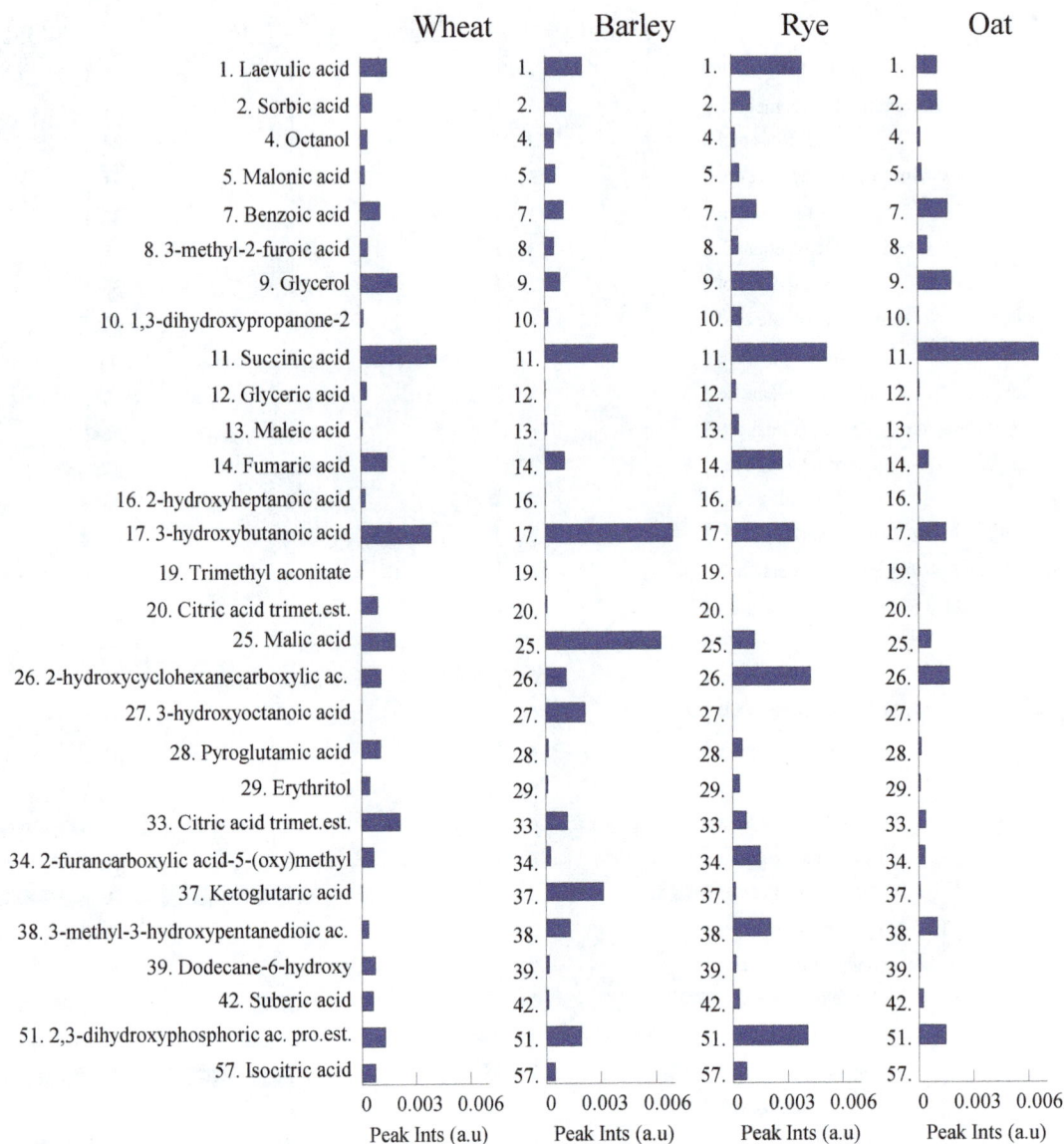

4. Conclusions

This paper outlines and demonstrates an optimized, relatively unbiased, comprehensive and high-throughput metabolomic profiling of whole-grain cereals based on new technologies developed within GC-MS metabolomics and chemometrics. A metabolite extraction protocol optimized towards phenolics and organic acids of whole-grains, and an unbiased and high-throughput protocol, was developed that allow processing of up to 60 samples per day. The hydrochloric acid based hydrolysis allowed extraction of all major cereal phenolics, free and conjugated, and enabled the detection of 32 phenolic and 30 organic acids from 50 mg of flour. A novel trimethylsilylation method based on TMSCN allowed the detection of up to 300 metabolites from the GC-MS profiles. The multi-way decomposition method PARAFAC2 facilitated deconvolution of overlapping, retention time shifted and low s/n ratio peaks with high precision and in a semi-automated manner. The resolved mass spectra of deconvoluted peaks allowed the identification of 89 metabolites using NIST and Golm metabolite databases. Multivariate and univariate analysis of phenolic profiles of cereals revealed that ferulic, caffeic and sinapinic acids and their esters were the main phenolics of whole-grain samples across the four cereals studied. Rye and oat showed higher concentrations of the most abundant phenolics acids compared to wheat and barley. Comparison of the relative concentrations of the nine most abundant phenolics of cereals with previously reported data showed that the acidic hydrolysis significantly improved detection of caffeic acid. However, metabolite profiles of cereals highly depend on several factors such as genotype, growth conditions, harvest time and storage. Thus, essential secondary metabolite profile comparisons of different cereals as well as different varieties require a strictly controlled experimental design. This paper has demonstrated a new methodology that is ready to be applied in a larger metabolomic profiling studies that may reveal biological information related to phenolic and organic acids of whole-grain cereals. Moreover, the protocol developed can easily be modified for polar metabolite fractions, including mono- and di-saccharides and amino acids, of cereals by altering metabolite extraction method and the additional of a methoximation step in GC-MS derivatization.

Acknowledgements

Faculty of Science is acknowledged for support to the elite-research area "metabolomics and bioactive compounds" with a PhD stipendium to B. Khakimov and The Ministry of Science and Technology is acknowledged for a grant to University of Copenhagen (S.B. Engelsen) with the title "metabolomics infrastructure" under which the GC-MS was acquired.

Author Contributions

B.K. B.M.J. and S.B.E. designed the study; B.K. conducted the GC-MS analysis. B.K. and S.B.E. performed the chemometric analysis and drafted the manuscript. All authors contributed to, read and approved the final manuscript.

Conflicts of Interest

The authors declare no conflict of interest.

References

1. Zilic, S.; Sukalovic, V.H.T.; Dodig, D.; Maksimovic, V.; Maksimovic, M.; Basic, Z. Antioxidant activity of small grain cereals caused by phenolics and lipid soluble antioxidants. *J. Cereal Sci* **2011**, *54*, 417–424.
2. Björck, I.; Östman, E.; Kristensen, M.; Anson, N.M.; Price, R.K.; Haenen, G.R.M.M.; Havenaar, R.; Knudsen, K.E.B.; Frid, A.; Mykkänen, H.; *et al.* Cereal grains for nutrition and health benefits: Overview of results from *in vitro*, animal and human studies in the HEALTHGRAIN project. *Trends Food Sci. Technol.* **2012**, *25*, 87–100.

3.   Andersson, A.A.M.; Andersson, R.; Piironen, V.; Lampi, A.M.; Nystrom, L.; Boros, D.; Fras, A.; Gebruers, K.; Courtin, C.M.; Delcour, J.A.; *et al.* Contents of dietary fibre components and their relation to associated bioactive components in whole grain wheat samples from the HEALTHGRAIN diversity screen. *Food Chem.* **2013**, *136*, 1243–1248.

4.   Amarowicz, R.; Zegarska, Z.; Pegg, R.B.; Karamac, M.; Kosinska, A. Antioxidant and radical scavenging activities of a barley crude extract and its fractions. *Czech J. Food Sci.* **2007**, *25*, 73–80.

5.   Wood, P.J. Cereal beta-glucans in diet and health. *J. Cereal Sci.* **2007**, *46*, 230–238.

6.   Mcintosh, G.H.; Whyte, J.; Mcarthur, R.; Nestel, P.J. Barley and wheat foods—Influence on plasma-cholesterol concentrations in hypercholesterolemic men. *Am. J. Clin. Nutr.* **1991**, *53*, 1205–1209.

7.   Madhujith, T.; Shahidi, F. Antioxidative and antiproliferative properties of selected barley (*Hordeum vulgarae* L.) cultivars and their potential for inhibition of low-density lipoprotein (LDL) cholesterol oxidation. *J. Agric. Food Chem.* **2007**, *55*, 5018–5024.

8.   Behall, K.M.; Scholfield, D.J.; Hallfrisch, J. Diets containing barley significantly reduce lipids in mildly hypercholesterolemic men and women. *Am. J. Clin. Nutr.* **2004**, *80*, 1185–1193.

9.   Gibson, S.M.; Strauss, G. Implication of phenolic-acids as texturizing agents during cooking-extrusion cereals. *Abstr. Pap. Am. Chem. Soc.* **1991**, *202*, 150.

10. Vinson, J.A.; Erk, K.M.; Wang, S.Y.; Marchegiani, J.Z.; Rose, M.F. Total polyphenol antioxidants in whole grain cereals and snacks: Surprising sources of antioxidants in the US diet. *Abstr. Pap. Am. Chem. Soc.* **2009**, *238*, 246.

11. Khakimov, B.; Bak, S.; Engelsen, S.B. High-throughput cereal metabolomics: Current analytical technologies, challenges and perspectives. *J. Cereal Sci.* **2014**, *59*, 393–418.

12. Soltesz, A.; Smedley, M.; Vashegyi, I.; Galiba, G.; Harwood, W.; Vagujfalvi, A. Transgenic barley lines prove the involvement of TaCBF14 and TaCBF15 in the cold acclimation process and in frost tolerance. *J. Exp. Bot.* **2013**, *64*, 1849–1862.

13. Widodo; Patterson, J.H.; Newbigin, E.; Tester, M.; Bacic, A.; Roessner, U. Metabolic responses to salt stress of barley (*Hordeum vulgare* L.) cultivars, Sahara and Clipper, which differ in salinity tolerance. *J. Exp. Bot.* **2009**, *60*, 4089–4103.

14. Manavalan, L.P.; Chen, X.; Clarke, J.; Salmeron, J.; Nguyen, H.T. RNAi-mediated disruption of squalene synthase improves drought tolerance and yield in rice. *J. Exp. Bot.* **2012**, *63*, 163–175.

15. Balmer, D.; Flors, V.; Glauser, G.; Mauch-Mani, B. Metabolomics of cereals under biotic stress: Current knowledge and techniques. *Front. Plant Sci.* **2013**, *4*, 82.

16. Fernie, A.R.; Schauer, N. Metabolomics-assisted breeding: A viable option for crop improvement? *Trends Genet.* **2009**, *25*, 39–48.

17. Bino, R.J.; Hall, R.D.; Fiehn, O.; Kopka, J.; Saito, K.; Draper, J.; Nikolau, B.J.; Mendes, P.; Roessner-Tunali, U.; Beale, M.H.; *et al.* Potential of metabolomics as a functional genomics tool. *Trends Plant Sci.* **2004**, *9*, 418–425.

18. Li, L.; Shewry, P.R.; Ward, J.L. Phenolic acids in wheat varieties in the HEALTHGRAIN diversity screen. *J. Agric. Food Chem.* **2008**, *56*, 9732–9739.

19. Fernandez-Orozco, R.; Li, L.; Harflett, C.; Shewry, P.R.; Ward, J.L. Effects of environment and genotype on phenolic acids in wheat in the HEALTHGRAIN diversity screen. *J. Agric. Food Chem.* **2010**, *58*, 9341–9352.

20. Shewry, P.R.; Piironen, V.; Lampi, A.M.; Edelmann, M.; Kariluoto, S.; Nurmi, T.; Fernandez-Orozco, R.; Ravel, C.; Charmet, G.; Andersson, A.A.M.; *et al.* The HEALTHGRAIN wheat diversity screen: Effects of genotype and environment on phytochemicals and dietary fiber components. *J. Agric. Food Chem.* **2010**, *58*, 9291–9298.

21. Andersson, A.A.M.; Lampi, A.M.; Nystrom, L.; Piironen, V.; Li, L.; Ward, J.L.; Gebruers, K.; Courtin, C.M.; Delcour, J.A.; Boros, D.; *et al.* Phytochemical and dietary fiber components in barley varieties in the HEALTHGRAIN diversity screen. *J. Agric. Food Chem.* **2008**, *56*, 9767–9776.

22. Shewry, P.R.; Piironen, V.; Lampi, A.M.; Nystrom, L.; Li, L.; Rakszegi, M.; Fras, A.; Boros, D.; Gebruers, K.; Courtin, C.M.; *et al.* Phytochemical and fiber components in oat varieties in the HEALTHGRAIN diversity screen. *J. Agric. Food Chem.* **2008**, *56*, 9777–9784.

23. Nyström, L.; Lampi, A.M.; Andersson, A.A.M.; Kamal-Eldin, A.; Gebruers, K.; Courtin, C.M.; Delcour, J.A.; Li, L.; Ward, J.L.; Fras, A.; *et al.* Phytochemicals and dietary fiber components in rye varieties in the HEALTHGRAIN diversity screen. *J. Agric. Food Chem.* **2008**, *56*, 9758–9766.

24. Ward, J.L.; Poutanen, K.; Gebruers, K.; Piironen, V.; Lampi, A.M.; Nystrom, L.; Andersson, A.A.M.; Aman, P.; Boros, D.; Rakszegi, M.; *et al.* The HEALTHGRAIN cereal diversity screen: Concept, results, and prospects. *J. Agric. Food Chem.* **2008**, *56*, 9699–9709.

25. Arranz, S.; Calixto, F.S. Analysis of polyphenols in cereals may be improved performing acidic hydrolysis: A study in wheat flour and wheat bran and cereals of the diet. *J. Cereal Sci.* **2010**, *51*, 313–318.

26. Sani, I.M.; Iqbal, S.; Chan, K.W.; Ismail, M. Effect of acid and base catalyzed hydrolysis on the yield of phenolics and antioxidant activity of extracts from germinated brown rice (GBR). *Molecules* **2012**, *17*, 7584–7594.

27. Khakimov, B.; Motawia, M.S.; Bak, S.; Engelsen, S.B. The use of trimethylsilyl cyanide derivatization for robust and broad-spectrum high-throughput gas chromatography-mass spectrometry based metabolomics. *Anal. Bioanal. Chem.* **2013**, *405*, 9193–9205.

28. Bro, R.; Andersson, C.A.; Kiers, H.A.L. PARAFAC2—Part II. Modeling chromatographic data with retention time shifts. *J. Chemom.* **1999**, *13*, 295–309.

29. Amigo, J.M.; Skov, T.; Coello, J.; Maspoch, S.; Bro, R. Solving GC-MS problems with PARAFAC2. *Trac-Trends Anal. Chem.* **2008**, *27*, 714–725.

30. Khakimov, B.; Amigo, J.M.; Bak, S.; Engelsen, S.B. Plant metabolomics: Resolution and quantification of elusive peaks in liquid chromatography-mass spectrometry profiles of complex plant extracts using multi-way decomposition methods. *J. Chromatogr. A* **2012**, *1266*, 84–94.

31. Vandendool, H.; Kratz, P.D. A generalization of retention index system including linear temperature programmed gas-liquid partition chromatography. *J. Chromatogr.* **1963**, *11*, 463.

32. Golm Metabolome Database. Available online: http://gmd.mpimp-golm.mpg.de/ (accessed on 5 November 2013).

33. Hotelling, H. Analysis of a complex of statistical variables into principal components. *J. Educ. Psychol.* **1933**, *24*, 417–441.

34. Sumner, L.; Amberg, A.; Barrett, D.; Beale, M.; Beger, R.; Daykin, C.; Fan, T.; Fiehn, O.; Goodacre, R.; Griffin, J.; *et al.* Proposed minimum reporting standards for chemical analysis. *Metabolomics* **2007**, *3*, 211–221.

14

Molecular Typing of *Campylobacter jejuni* and *Campylobacter coli* Isolated from Various Retail Meats by MLST and PFGE

Aneesa Noormohamed and Mohamed K. Fakhr *

Department of Biological Science, The University of Tulsa, Tulsa, OK 74104, USA;
E-Mail: aneesa-noormohamed@utulsa.edu

* Author to whom correspondence should be addressed; E-Mail: Mohamed-fakhr@utulsa.edu

Abstract: *Campylobacter* species are one of the leading causes of foodborne disease in the United States. *Campylobacter jejuni* and *Campylobacter coli* are the two main species of concern to human health and cause approximately 95% of human infections. Molecular typing methods, such as pulsed-field gel electrophoresis (PFGE) and multilocus sequence typing (MLST) are often used to source track foodborne bacterial pathogens. The aim of the present study was to compare PFGE and MLST in typing strains of *C. jejuni* and *C. coli* that were isolated from different Oklahoma retail meat sources. A total of 47 *Campylobacter* isolates (28 *C. jejuni* and 19 *C. coli*) isolated from various retail meat samples (beef, beef livers, pork, chicken, turkey, chicken livers, and chicken gizzards) were subjected to pulsed-field gel electrophoresis (PFGE) and multilocus sequence typing (MLST). PFGE was able to group the 47 *Campylobacter* isolates into two major clusters (one for *C. jejuni* and one for *C. coli*) but failed to differentiate the isolates according to their source. MLST revealed 21 different sequence types (STs) that belonged to eight different clonal complexes. Twelve of the screened *Campylobacter* isolates (8 *C. jejuni* and 4 *C. coli*) did not show any defined STs. All the defined STs of *C. coli* isolates belonged to ST-828 complex. The majority of *C. jejuni* isolates belonged to ST-353, ST-607, ST-52, ST-61, and ST-21 complexes. It is worthy to mention that, while the majority of *Campylobacter* isolates in this study showed STs that are commonly associated with human infections along with other sources, most of the STs from chicken livers were solely reported in human cases. In conclusion, retail meat *Campylobacter* isolates tested in this study particularly those from chicken livers showed relatedness to STs commonly associated with humans. Molecular typing, particularly MLST, proved to be a helpful tool in suggesting this relatedness to *Campylobacter* human isolates.

Keywords: *Campylobacter*; MLST; PFGE; molecular typing; retail meats; poultry; beef; pork; livers; foodborne pathogens

---

1. Introduction

Campylobacter is a foodborne pathogen that is one of the leading causes of bacterial gastroenteritis [1]. It causes an estimated 1.3 million infections a year [1]. It is the third most common cause of bacterial foodborne illness in the United States, after *Salmonella* [1]. The most common species isolated are *Campylobacter jejuni* and *Campylobacter coli*, which, together, cause around 95% of all *Campylobacter* infections [2,3]. Contaminated food is the most common mode of infection with *Campylobacter*. The most common food source is poultry [4].

Molecular typing is used to differentiate between isolates of the same species of bacteria [5]. Genotyping methods can be used to identify the genetic relatedness between different strains of bacteria. In order to track *Campylobacter* infections, various genotyping methods are used, such as pulsed-field gel electrophoresis (PFGE) and multilocus sequence typing (MLST). PFGE is based on gel electrophoresis of restriction digested genomic DNA. Traditional gel electrophoresis has a constant current in one direction so only small fragments can enter the gel and be separated. In PFGE, the direction of current changes regularly (pulsed) and, thus, large fragments twist and move slowly through the gel [6,7]. The pattern of the bands determines the relatedness of the isolates. PFGE is considered the "gold standard" in molecular typing for most bacteria, including foodborne pathogens, as the entire genome of the microbe is analyzed to create restriction profiles [8,9], however, it has its disadvantages in that it requires expensive equipment and complicated protocols, in addition to which, there are no standard methods for the interpretation of data, or sharing of this data with other scientists [9–11]. In fact, the genetic variation among *Campylobacter* becomes a concern when using PFGE for genotyping [12]. Some strains are not typable using either of the commonly used restriction enzymes *Sma*I or *Kpn*I, which bring about questions as to the usefulness of PFGE with *Campylobacter* species [11,13].

MLST typing is based on gene sequences of seven selected genes, which are considered "housekeeping" genes. These genes are selected as they are fairly conserved. For each gene, each recorded sequence is given a number. The resulting seven-digit number defines the isolate. The isolates with related sequence types (STs) can also be grouped together into clonal complexes. There may be minor differences between "identical" MLST isolates (e.g., from the PFGE pattern). In the case of MLST, there is a useful website that has, not only *Campylobacter* MLST primers, but it also allows scientists to input the gene sequences, which can then be accessed by scientists worldwide to compare, and the protocols are not as involved [14]. MLST also has the discriminatory power to characterize hypervariable genomes, such as those of *Campylobacter* [15], although it isn't able to separate closely-related isolates [16,17]. More recently, developed MLST protocols to study several different bacterial pathogens became available [18,19]. The MLST schemes for *C. jejuni* and *C. coli* have previously been determined [14,20] and are available on the MLST website [21]. The MLST website also carries information on several different sequence types and is used to share this information with scientists worldwide. Alternate schemes that do not use all the same genes are also available [22,23].

The objective of this study was to determine the genetic relatedness among 47 strains of *C. jejuni* and *C. coli*, isolated from different retail meat sources and to determine if one typing method is superior to the other one in determining such relatedness.

2. Experimental Section

2.1. Bacterial Isolates

Forty-seven *C. jejuni* and *C. coli*, previously isolated from retail meat samples, were used in this study for MLST and PFGE [24,25]. The isolates were selected to represent both species (*C. jejuni* and *C. coli*), several meat brands, as well as different retail meat sources, such as chicken (breast and thighs), turkey (breast, thighs, neck pieces, and ground), beef livers, pork (tongue), chicken livers, and chicken gizzards (Table 1). All isolates were kept frozen at −80 °C in Brucella broth (Becton Dickinson, Sparks, MD, USA) with 20% glycerol.

Table 1. Number and sources of isolates for each *Campylobacter* species used in this study.

Source	C. jejuni	C. coli
Chicken	7	3
Chicken Livers	4	5
Chicken Gizzards	7	2
Turkey	5	2
Beef Livers	5	5
Pork	0	2
Total	28	19

2.2. Pulsed-Field Gel Electrophoresis Typing

The isolates were typed by PFGE following the PulseNet protocol for *Campylobacter* [26]. Briefly, isolates were grown on MH agar with 5% laked-horse blood and then diluted to the required concentration and agarose-embedded plugs were made and washed. They were then digested with *SmaI* restriction enzyme (Promega, Madison, WI, USA). The digested plugs were run in Seakem agarose gel (Lonza, Allendale, NJ, USA) with 0.5× Tris-Borate EDTA (TBE) buffer (Amresco, Solon, OH, USA) to separate the bands on the CHEF Mapper PFGE system (Bio-Rad) by running for 16 h at 14 °C switching directions every 6.76 s and ending with 35.38 s (25). *Salmonella enterica* serovar Braenderup digested with *XbaI* (Promega, Madison, WI, USA) was used as the molecular reference marker. Gels were stained with ethidium bromide and viewed and recorded under UV transillumination (UVP, Upland, CA, USA). Gel images were analyzed using BioNumerics software (Applied Maths, Austin, TX, USA). The banding patterns were clustered using Dice coefficients using unweighted pair group method, with arithmetic mean (UPGMA), and a 3% band tolerance.

2.3. Multilocus Sequence Typing

MLST was performed for the same 47 isolates that were typed in the PFGE study. PCR for each of the following seven housekeeping genes was performed: *aspA* (aspartase A), *glnA* (glutamine synthetase), *gltA* (citrate synthase), *glyA* (serine hydroxymethyltransferase), *pgm* (phosphoglucomutase), *tkt* (transketolase), and *uncA* (ATP synthase α subunit) [14,27]. Bacterial DNA extracts used in polymerase chain reaction (PCR) were prepared from *Campylobacter* cultures using the single cell lysing buffer (SCLB) method [28].

The selected isolates were tested for the presence of the seven different housekeeping genes used in the MLST scheme for *C. jejuni* by PCR reactions. The primers used were available at the MLST website [29,30] and are shown in Table 2. The PCR was carried out in 25 μL reactions. Each 25 μL reaction contained 12.5 μL GoTaq® Green Master Mix (Promega, Madison, WI, USA), 3.5 μL sterile water (Promega, Madison, WI, USA), 1 μL (25 pmol) each primer (IDT, Coralville, IA, USA), and 3 μL of template DNA. The cycling conditions were set as follows: (1) 95 °C for 5 min; (2) 94 °C for 1 min; (3) 50 °C for 1 min; (4) 72 °C for 1 min; and (5) 72 °C for 10 min. Steps 2 through 4 were repeated for 35 cycles. Once the cycles were complete, reactions were held at 4 °C until gel electrophoresis. Ten microliters of PCR product was subjected to horizontal electrophoresis in a 1% agarose gel in 1× Tris-acetate-EDTA (TAE) buffer. A 1 kb plus ladder (Bioneer, Alameda, CA, USA) was used as the molecular marker. Gels were viewed and recorded by ultraviolet transillumination, using a UV imager (UVP). Sterile water was used as the negative control.

The PCR products were purified using ExoSAP-IT enzyme (Affymetrix, Santa Clara, CA, USA). The sequencing PCR reaction was prepared according to a modified ABI 3130*xl* manufacturer′s sequencing protocol (Applied Biosystems, Foster City, CA, USA). Briefly, sequencing reactions were prepared to a 15 μL volume containing 3.5 μL purified PCR product, 1.5 μL primer, 0.5 μL sequencing buffer, 2 μL betaine, 0.5 μL BigDye, and 2 μL RNase-free water. The cycling conditions were set up according to the ABI capillary sequencer instructions (Applied Biosystems, Foster City, CA, USA). The sequenced products were then read using the ABI 3130*xl* (Applied Biosystems) and analyzed using BioNumerics software (Applied Maths, Austin, TX, USA), which has a function to determine STs and clonal complexes by directly submitting the sequences to the MLST website.

Table 2. Polymerase chain reaction (PCR) and sequencing primer sets for the multilocus sequence typing (MLST) scheme for *C. jejuni* and *C. coli*.

Genes	Primer Sequences	Use	References
aspA	5′-AGTACTAATGATGCTTATCC-3′ 5′-ATTTCATCAATTTGTTCTTTGC-3′	*C. jejuni* PCR	[14]
glnA	5′-TAGGAACTTGGCATCATATTACC-3′ 5′-TTGGACGAGCTTCTACTGGC-3′	*C. jejuni* PCR	[14]
gltA	5′-GGGCTTGACTTCTACAGCTACTTG-3′ 5′-CCAAATAAAGTTGTCTTGGACGG-3′	*C. jejuni* PCR	[14]
glyA	5′-GAGTTAGAGCGTCAATGTGAAGG-3′ 5′-AAACCTCTGGCAGTAAGGGC-3′	*C. jejuni* PCR	[14]
pgm	5′-TACTAATAATATCTTAGTAGG-3′ 5′-CACAACATTTTTCATTTCTTTTTC-3′	*C. jejuni* PCR	[14]

Table 2. *Cont.*

Genes	Primer Sequences	Use	References
tkt	5'-GCAAACTCAGGACACCCAGG-3' 5'-AAAGCATTGTTAATGGCTGC-3'	*C. jejuni* PCR	[14]
uncA	5'-ATGGACTTAAGAATATTATGGC-3' 5'-GCTAAGCGGAGAATAAGGTGG-3'	*C. jejuni* PCR	[14]
aspA	5'-AGTACTAATGATGCTTATCC-3' 5'-ATTTCATCAATTTGTTCTTTGC-3'	*C. jejuni* sequencing	[14]
glnA	5'-TAGGAACTTGGCATCATATTACC-3' 5'-TTGGACGAGCTTCTACTGGC-3'	*C. jejuni* sequencing	[14]
gltA	5'-GGGCTTGACTTCTACAGCTACTTG-3' 5'-CCAAATAAAGTTGTCTTGGACGG-3'	*C. jejuni* sequencing	[14]
glyA	5'-GAGTTAGAGCGTCAATGTGAAGG-3' 5'-AAACCTCTGGCAGTAAGGGC-3'	*C. jejuni* sequencing	[14]
pgm	5'-TACTAATAATATCTTAGTAGG-3' 5'-CACAACATTTTTCATTTCTTTTTC-3'	*C. jejuni* sequencing	[14]
tkt	5'-GCAAACTCAGGACACCCAGG-3' 5'-AAAGCATTGTTAATGGCTGC-3'	*C. jejuni* sequencing	[14]
uncA	5'-ATGGACTTAAGAATATTATGGC-3' 5'-GCTAAGCGGAGAATAAGGTGG-3'	*C. jejuni* sequencing	[14]
aspA	5'-CCAACTGCAAGATGCTGTACC-3' 5'-TTCATTTGCGGTAATACCATC-3'	*C. coli* PCR and sequencing	[27]
glnA	5'-CATGCAATCAATGAAGAAAC-3' 5'-TTCCATAAGCTCATATGAAC-3'	*C. coli* PCR and sequencing	[27]
gltA	5'-CTTATATTGATGGAGAAAATGG-3' 5'-CCAAAGCGCACCAATACCTG-3'	*C. coli* PCR and sequencing	[27]
glyA	5'-AGCTAATCAAGGTGTTTATGCGG-3' 5'-AGGTGATTATCCGTTCCATCGC-3'	*C. coli* PCR and sequencing	[27]
pgm	5'-GGTTTTAGATGTGGCTCATG-3' 5'-TCCAGAATAGCGAAATAAGG-3'	*C. coli* PCR and sequencing	[27]
tkt	5'-GCTTAGCAGATATTTTAAGTG-3' 5'-AAGCCTGCTTGTTCTTTGGC-3'	*C. coli* PCR and sequencing	[27]
uncA	5'-AAAGTACAGTGGCACAAGTGG-3' 5'-TGCCTCATCTAAATCACTAGC-3'	*C. coli* PCR and sequencing	[27]

3. Results and Discussion

3.1. Pulsed-Field Gel Electrophoresis

The results of the PFGE showed that the isolates studied were separated into two major groups according to their species (*C. jejuni* and *C. coli*) (Figure 1). The isolates were also able to be separated by their ST-complexes within the species groups. PFGE was able to group the 47 isolates into 2 major clusters (one for *C. jejuni* and one for *C. coli*) but wasn't able to differentiate the isolates by meat source within the species (Figure 1). By PFGE separation, the isolates were found to also cluster into their ST-complexes, such as for ST-61, ST-21, ST-52 for *C. jejuni*, and ST-828 for *C. coli* (Figure 1). Among the *C. jejuni* isolates, ST-61 was related to beef liver isolates and ST-21 and ST-52 related to chicken sources. ST-607 isolates did not all cluster closely together but they all belonged to poultry sources and to the same ST (1212). ST-353 isolates also did not cluster together, but they all belonged to poultry sources (Figure 1). When using PFGE for genotyping, genetic variation among *Campylobacter* strains becomes a concern [12]. Some strains are not typable, using either of the commonly used restriction enzymes *Sma*I or *Kpn*I, which creates questions about the usefulness of PFGE with *Campylobacter* species [11,13]. In our study PFGE was able to group the 47 *Campylobacter* isolates into two major clusters (one for *C. jejuni* and one for *C. coli*) but failed to differentiate the isolates according to their source.

Figure 1. The MLST and PFGE profile comparison of the *Campylobacter jejuni* and *Campylobacter coli* isolates. Symbols represent the different sources of the isolates. No ST defined means that no sequence type was identified for that particular isolate. ST, sequence type; ST complex, MLST clonal complex.

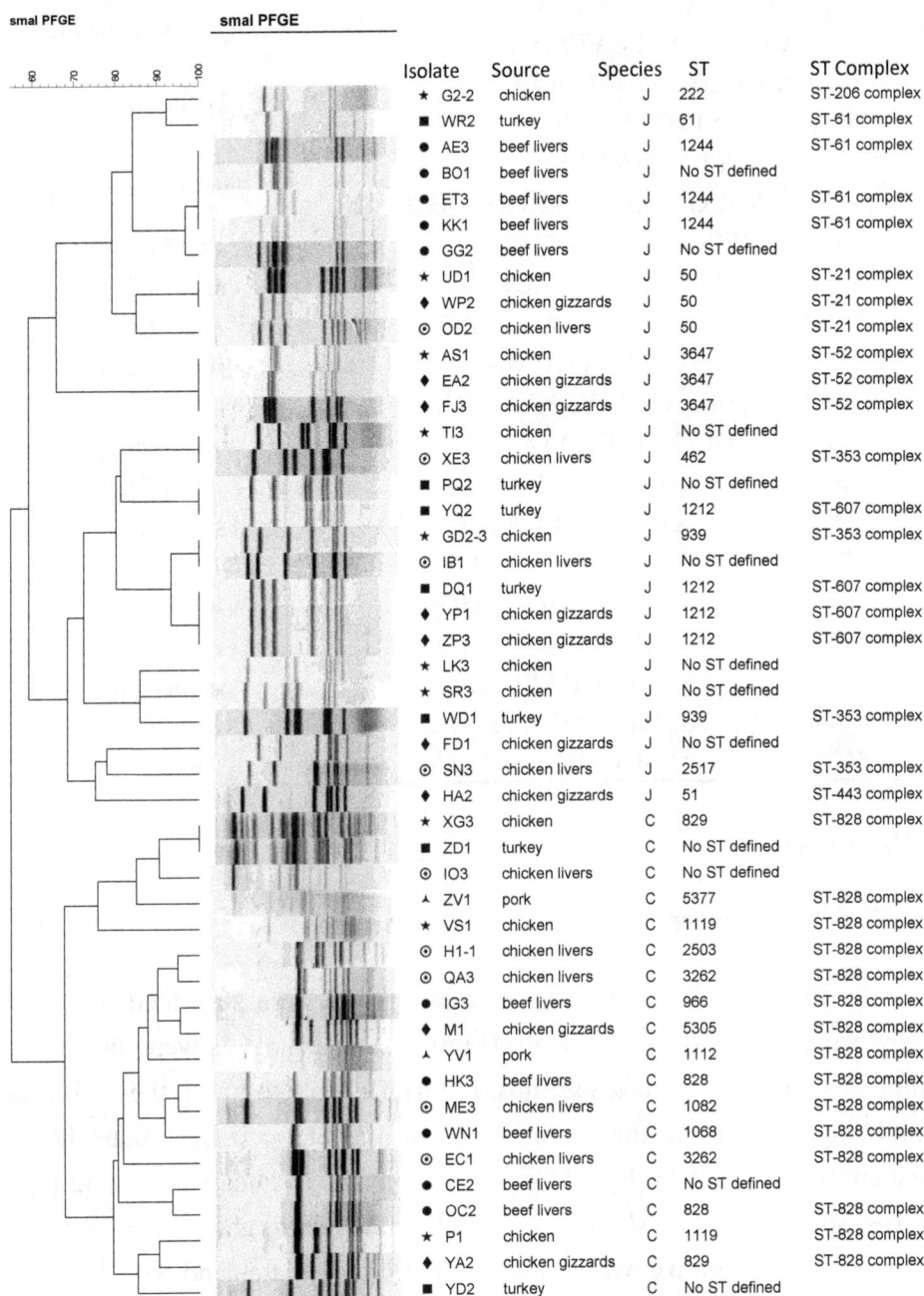

Isolate	Source	Species	ST	ST Complex
★ G2-2	chicken	J	222	ST-206 complex
■ WR2	turkey	J	61	ST-61 complex
● AE3	beef livers	J	1244	ST-61 complex
● BO1	beef livers	J	No ST defined	
● ET3	beef livers	J	1244	ST-61 complex
● KK1	beef livers	J	1244	ST-61 complex
● GG2	beef livers	J	No ST defined	
★ UD1	chicken	J	50	ST-21 complex
◆ WP2	chicken gizzards	J	50	ST-21 complex
⊙ OD2	chicken livers	J	50	ST-21 complex
★ AS1	chicken	J	3647	ST-52 complex
◆ EA2	chicken gizzards	J	3647	ST-52 complex
◆ FJ3	chicken gizzards	J	3647	ST-52 complex
★ TI3	chicken	J	No ST defined	
⊙ XE3	chicken livers	J	462	ST-353 complex
■ PQ2	turkey	J	No ST defined	
■ YQ2	turkey	J	1212	ST-607 complex
★ GD2-3	chicken	J	939	ST-353 complex
⊙ IB1	chicken livers	J	No ST defined	
■ DQ1	turkey	J	1212	ST-607 complex
◆ YP1	chicken gizzards	J	1212	ST-607 complex
◆ ZP3	chicken gizzards	J	1212	ST-607 complex
★ LK3	chicken	J	No ST defined	
★ SR3	chicken	J	No ST defined	
■ WD1	turkey	J	939	ST-353 complex
◆ FD1	chicken gizzards	J	No ST defined	
⊙ SN3	chicken livers	J	2517	ST-353 complex
◆ HA2	chicken gizzards	J	51	ST-443 complex
★ XG3	chicken	C	829	ST-828 complex
■ ZD1	turkey	C	No ST defined	
⊙ IO3	chicken livers	C	No ST defined	
⋏ ZV1	pork	C	5377	ST-828 complex
★ VS1	chicken	C	1119	ST-828 complex
⊙ H1-1	chicken livers	C	2503	ST-828 complex
⊙ QA3	chicken livers	C	3262	ST-828 complex
● IG3	beef livers	C	966	ST-828 complex
◆ M1	chicken gizzards	C	5305	ST-828 complex
⋏ YV1	pork	C	1112	ST-828 complex
● HK3	beef livers	C	828	ST-828 complex
⊙ ME3	chicken livers	C	1082	ST-828 complex
● WN1	beef livers	C	1068	ST-828 complex
⊙ EC1	chicken livers	C	3262	ST-828 complex
● CE2	beef livers	C	No ST defined	
● OC2	beef livers	C	828	ST-828 complex
★ P1	chicken	C	1119	ST-828 complex
◆ YA2	chicken gizzards	C	829	ST-828 complex
■ YD2	turkey	C	No ST defined	

3.2. Multilocus Sequence Typing

MLST was able to separate the isolates into 21 different STs, which belonged to eight different clonal complexes (Figure 1). Twelve of the isolates (eight *C. jejuni* and four *C. coli*) were not assigned STs (Figure 1). All the defined STs of *C. coli* isolates belonged to ST-828 complex. The majority of the *C. jejuni* isolates grouped into ST-353, ST-607, ST-52, and ST-61 clonal complexes (Figure 1). The most common clonal complex observed in this study was the ST-828 complex, which consists mostly of *C. coli* isolates. Other clonal complexes identified were ST-21, ST-61, ST-52, ST-206, ST-353, ST-443, and ST-607. Cluster analysis by Minimum Spanning Tree created using the BioNumerics software also shows that the isolates clustered into two distinct groups according to their species, which are *C. jejuni* and *C. coli* (Figure 2). *Campylobacter jejuni* appears to be more diverse than *Campylobacter coli* in regards to their STs and clonal complexes distributions (Figure 2).

Figure 2. Minimum spanning tree showing the clustering of the *Campylobacter* STs showing the two species type clusters. Each circle represents a clonal complex. The number inside each circle is the clonal complex. Red circles denote isolates that were not defined. The size of the circle is proportional to the number of strains represented. Thick lines denote closer association between the groups and thin lines denote less. The dashed lines denote least association between members.

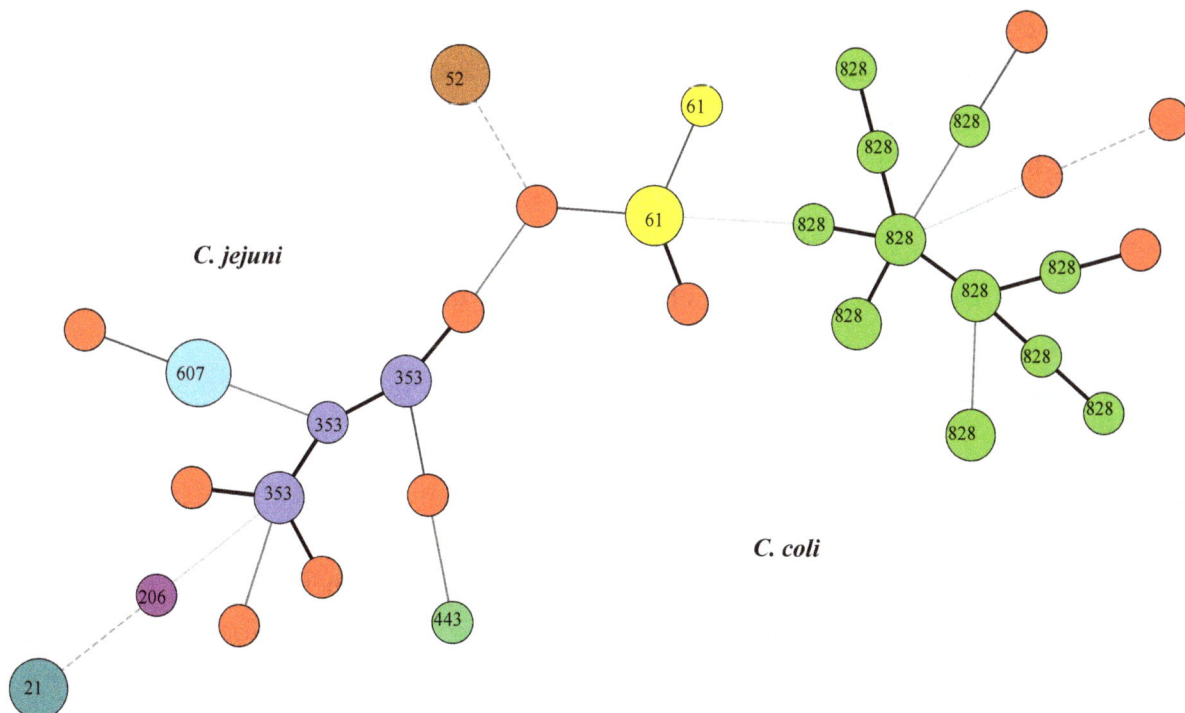

Figure 2 shows the separation of the isolates by MLST according to their ST-complexes. The isolates were separated according to the species of *Campylobacter* they belong to and, within those groups, there is a separation into ST-complexes. The unidentified isolates were also added and show affiliation with one group or the other. The most common clonal complex was ST-828 consisting of *C. coli* isolates, which has been previously observed [11,17,27,31,32]. ST-828 was also reported as the most common complex among *C. coli* isolates by other studies [11,27,31–34]. This was also the case with the *C. coli* isolates in this study for all of our defined isolates. Most human infections are caused by complexes ST-828 and ST-1150 [17]. In fact, the ST-828 complex is commonly associated with human isolates, as well as chicken meat or offal, according to data from the MLST website [17]. ST-21 has also been reported to be the most commonly detected complex among the isolates that were published on the MLST website with the most isolates from human origins [17], and the next most common belong to chicken isolates [17,35,36]. ST-21 has also been found in bovine and ovine isolates [37,38]. Chicken is also the source of ST-52, ST-61, ST-206, ST-353, and ST-828 in other studies [5]. In our study, all of these ST complexes belonged to various poultry isolates. ST-353 and ST-21 were also previously reported among *C. jejuni* isolates [5,34,37,38]. Colles *et al.*, 2003 [39], reported in their study that the ST-61 complex was associated with sheep isolates. Data collected from the MLST website in 2012 by Colles and Maiden [17], found that ST-61 was most commonly associated with human isolates and the next most common source was beef offal or meat. ST-206 was also most associated with human isolates in that study [17].

The fact that all the *C. coli* isolates belong to the same complex could be due to the more conservative nature of the *C. coli* genome and the *C. jejuni* genome being more variable [27,40]. Colles *et al.* [39], and Manning *et al.* [22], found that there was no association of the STs with the host, inferring that this could be due to the lack of diversity among *C. coli* or possibly due to their sample not being very diverse. Miller *et al.* [31], reported that, in their MLST studies, there was some association among *C. coli* and their STs to specific hosts suggesting that source tracking would be possible with *C. coli*.

Most of the STs identified in this study were found to be associated with human and other sources of infection. Most of the STs found in the chicken livers were STs associated with human infection only according to the MLST website [29]. Adding to the importance of chicken livers as a public health risk is the recent discovery by Strachan *et al.* [41], that molecular source attribution by MLST demonstrated that *Campylobacter* strains from chicken livers were most similar to those found commonly in humans, which provides further evidence that chicken liver is a probable source of human infection.

4. Conclusions

In conclusion, retail meat *Campylobacter* isolates tested in this study particularly those from chicken livers showed relatedness to STs commonly associated with humans. Molecular typing particularly MLST proved to be a helpful tool in suggesting this relatedness to *Campylobacter* human isolates and can be regarded as superior to PFGE in this regard. The genetic variation among *C. jejuni* strains appeared higher than that among *C. coli* strains using MLST in our study.

Acknowledgments

The authors would like to acknowledge financial support from the Graduate School of The University of Tulsa (Tulsa, OK, USA) for granting Aneesa Noormohamed a Bellwether Doctoral Dissertation Fellowship.

Conflicts of Interest

The authors declare no conflict of interest.

References

1. Centers for Disease Control (CDC). CDC Estimates of Foodborne Illness in the United States, 2011. Available online: http://www.cdc.gov/foodborneburden/2011-foodborne-estimates.html (accessed on 16 May 2013).
2. Debruyne, L.; Gevers, D.; Vandamme, P. Taxonomy of the Family *Campylobacteraceae*. In *Campylobacter*, 3rd ed.; Nachamkin, I., Szymanski, C.M., Blaser, M.J., Eds.; American Society for Microbiology: Washington, DC, USA, 2008; pp. 3–26.
3. Lastovica, A.J.; Allos, B.M. Clinical Significance of *Campylobacter* and Related Species Other than *Campylobacter jejuni* and *Campylobacter coli*. In *Campylobacter*, 3rd ed.; Szymanski, C.M., Blaser, M.J., Eds.; American Society for Microbiology: Washington, DC, USA, 2008; pp. 123–149.
4. Zhao, S.; Young, S.R.; Tong, E.; Abbott, J.W.; Womack, N.; Friedman, S.L.; McDermott, P.F. Antimicrobial resistance of *Campylobacter* isolates from retail meat in the United States: 2002–2007. *Appl. Environ. Microbiol.* **2010**, *76*, 7949–7956.
5. Behringer, M.; Miller, W.G.; Oyarzabal, O.A. Typing of *Campylobacter jejuni* and *Campylobacter coli* isolated from live broilers and retail broiler meat by flaA-RFLP, MLST, PFGE and REP-PCR. *J. Microbiol. Methods* **2011**, *84*, 194–201.
6. Lukinmaa, S.; Nakari, U.M.; Eklund, M.; Siitonen, A. Application of molecular genetic methods in diagnostics and epidemiology of food-borne bacterial pathogens. *APMIS* **2004**, *112*, 908–929.
7. Van Belkum, A.; Tassios, P.T.; Dijkshoorn, L.; Haeggman, S.; Cookson, B.; Fry, N.K.; Fussing, V.; Green, J.; Feil, E.; Gerner-Smidt, P.; *et al.* Guidelines for the validation and application of typing methods for use in bacterial epidemiology. *Clin. Microbiol. Infect.* **2007**, *13*, 1–46.
8. Maslow, J.N.; Slutsky, A.M.; Arbeit, R.D. The Application of Pulsed-Field Gel Electrophoresis to Molecular Epidemiology. In *Diagnostic Molecular Microbiology*; Persing, D.H., Tenover, F.C., Smith, T.F., White, T.J., Eds.; ASM Press: Washington, DC, USA, 1993; pp. 563–572.
9. Olive, D.M.; Bean, P. Principles and applications of methods for DNA-based typing of microbial organisms. *J. Clin. Microbiol.* **1999**, *37*, 1661–1669.
10. Wassenaar, T.M.; Newell, D.G. Genotyping of *Campylobacter* spp. *Appl. Environ. Microbiol.* **2000**, *66*, 1–9.
11. Thakur, S.; White, D.G.; McDermott, P.F.; Zhao, S.; Kroft, B.; Gebreyes, W.; Abbott, J.; Cullen, P.; English, L.; Carter, P.; *et al.* Genotyping of *Campylobacter coli* isolated from humans and retail meats using multilocus sequence typing and pulsed-field gel electrophoresis. *J. Appl. Microbiol.* **2009**, *106*, 1722–1733.
12. On, S.L.; Nielsen, E.M.; Engberg, J.; Madsen, M. Validity of SmaI-defined genotypes of *Campylobacter jejuni* examined by *Sal*I, *Kpn*I, and *Bam*HI polymorphisms: Evidence of identical clones infecting humans, poultry, and cattle. *Epidemiol. Infect.* **1998**, *120*, 231–237.

13. Oyarzabal, O.A.; Backert, S.; Williams, L.L.; Lastovica, A.J.; Miller, R.S.; Pierce, S.J.; Vieira, S.L.; Rebollo-Carrato, F. Molecular typing of *Campylobacter jejuni* strains isolated from commercial broilers in Puerto Rico. *J. Appl. Microbiol.* **2008**, *105*, 800–812.

14. Dingle, K.E.; Colles, F.M.; Wareing, D.R.A.; Ure, R.; Fox, A.J.; Bolton, F.J.; Bootsma, R.J.L.; Willems, R.; Urwin, R.; Maiden, M.C.J. Multilocus sequence typing system for *Campylobacter jejuni*. *J. Clin. Microbiol.* **2001**, *39*, 14–23.

15. Dingle, K.E.; Colles, F.M.; Ure, R.; Wagenaar, J.A.; Duim, B.; Bolton, F.J.; Fox, A.J.; Wareing, D.R.; Maiden, M.C. Molecular characterization of *Campylobacter jejuni* clones: A basis for epidemiologic investigation. *Emerg. Infect. Dis.* **2002**, *8*, 949–955.

16. Clark, C.G.; Taboada, E.; Grant, C.C.R.; Blakeston, C.; Pollari, F.; Marshall, B.; Rahn, K.; Mackinnon, J.; Daignault, D.; Pillai, D.; *et al.* Comparison of molecular typing methods useful for detecting clusters of *Campylobacter jejuni* and *C. coli* isolates through routine surveillance. *J. Clin. Microbiol.* **2012**, *50*, 798–809.

17. Colles, F.M.; Maiden, M.C.J. *Campylobacter* sequence typing databases: Applications and future prospects. *Microbiology* **2012**, *158*, 2695–2709.

18. Kotetishvili, M.; Stine, O.C.; Chen, Y.; Kreger, A.; Sulakvelidze, A.; Sozhamannan, S.; Morris, J.G., Jr. Multilocus sequence typing has better discriminatory ability for typing *Vibrio cholerae* than does pulsed-field gel electrophoresis and provides a measure of phylogenetic relatedness. *J. Clin. Microbiol.* **2003**, *41*, 2191–2196.

19. Alcaine, S.D.; Soyer, Y.; Warnick, L.D.; Su, W.L.; Sukhnanand, S.; Richards, J.; Fortes, E.D.; McDonough, P.; Root, T.P.; Dumas, N.B.; *et al.* Multilocus sequence typing supports the hypothesis that cow- and human-associated *Salmonella* isolates represent distinct and overlapping populations. *Appl. Environ. Microbiol.* **2006**, *72*, 7575–7585.

20. Miller, W.G.; On, S.L.; Wang, G.; Fontanoz, S.; Lastovica, A.J.; Mandrell, R.E. Extended multilocus sequence typing system for *Campylobacter coli*, *C. lari*, *C. upsaliensis*, and *C. helveticus. J. Clin. Microbiol.* **2005**, *43*, 2315–2329.

21. Campylobacter MLST Home Page. Available online: www.pubmlst.org/campylobacter/ (accessed on 15 March 2012).

22. Manning, G.; Dowson, C.G.; Bagnall, M.C.; Ahmed, I.H.; West, M.; Newell, D.G. Multilocus sequence typing for comparison of veterinary and human isolates of *Campylobacter jejuni*. *Appl. Environ. Microbiol.* **2003**, *69*, 6370–6379.

23. Suerbaum, S.; Lohrengel, M.; Sonnevend, A.; Ruberg, F.; Kist, M. Allelic diversity and recombination in *Campylobacter jejuni*. *J. Bacteriol.* **2001**, *183*, 2553–2559.

24. Noormohamed, A.; Fakhr, M.K. Incidence and antimicrobial resistance profiling of *Campylobacter* in retail chicken livers and gizzards. *Foodborne Pathog. Dis.* **2012**, *9*, 617–624.

25. Noormohamed, A.; Fakhr, M.K. A higher prevalence rate of *Campylobacter* in retail beef livers compared to other beef and pork meat cuts. *Int. J. Environ. Res. Public Health* **2013**, *10*, 2058–2068.

26. Centers for Disease Control. Standard Operating Procedure for PulseNet PFGE of *Campylobacter jejuni*, 2011. Available online: http://www.cdc.gov/pulsenet/PDF/campylobacter-pfge-protocol-508c.pdf (accessed on 15 August 2011).

27. Dingle, K.E.; Colles, F.M.; Falush, D.; Maiden, M.C. Sequence typing and comparison of population biology of *Campylobacter coli* and *Campylobacter jejuni*. *J. Clin. Microbiol.* **2005**, *43*, 340–347.

28. Marmur, J. A procedure for the isolation of deoxyribonucleic acid from microorganisms. *J. Mol. Biol.* **1961**, *3*, 208–218.

29. Jolley, K.A.; Maiden, M.C. BIGSdb: Scalable analysis of bacterial genome variation at the population level. *BMC Bioinformatics* **2010**, *11*, 595.

30. Primers Used for MLST of *Campylobacter*. Avalaible online: pubmlst.org/campylobacter/info/primers.shtml (accessed on 15 March 2012).

31. Miller, W.G.; Englen, M.D.; Kathariou, S.; Wesley, I.V.; Wang, G.; Pittenger-Alley, L.; Siletz, R.M.; Muraoka, W.; Fedorka-Cray, P.J.; Mandrell, R.E. Identification of host-associated alleles by multilocus sequence typing of *Campylobacter coli* strains from food animals. *Microbiology* **2006**, *152*, 245–255.

32. Thakur, S.; Morrow, W.E.; Funk, J.A.; Bahnson, P.B.; Gebreyes, W.A. Molecular epidemiologic investigation of *Campylobacter coli* in swine production systems, using multilocus sequence typing. *Appl. Environ. Microbiol.* **2006**, *72*, 5666–5669.

33. Abley, M.J.; Wittum, T.E.; Funk, J.A.; Gebreyes, W.A. Antimicrobial susceptibility, pulsed-field gel electrophoresis, and multi-locus sequence typing of *Campylobacter coli* in swine before, during, and after the slaughter process. *Foodborne Pathog. Dis.* **2012**, *9*, 506–512.

34. Carrillo, C.D.; Kruczkiewicz, P.; Mutschall, S.; Tudor, A.; Clark, C.; Taboada, E.N. A framework for assessing the concordance of molecular typing methods and the true strain phylogeny of *Campylobacter jejuni* and *C. coli* using draft genome sequence data. *Front. Cell. Infect. Microbiol.* **2012**, *2*, 57.

35. Magnússon, S.H.; Guðmundsdóttir, S.; Reynisson, E.; Rúnarsson, A.R.; Harðardóttir, H.; Gunnarson, E.; Georgsson, F.; Reiersen, J.; Marteinsson, V.T. Comparison of *Campylobacter jejuni* isolates from human, food, veterinary and environmental sources in Iceland using PFGE, MLST and fla-SVR sequencing. *J. Appl. Microbiol.* **2011**, *111*, 971–981.

36. Griekspoor, P.; Engvall, E.O.; Olsen, B.; Waldenstrom, J. Multilocus sequence typing of *Campylobacter jejuni* from broilers. *Vet. Microbiol.* **2010**, *140*, 180–185.

37. Cornelius, A.J.; Gilpin, B.; Carter, P.; Nicol, C.; On, S.L. Comparison of PCR binary typing (P-BIT), a new approach to epidemiological subtyping of *Campylobacter jejuni*, with serotyping, pulsed-field gel electrophoresis, and multilocus sequence typing methods. *Appl. Environ. Microbiol.* **2010**, *76*, 1533–1544.

38. Nielsen, L.N.; Sheppard, S.K.; McCarthy, N.D.; Maiden, M.C.; Ingmer, H.; Krogfelt, K.A. MLST clustering of *Campylobacter jejuni* isolates from patients with gastroenteritis, reactive arthritis and Guillain-Barré syndrome. *J. Appl. Microbiol.* **2010**, *108*, 591–599.

39. Colles, F.M.; Jones, K.; Harding, R.M.; Maiden, M.C. Genetic diversity of *Campylobacter jejuni* isolates from farm animals and the farm environment. *Appl. Environ. Microbiol.* **2003**, *69*, 7409–7413.

40. Duim, B.; Wassenaar, T.M.; Rigter, A.; Wagenaar, J. High-resolution genotyping of *Campylobacter* strains isolated from poultry and humans with amplified fragment length polymorphism fingerprinting. *Appl. Environ. Microbiol.* **1999**, *65*, 2369–2375.

41. Strachan, N.J.C.; MacRae, M.; Thomson, A.; Rotariu, O.; Ogden, I.D.; Forbes, K.J. Source attribution, prevalence and enumeration of *Campylobacter* spp. from retail liver. *Int. J. Food Microbiol.* **2012,** *153,* 234–236.

Permissions

The contributors of this book come from diverse backgrounds, making this book a truly international effort. This book will bring forth new frontiers with its revolutionizing research information and detailed analysis of the nascent developments around the world.

We would like to thank all the contributing authors for lending their expertise to make the book truly unique. They have played a crucial role in the development of this book. Without their invaluable contributions this book wouldn't have been possible. They have made vital efforts to compile up to date information on the varied aspects of this subject to make this book a valuable addition to the collection of many professionals and students.

This book was conceptualized with the vision of imparting up-to-date information and advanced data in this field. To ensure the same, a matchless editorial board was set up. Every individual on the board went through rigorous rounds of assessment to prove their worth. After which they invested a large part of their time researching and compiling the most relevant data for our readers.

The editorial board has been involved in producing this book since its inception. They have spent rigorous hours researching and exploring the diverse topics which have resulted in the successful publishing of this book. They have passed on their knowledge of decades through this book. To expedite this challenging task, the publisher supported the team at every step. A small team of assistant editors was also appointed to further simplify the editing procedure and attain best results for the readers.

Apart from the editorial board, the designing team has also invested a significant amount of their time in understanding the subject and creating the most relevant covers. They scrutinized every image to scout for the most suitable representation of the subject and create an appropriate cover for the book.

The publishing team has been an ardent support to the editorial, designing and production team. Their endless efforts to recruit the best for this project, has resulted in the accomplishment of this book. They are a veteran in the field of academics and their pool of knowledge is as vast as their experience in printing. Their expertise and guidance has proved useful at every step. Their uncompromising quality standards have made this book an exceptional effort. Their encouragement from time to time has been an inspiration for everyone.

The publisher and the editorial board hope that this book will prove to be a valuable piece of knowledge for researchers, students, practitioners and scholars across the globe.

List of Contributors

Stephen Lock
AB SCIEX, Pheonix House, Centre Park, Warrington, WA1 1RX, UK

Indra Prakash
The Coca-Cola Company, Atlanta, GA 30313, USA

Avetik Markosyan
PureCircle Limited, Lengkuk Teknologi, 71760 Bandar Enstek, Negeri Sembilan, Malaysia

Cynthia Bunders
The Coca-Cola Company, Atlanta, GA 30313, USA

Yukiko Wadamori
Food Group, Department of Wine, Food and Molecular Biosciences, Lincoln University, Lincoln 7647, Canterbury, New Zealand

Leo Vanhanen
Food Group, Department of Wine, Food and Molecular Biosciences, Lincoln University, Lincoln 7647, Canterbury, New Zealand

Geoffrey P. Savage
Food Group, Department of Wine, Food and Molecular Biosciences, Lincoln University, Lincoln 7647, Canterbury, New Zealand

Yin Yang
Laboratory of Natural Product Chemistry, UMR CNRS 6134, Grimaldi Campus, Corsican University, BP 52, Corte 20250, France

Marie-José Battesti
Laboratory of Natural Product Chemistry, UMR CNRS 6134, Grimaldi Campus, Corsican University, BP 52, Corte 20250, France

Jean Costa
Laboratory of Natural Product Chemistry, UMR CNRS 6134, Grimaldi Campus, Corsican University, BP 52, Corte 20250, France

Julien Paolini
Laboratory of Natural Product Chemistry, UMR CNRS 6134, Grimaldi Campus, Corsican University, BP 52, Corte 20250, France

Elena Gonzalez-Fandos
Food Technology Department, CIVA Research Center, University of La Rioja, Madre de Dios 51, 26006 Logroño, La Rioja, Spain

Barbara Herrera
Food Technology Department, CIVA Research Center, University of La Rioja, Madre de Dios 51, 26006 Logroño, La Rioja, Spain

Douglas S. Kalman
Nutrition/Endocrinology Department, Miami Research Associates, 6141 Sunset Drive, Suite 301, Miami, FL 33143, USA

Soheila J. Maleki
Southern Regional Research Center, Agricultural Research Service, U.S. Department of Agriculture, 1100 Robert E. Lee Blvd, New Orleans, LA 70124, USA

David A. Schmitt
Southern Regional Research Center, Agricultural Research Service, U.S. Department of Agriculture, 1100 Robert E. Lee Blvd, New Orleans, LA 70124, USA
Marathon Petroleum, Texas City, TX 77590, USA

Maria Galeano
Southern Regional Research Center, Agricultural Research Service, U.S. Department of Agriculture, 1100 Robert E. Lee Blvd, New Orleans, LA 70124, USA
Vital Source Technologies, Raleigh, NC 27601, USA

Barry K. Hurlburt
Southern Regional Research Center, Agricultural Research Service, U.S. Department of Agriculture, 1100 Robert E. Lee Blvd, New Orleans, LA 70124, USA

Nuria Prieto
Genetics Department, Biology Faculty, Complutense University of Madrid, Madrid 28040, Spain

Carmen Burbano
Food Technology Department, National Institute for Agricultural and Food Research and Technology (INIA), Ctra. La Coruña Km. 7.5, Madrid 28040, Spain

Elisa Iniesto
Genetics Department, Biology Faculty, Complutense University of Madrid, Madrid 28040, Spain

Julia Rodríguez
Allergy Service, Research Institute Hospital 12 de Octubre (i+12), Avenida de Córdoba s/n, Madrid 28041, Spain

Beatriz Cabanillas
Allergy Service, Research Institute Hospital 12 de Octubre (i+12), Avenida de Córdoba s/n, Madrid 28041, Spain

Jesus F. Crespo
Allergy Service, Research Institute Hospital 12 de Octubre (i+12), Avenida de Córdoba s/n, Madrid 28041, Spain

Mercedes M. Pedrosa
Food Technology Department, National Institute for Agricultural and Food Research and Technology (INIA), Ctra. La Coruña Km. 7.5, Madrid 28040, Spain

Mercedes Muzquiz
Food Technology Department, National Institute for Agricultural and Food Research and Technology (INIA), Ctra. La Coruña Km. 7.5, Madrid 28040, Spain

Juan Carlos del Pozo
Center for Biotechnology and Plant Genomic, Polytechnic University of Madrid-National Institute for Agricultural and Food Research and Technology (UPM-INIA), Montegancedo Campus, Boadilla del Monte, Madrid 28660, Spain

Rosario Linacero
Genetics Department, Biology Faculty, Complutense University of Madrid, Madrid 28040, Spain

Carmen Cuadrado
Food Technology Department, National Institute for Agricultural and Food Research and Technology (INIA), Ctra. La Coruña Km. 7.5, Madrid 28040, Spain

Agnieszka Golon
School of Engineering and Science, Jacobs University Bremen, Campus Ring 1, 28759 Bremen, Germany

Christian Kropf
Henkel AG & Co. KGaA, Henkelstr. 67, 40589 Düsseldorf, Germany

Inga Vockenroth
Henkel AG & Co. KGaA, Henkelstr. 67, 40589 Düsseldorf, Germany

Nikolai Kuhnert
School of Engineering and Science, Jacobs University Bremen, Campus Ring 1, 28759 Bremen, Germany

Senay Simsek
Department of Plant Sciences, North Dakota State University, P.O. Box 6050, Department #7670, Fargo, ND 58108-6050, USA

Jae-Bom Ohm
USDA-ARS Hard Red Spring and Durum Wheat Quality Laboratory, Harris Hall, North Dakota State University, Fargo, ND 58108, USA

Haiyan Lu
Department of Plant Sciences, North Dakota State University, P.O. Box 6050, Department #7670, Fargo, ND 58108-6050, USA

Mory Rugg
Department of Plant Sciences, North Dakota State University, P.O. Box 6050, Department #7670, Fargo, ND 58108-6050, USA

William Berzonsky
Department of Plant Sciences, South Dakota State University, Brookings, SD 57007-2141, USA

Mohammed S. Alamri
Nutrition and Food Sciences Department, College of Food and Agricultural Sciences; King Saud University, P.O. Box 2460, Riyadh 11451, Saudi Arabia

Mohamed Mergoum
Department of Plant Sciences, North Dakota State University, P.O. Box 6050, Department #7670, Fargo, ND 58108-6050, USA

D. Cozzolino
School of Agriculture, Food and Wine, The University of Adelaide, Waite Campus, PMB 1 Glen Osmond, SA, 5064, Australia

S. Degner
School of Agriculture, Food and Wine, The University of Adelaide, Waite Campus, PMB 1 Glen Osmond, SA, 5064, Australia

J. Eglinton
School of Agriculture, Food and Wine, The University of Adelaide, Waite Campus, PMB 1 Glen Osmond, SA, 5064, Australia

Bekzod Khakimov
Department of Food Science, Faculty of Science, University of Copenhagen, Rolighedsvej 30, Frederiksberg C, 1958 Copenhagen, Denmark

Birthe Møller Jespersen
Department of Food Science, Faculty of Science, University of Copenhagen, Rolighedsvej 30, Frederiksberg C, 1958 Copenhagen, Denmark

Søren Balling Engelsen
Department of Food Science, Faculty of Science, University of Copenhagen, Rolighedsvej 30, Frederiksberg C, 1958 Copenhagen, Denmark

Aneesa Noormohamed
Department of Biological Science, The University of Tulsa, Tulsa, OK 74104, USA

Mohamed K. Fakhr
Department of Biological Science, The University of Tulsa, Tulsa, OK 74104, USA

www.ingramcontent.com/pod-product-compliance
Lightning Source LLC
Chambersburg PA
CBHW080700200326
41458CB00013B/4923